STOCHASTIC CONTROL OF PARTIALLY OBSERVABLE SYSTEMS

STOCHASTIC CONTROL OF PARTIALLY OBSERVABLE SYSTEMS

ALAIN BENSOUSSAN
University Paris Dauphine

CAMBRIDGE
UNIVERSITY PRESS

PUBLISHED BY THE PRESS SYNDICATE OF THE UNIVERSITY OF CAMBRIDGE
The Pitt Building, Trumpington Street, Cambridge, United Kingdom

CAMBRIDGE UNIVERSITY PRESS
The Edinburgh Building, Cambridge CB2 2RU, UK
40 West 20th Street, New York NY 10011–4211, USA
477 Williamstown Road, Port Melbourne, VIC 3207, Australia
Ruiz de Alarcón 13, 28014 Madrid, Spain
Dock House, The Waterfront, Cape Town 8001, South Africa

http://www.cambridge.org

First published 1992
First paperback edition 2004

A catalogue record for this book is available from the British Library

ISBN 0 521 35403 X hardback
ISBN 0 521 61197 0 paperback

Contents

Preface

The problem of stochastic control of partially observable systems plays an important role in many applications. All real problems are in fact of this type, and deterministic control as well as stochastic control with full observation can only be approximations of the real world. This justifies the importance of having a theory as complete as possible, which can be used for numerical implementation.

In the first three chapters of this book we study problems which can be dealt with directly by algebraic manipulations, without using the complete theory. This is because the system and the observation have linear dynamics, and the cost is either quadratic or the exponential of a quadratic functional.

In Chapters 4 to 6, we present the theory of non linear filtering, which is the basic step in formulating the control problem adequately. The main difficulty, especially from the point of view of numerical applications, is that there are no statistics which are finite dimensional, and the basic object to be computed is the conditional probability. This is the solution of a stochastic partial differential equation (PDE) studied in Chapter 4. Chapters 5 and 6 are devoted to approximations, when perturbation methods are applicable, or to particular cases when simplification occurs, and sufficient statistics exist.

In Chapter 7, we study stochastic control problems with partial information, in an intermediate case, namely when the direct methods of Chapters 1, 2, 3 are not applicable yet the full theory is not necessary, either because finite dimensional sufficient statistics are available, or approximations are possible. In Chapter 8, we present the stochastic maximum principle and dynamic programming approach to the problem of stochastic control with partial information in the general case, which implies infinite dimensionality. In Chapter 9 we state, in a limited framework, some existence results.

Needless to say, I make no pretence to be exhaustive, and in this presentation have favoured analytic techniques, rather than algebraic and probabilistic ones. Reference to relevant literature is given throughout the book, for further study of these aspects. Numerical methods are not considered at all (see Legrand (1989) and Bensoussan and Runggaldier (1987) for more in this direction).

1 Linear filtering theory

Introduction

The objective of this chapter is to derive the Kalman filter in discrete time as well as in continuous time.

There are several ways to derive these formulas, especially in discrete time where an elementary direct approach is feasible. A more systematic way to proceed is to use the connection between the optimal filter and the least square or maximum likelihood estimate. This approach is convenient in continuous time.

1.1   Filtering theory in discrete time

1.1.1   The model

We consider the following dynamic system in discrete time:

$$x_{k+1} = F_k x_k + f_k + w_k, \quad k = 0, 1, \ldots, N-1,$$
$$x_0 = \xi \tag{1.1.1}$$
$$y_k = H_k x_k + h_k + b_k, \quad k = 0, 1, \ldots, N-1 \tag{1.1.2}$$

where

$$x_k \in R^n, \quad y_k \in R^m; \tag{1.1.3}$$

F_k, H_k are matrices with appropriate sizes; f_k is a sequence of vectors in R^n, h_k a sequence of vectors in R^m.

ξ is a random variable with gaussian probability law, with mean
x_0 and covariance matrix P_0 $\tag{1.1.4}$

w_k, b_k are random gaussian variables with mean 0 and covariance matrices Q_k, P_k respectively; the matrix R_k is positive definite. The variables, ξ, w_k, b_k are mutually independent.

Let us consider the sequence of σ-algebras,

$$\mathcal{Y}^k = \sigma(y_0, \ldots, y_{k-1}), \quad k = 1, \ldots, N.$$

The problem is to compute

$$\hat{x}_N = E[x_N | \mathcal{Y}^N]. \tag{1.1.5}$$

1.1.2   The best linear estimate

A linear estimate is defined as follows:

$$\mathcal{F}_S = \bar{x}_N + \sum_{k=0}^{N-1} S_k(y_k - \bar{y}_k)$$

where $S_1, \ldots, S_N$ are matrices in $L(R^m; R^n)$ defining the filter $\mathcal{F}$, and $\bar{x}_N$, $\bar{y}_k$ represent the means of x_N, y_k respectively. The set $S_1, \ldots, S_N$ is represented globally by S.

Note that the sequences $\bar{x}_k$, $\bar{y}_k$ are defined by the formulas

$$\bar{x}_{k+1} = F_k \bar{x}_k + f_k, \quad k = 0, \ldots, N-1 \tag{1.1.6}$$

$$\bar{y}_k = H_k \bar{x}_k + h_k.$$

The best linear filter is obtained by choosing S in order to minimize the functional

$$\mathcal{L}(S) = E(x_N - \mathcal{F}_S)^*(x_N - \mathcal{F}_S) \tag{1.1.7}$$

where * denotes the transpose.

We need some notation. Let Λ_{kl} denote the correlation matrix of the process x_k, namely

$$\Lambda_{kl} \in L(R^n; R^n)$$

and

$$\Lambda_{kl} = E(x_k - \bar{x}_k)(x_l - \bar{x}_l)^*.$$

Note that $\Lambda_{kl}^* = \Lambda_{lk}$, and if $M \in L(R^n; R^n)$

$$E(x_k - \bar{x}_k)^* M(x_l - \bar{x}_l) = \mathrm{tr}\,\Lambda_{lk} M$$

we then deduce easily that

$$\begin{aligned}
\mathcal{L}(S) = &\mathrm{tr}\,\Lambda_{NN} + \sum_{k=0}^{N-1} \mathrm{tr}\, R_k S_k^* S_k \\
&+ \sum_{k,l=0}^{N-1} \mathrm{tr}\, \Lambda_{lk} H_k^* S_k^* S_l H_l \\
&- 2\sum_{k=0}^{N-1} \mathrm{tr}\, \Lambda_{kN} S_k H_k.
\end{aligned} \tag{1.1.8}$$

We then deduce easily

Proposition 1.1.1 *There exists a unique S minimizing the functional $\mathcal{L}(S)$.*

Proof

The set of S is clearly a finite dimensional vector space. We equip it with the scalar product

$$(S, \tilde{S}) = \sum_k \mathrm{tr} S_k^* \tilde{S}_k. \tag{1.1.9}$$

The functional $\mathcal{L}(S)$ is a quadratic form and

$$\sum_{k,l} \mathrm{tr}\Lambda_{lk} H_k^* S_k^* S_l H_l \geq 0$$

$$\begin{aligned}
\sum_k \mathrm{tr} R_k S_k^* S_k &= \sum_{k=0}^{N-1} \sum_{h=1}^{n} \sum_{i,j=1}^{m} R_{k,ij} S_{k,hi} S_{k,hj} \\
&\geq \alpha \sum_{k=0}^{N-1} \sum_{h=1}^{n} \sum_{i=1}^{m} (S_{k,hi})^2 \\
&= \alpha \|S\|^2.
\end{aligned}$$

Therefore $\mathcal{L}(S)$ has a unique minimum $\hat{S}$ and $\hat{S}$ is uniquely defined by the equation

$$
\sum_k \mathrm{tr}(\hat{S}_k S_k R_k + R_k \hat{S}_k^* S_k)
$$

$$
+ \sum_{k,l} \mathrm{tr}(\Lambda_{lk} H_k^* \hat{S}_k^* S_l H_l + H_l^* \hat{S}_l^* S_k H_k \Lambda_{kl}) \qquad (1.1.10)
$$

$$
- 2 \sum_k \mathrm{tr}\, \Lambda_{kN} S_k H_k = 0, \quad \forall S.
$$

∎

In fact the best linear estimate is also the best possible estimate, by virtue of the gaussian properties. We can state the following:

Proposition 1.1.2 *We have*

$$
\hat{x}_N = \mathcal{F}_{\hat{S}} = \bar{x}_N + \sum_{k=0}^{N-1} \hat{S}_k (y_k - \bar{y}_k). \qquad (1.1.11)
$$

Proof

Define

$$
\epsilon_N = x_N - \mathcal{F}_{\hat{S}};
$$

then from the definition (1.1.7) one has

$$
\sum_k E[\epsilon_N^* S_k y_k + y_k^* S_k^* \epsilon_N] = 0 \quad \forall S_0, \ldots, S_{N-1}.
$$

Since $S_0, \ldots, S_{N-1}$ are arbitrary, it follows that ϵ_N^* is not correlated with $y_0, \ldots, y_{N-1}$ thus is also independent of them. Hence the desired result. ∎

1.1.3   The discrete time Kalman filter

In equation (1.1.10) the explicit relations defining the process x_k do not play any role. On the other hand the best estimate $\hat{S}$ depends on N and there is no recursive property as N varies.

A recursive form of the filter can be obtained using the explicit relations (1.1.1), and this form is called the Kalman filter, since the seminal papers of Kalman (Kalman 1960; Kalman and Bucy, 1961).

Let us set

$$
P_N = E(x_N - \mathcal{F}_{\hat{S}})(x_N^* - F_{\hat{S}}^*) = E\epsilon_N \epsilon_N^* \qquad (1.1.12)
$$

which represents the covariance of the error. Clearly one has

$$
\mathcal{L}(\hat{S}) = \mathrm{tr}\, P_N. \qquad (1.1.13)
$$

Let us also define

$$
\hat{x}_N^+ = E[x_N | y^{N+1}]
$$

$$
= E[x_N | y_0, \ldots, y_N] \qquad (1.1.14)
$$

$$
\epsilon_N^+ = x_N - \hat{x}_N^+ \qquad (1.1.15)
$$

$$
P_N^+ = E\epsilon_N^+ (\epsilon_N^+)^*. \qquad (1.1.16)
$$

We shall prove the following formulae.

Theorem 1.1.1 *For the model (1.1.1), (1.1.2) the following formulae hold:*

$$\hat{x}_{N+1} = F_N \hat{x}_N^+ + f_N \tag{1.1.17}$$

$$\hat{x}_N^+ = \hat{x}_N + P_N H_N^* (R_N + H_N P_N H_N^*)^{-1}(y_N - H_N \hat{x}_N - h_N)$$

$$P_{N+1} = Q_N + F_N P_N^+ F_N^*$$

$$P_N^+ = P_N - P_N H_N^* (H_N P_N H_N^* + R_N)^{-1} H_N P_N \tag{1.1.18}$$

$$\hat{x}_0 = x_0$$

$$P_0 = P_0.$$

The proof requires the concept of innovation, which we develop now.

1.1.4   Innovation process

Let us consider the process

$$\nu_N = y_N - (H_N \hat{x}_N + h_N), \tag{1.1.19}$$

then we have the following:

Proposition 1.1.3 *The process ν_N is a gaussian process such that*

$$E\nu_k = 0 \tag{1.1.20}$$

$$E\nu_k \nu_l^* = \delta_{kl}(R_k + H_k P_k H_k^*), \quad \forall k, l.$$

Moreover ν_N is independent of $\mathcal{Y}^N$.

Proof

We have

$$\nu_k = H_k \epsilon_k + b_k, \quad k = 0, 1, \dots \tag{1.1.21}$$

and ν_k is a gaussian variable such that $E\nu_k = 0$.

Moreover

$$E\nu_k \nu_k^* = E(H_k \epsilon_k + b_k)(\epsilon_k^* H_k^* + b_k^*).$$

Since b_k is independent of ϵ_k, we deduce immediately

$$E\nu_k \nu_k^* = R_k + H_k P_k H_k^*.$$

Moreover for $l = 0, \dots, k-1$,

$$E[\nu_k y_l^* | \mathcal{Y}^k] = E[\nu_k | \mathcal{Y}^k]y_l^* = 0$$

hence also

$$E[\nu_k \nu_l^*] + E[\nu_k \hat{x}_l^*]H_l^* = 0.$$

Using

$$E[\nu_k \hat{x}_l^*] = E\{E[\nu_k | \mathcal{Y}^k]\hat{x}_l^*\} = 0$$

we deduce

$$E\nu_k \nu_l^* = 0, \quad \text{for } l = 0, \dots, k-1.$$

The proof has been completed.   ∎

1.1.5   Proof of Theorem 1.1.1

The first relation of (1.1.17) as well as the first relation of (1.1.18) are immediate.

Moreover using the innovation we can assert that

$$\hat{x}_N^+ = E[x_N | y_0, \ldots, y_{N-1}, \nu_N].$$

Using the fact that $\hat{x}_N^+$ is also the best linear estimate of x_N, given $y_0, \ldots, y_{N-1}, \nu_N$ (by analogy with Proposition 1.1.2), we can write the formula

$$\hat{x}_N^+ = \hat{x}_N + K_N \nu_N \qquad (1.1.22)$$

where K_N is a gain factor to be determined. Note that the independence of ν_N from $y_0, \ldots, y_{N-1}$ has been used to derive (1.1.22). It remains to fix K_N, knowing that it minimizes the covariance of the error.

The error can be written as

$$\epsilon_N^+ = \epsilon_N - K_N \nu_N$$

hence the covariance $E\epsilon_N^+ (\epsilon_N^+)^*$ is

$$P_N^+ = P_N + K_N(H_N P_N H_N^* + R_N)K_N^* - K_N E\nu_N \epsilon_N^* - E\epsilon_N \nu_N^* K_N.$$

But

$$\begin{aligned} E\nu_N \epsilon_N^* &= E y_N \epsilon_N^* \\ &= E(H_N x_N + h_N)\epsilon_N^* \\ &= E H_N (x_N - \hat{x}_N)\epsilon_N^* = H_N P_N. \end{aligned}$$

Therefore we have

$$\begin{aligned} P_N^+ &= P_N + K_N(H_N P_N H_N^* + R_N)K_N^* - K_N H_N P_N - P_N H_N^* K_N^* \\ &= P_N + [K_N - P_N H_N^*(H_N P_N H_N^* + R_N)^{-1}](H_N P_N H_N^* + R_N) \\ &\quad \times [K_N^* - (H_N P_N H_N^* + R_N)^{-1} H_N P_N] \\ &\quad - P_N H_N^*(H_N P_N H_N^* + R_N)^{-1} H_N P_N \end{aligned}$$

by completing the square.

It follows immediately that the best value of K_N is

$$K_N = P_N H_N^*(H_N P_N H_N^* + R_N)^{-1}$$

and the second formulae in (1.1.17) and (1.1.18) follow. The proof is now complete. ∎

1.1.6   Least squares estimate

There is another approach to derive the Kalman filter, which is to some extent very intuitive but at the same time is more a lucky consequence of the linear gaussian character of the model than based on theoretical reasons.

The idea consists in looking at ξ, w_k in (1.1.1) as decision variables to be chosen in order to minimize the cost

$$
\begin{aligned}
K(\xi, w) =& (\xi - x_0)^* P_c^{-1}(\xi - x_0) + \sum_{k=0}^{N-1} w_k^* Q_k^{-1} w_k \\
& + \sum_{k=0}^{N-1} [y_k - (H_k x_k + h_k)]^* R_k^{-1} [y_k - (H_k x_k + h_k)].
\end{aligned}
\tag{1.1.23}
$$

In (1.1.23) the quantities $y_0, \ldots, y_{N-1}$ are considered as given. This is a control problem in which the state equations are (1.1.1) and ξ, w_k are the control variables. The functional K is called a likelihood function. This is because we can express

$$
\text{Prob}(\xi = \xi_0, w_k = w_{k_0}, y_k = y_{k_0}) = C \exp\left(-\frac{1}{2} K_0(\xi_0, w_0)\right)
\tag{1.1.24}
$$

in which $K_0(\xi, w)$ represents the expression (1.1.23) with $y_k = y_{k_0}$. Therefore if we minimize (1.1.23) we maximize the conditional probability of ξ, w given y. The control problem (1.1.83) is solved by standard techniques. Denoting the optimal controls by $\hat{\xi}, \hat{w}$ and the corresponding optimal state by $\hat{p}$, we have the relations

$$
\begin{aligned}
\hat{p}_{k+1} =& F_k \hat{p}_k + f_k + \hat{w}_k, \quad k = 0, \ldots, N-1 \\
\hat{p}_0 =& \hat{\xi}
\end{aligned}
\tag{1.1.25}
$$

and

$$
(\hat{\xi} - \bar{x}_0)^* P_0^{-1} \xi + \sum_{k=0}^{N-1} \hat{w}_k^* Q_k^{-1} w_k - \sum_{k=0}^{N-1} (y_k - H_k \hat{p}_k - h_k)^* R_k^{-1} H_k \tilde{x}_k = 0
\tag{1.1.26}
$$

for any ξ, w and $\tilde{x}_k$ given by

$$
\begin{aligned}
\tilde{x}_{k+1} =& F_k \tilde{x}_k + w_k \\
\tilde{x}_0 =& \xi.
\end{aligned}
\tag{1.1.27}
$$

Introducing the adjoint variable $\hat{q}_k$ as the solutions of

$$
\begin{aligned}
\hat{q}_k =& F_k^* \hat{q}_{k+1} - H_k^* R_k^{-1}(y_k - H_k \hat{p}_k - h_k) \\
\hat{q}_N =& 0
\end{aligned}
\tag{1.1.28}
$$

we deduce from (1.1.26) that

$$
(\hat{\xi} - x_0) P_0^{-1} \xi + \sum_{k=0}^{N-1} \hat{w}_k^* Q_k^{-1} w_k + \sum_{k=0}^{N-1} (\hat{q}_k^* - \hat{q}_{k+1}^* F_k) \tilde{x}_k = 0
$$

and using (1.1.27)

$$
(\hat{\xi} - x_0) P_0^{-1} \xi + \sum_{k=0}^{N-1} \hat{w}_k^* Q_k^{-1} w_k + \sum_{k=0}^{N-1} \hat{q}_k^* \tilde{x}_k - \sum_{k=0}^{N-1} \hat{q}_{k+1}^* \tilde{x}_{k+1} + \sum_{k=0}^{N-1} \hat{q}_{k+1}^* w_k = 0
$$

or finally

$$[(\hat{\xi} - x_0)^* P_0^{-1} + \hat{q}_0^*]\xi + \sum_{k=0}^{N-1} (\hat{w}_k^* Q_k^{-1} + \hat{q}_{k+1}^*) w_k = 0$$

which implies

$$\hat{\xi} = x_0 - P_0 \hat{q}_0,$$
$$\hat{w}_k = -Q_k \hat{q}_{k+1}$$

which used in (1.1.25) yields

$$\hat{p}_{k+1} = F_k \hat{p}_k + f_k - Q_k \hat{q}_{k+1}$$
$$\hat{p}_0 = x_0 - P_0 \hat{q}_0. \tag{1.1.29}$$

Let us check that the pair $\hat{p}_k$, $\hat{q}_k$ satisfies the following affine relation

$$\hat{p}_k = r_k - \Sigma_k \hat{q}_k \tag{1.1.30}$$

where r_k, Σ_k are to be determined. Using (1.1.30) in (1.1.28), (1.1.29) yields

$$r_{k+1} - \Sigma_{k+1} \hat{q}_{k+1} = F_k(r_k - \Sigma_k \hat{q}_k) - Q_k \hat{q}_{k+1} + f_k$$
$$(I + H_k^* R_k^{-1} H_k \Sigma_k)\hat{q}_k = F_k^* \hat{q}_{k+1} - H_k^* R_k^{-1}(y_k - H_k r_k - h_k).$$

Assuming for a while that Σ_k is positive definite (it is also assumed symmetric), we deduce

$$r_{k+1} - \Sigma_{k+1} \hat{q}_{k+1} = F_k r_k - F_k(\Sigma_k^{-1} + H_k^* R_k^{-1} H_k)^{-1}$$
$$\times [F_k^* \hat{q}_{k+1} - H_k^* R_k^{-1}(y_k - H_k r_k - h_k)] - Q_k \hat{q}_{k+1} + f_k.$$

Identifying terms so that this relation holds whatever the value of $\hat{q}_{k+1}$, we write

$$\Sigma_{k+1} = F_k(\Sigma_k^{-1} + H_k^* R_k^{-1} H_k)^{-1} F_k^* + Q_k$$
$$r_{k+1} = F_k r_k + f_k + F_k(\Sigma_k^{-1} + H_k^* R_k^{-1} H_k)^{-1} H_k^* R_k^{-1}(y_k - H_k r_k - h_k). \tag{1.1.31}$$

By virtue of the initial condition appearing in (1.1.29), we take

$$r_0 = x_0, \quad \Sigma_0 = P_0.$$

We note the algebraic relations

$$H_k(\Sigma_k^{-1} + H_k^* R_k^{-1} H_k)^{-1} = R_k(H_k \Sigma_k H_k^* + R_k)^{-1} H_k \Sigma_k \tag{1.1.32}$$

$$\Sigma_k - \Sigma_k H_k^*(R_k + H_k \Sigma_k H_k^*)^{-1} H_k \Sigma_k = (\Sigma_k^{-1} + H_k^* R_k^{-1} H_k)^{-1}$$

and hence we can rewrite (1.1.31) as follows:

$$\Sigma_{k+1} = F_k \Sigma_k F_k^* + Q_k - F_k \Sigma_k H_k^*(R_k + H_k \Sigma_k H_k^*)^{-1} H_k \Sigma_k F_k^* \tag{1.1.33}$$

$$r_{k+1} = F_k r_k + f_k + F_k \Sigma_k H_k^*(H_k \Sigma_k H_k^* + R_k)^{-1}(y_k - H_k r_k - h_k).$$

Comparing these formulae with those of Theorem 1.1, we deduce easily that

$$\Sigma_k = P_k, \quad r_k = \hat{x}_k.$$

Now from (1.1.30) we see

$$\hat{p}_N = r_N = \hat{x}_N$$

which proves that the optimal state at time N, corresponding to minimizing the likelihood function, coincides with the best estimate, the conditional mean.

This fact is not general. The maximum likelihood estimate does not coincides with the conditional mean in non linear, non gaussian situations. It remains an interesting suboptimal estimate to consider in all cases, in view of its nice computational features.

To illustrate further the interest of the maximum likelihood approach to guess intuitively the form of a desired estimate, let us suppose that we want to compute $\hat{x}_N$. By analogy with (1.1.23) we introduce this time the likelihood function

$$K^+(\xi, w) = (\xi - x_0)^* P_0^{-1}(\xi - x_0) + \sum_{k=0}^{N-1} w_k^* Q_k^{-1} w_k$$

$$+ \sum_{k=0}^{N} (y_k - H_k x_k - h_k)^* R_k^{-1} (y_k - H_k x_k - h_k). \qquad (1.1.34)$$

Denoting again (to save notation) by $\hat{\xi}$, $\hat{w}_k$, $\hat{r}_k$ the corresponding optimal control and state, we get

$$\hat{p}_{k+1} = F_k \hat{p}_k + f_k + \hat{w}_k, \quad k = 0, \ldots, N-1$$
$$\hat{p}_0 = \hat{\xi} \qquad (1.1.35)$$

and

$$(\hat{\xi} - \bar{x}_0)^* P_0^{-1} \xi + \sum_{k=0}^{N-1} \hat{w}_k^* Q_k^{-1} w_k - \sum_{k=0}^{N} (y_k - H_k \hat{p}_k - h_k)^* R_k^{-1} H_k \tilde{x}_k = 0 \qquad (1.1.36)$$

$$\tilde{x}_{k+1} = F_k \tilde{x}_k + w_k$$
$$\tilde{x}_0 = \xi.$$

Defining $\hat{q}_k$ by

$$\hat{q}_k = F_k^* \hat{q}_{k+1} - H_k^* R_k^{-1} (y_k - H_k \hat{p}_k - h_k), \quad k = 0, \ldots, N$$
$$\hat{q}_{N+1} = 0 \qquad (1.1.37)$$

we deduce from (1.1.36)

$$(\hat{\xi} - \bar{x}_0)^* P_0^{-1} \xi + \sum_{k=0}^{N-1} \hat{w}_k^* Q_k^{-1} w_k + \sum_{k=0}^{N} (\hat{q}_k^* - \hat{q}_{k+1}^* F_k) \tilde{x}_k = 0$$

that is

$$(\hat{\xi} - \bar{x}_0)^* P_0^{-1} \xi + \sum_{k=0}^{N-1} \hat{w}_k^* Q_k^{-1} w_k + \hat{q}_0^* \xi + \sum_{k=0}^{N-1} \hat{q}_{k+1}^* w_k = 0, \quad \forall \xi, \forall w_k$$

from which $\hat{\xi}$, $\hat{w}_k$ follow. We thus get

$$\hat{p}_{k+1} = F_k \hat{p}_k + f_k - Q_k \hat{q}_{k+1}, \quad k = 0, \ldots, N-1$$
$$\hat{p}_0 = x_0 - P_0 \hat{q}_0. \qquad (1.1.38)$$

Note that (1.1.37) can be written as

$$\hat{q}_k = F_k^* \hat{q}_{k+1} - H_k^* R_k^{-1}(y_k - H_k \hat{p}_k - h_k), \quad k = 0, \ldots, N-1$$
$$\hat{q}_N = -H_N^* R_N^{-1}(y_N - H_N \hat{p}_N - h_N). \tag{1.1.39}$$

One can easily check that

$$\hat{p}_k = \hat{x}_k - P_k \hat{q}_k$$

hence in particular, recalling that $\hat{p}_N = \hat{x}_N^+$, we deduce

$$\hat{x}_N^+ = \hat{x}_N + P_N H_N^* R_N^{-1}(y_N - H_N \hat{x}_N^+ - h_N)$$

hence

$$\hat{x}_N^+ = \hat{x}_N + (P_N^{-1} + H_N^* R_N^{-1} H_N)^{-1} H_N^* R_N^{-1}(y_N - H_N \hat{x}_N - h_N)$$

i.e.

$$\hat{x}_N^+ = \hat{x}_N + P_N H_N^* (H_N P_N H_N^* + R_N)^{-1}(y_N - H_N \hat{x}_N - h_N). \tag{1.1.40}$$

We can in particular note the relation

$$\hat{x}_N = F_{N-1} \hat{x}_{N-1}^+ + f_{N-1}. \tag{1.1.41}$$

It also follows from (1.1.40) that defining

$$\epsilon_N^+ = x_N - \hat{x}_N^+$$

$$P_N^+ = E\epsilon_N^+(\epsilon_N^+)^*$$

we have

$$\epsilon_N^+ = [I - P_N H_N^* (H_N P_N H_N^* + R_N)^{-1} H_N]\epsilon_N$$
$$- P_N H_N^* (H_N P_N H_N^* + R_N)^{-1} b_N$$

hence

$$P_N^* = [I - P_N H_N^* (H_N P_N H_N^* + R_N)^{-1} H_N]P_N[I - H_N^* (H_N P_N H_N^* + R_N)^{-1} H_N P_N]$$
$$+ P_N H_N^* (H_N P_N H_N^* + R_N)^{-1} R_N (H_N P_N H_N^* + R_N)^{-1} H_N P_N.$$

Simplifying, we deduce

$$P_N^+ = P_N - P_N H_N^* (H_N P_N H_N^* + R_N)^{-1} H_N P_N. \tag{1.1.42}$$

We recover in this manner all the formulae of Theorem 1.1.1.

1.2   Filtering theory in continuous time

1.2.1   The model

We develop here the continuous time analogue of the situation of section 1.1. The model is described as follows:

$$
\begin{aligned}
\mathrm{d}x &= [F(t)x + f(t)]\mathrm{d}t + \mathrm{d}w \\
x(0) &= \xi
\end{aligned}
\tag{1.2.1}
$$

$$
\begin{aligned}
\mathrm{d}y &= [H(t)x(t) + h(t)]\mathrm{d}t + \mathrm{d}b(t) \\
y(0) &= 0
\end{aligned}
\tag{1.2.2}
$$

where $x(\cdot) \in R^n$, $y(\cdot) \in R^m$.

ξ is a random variable with gaussian probability law with mean x_0 and covariance matrix P_0. $w(\cdot)$, $b(\cdot)$ are Wiener processes with covariance matrices $Q(\cdot)$, $R(\cdot)$ respectively. The matrix $R(\cdot)$ is uniformly positive definite. The variable ξ and the processes $w(\cdot)$, $b(\cdot)$ are mutually independent. $\tag{1.2.3}$

$$
F(\cdot),\ f(\cdot),\ H(\cdot),\ h(\cdot) \text{ are bounded.}
\tag{1.2.3$'$}
$$

Let us define the family of σ-algebras

$$
\mathcal{Y}^t = \sigma[y(s),\ 0 \le s \le t].
$$

The problem consists in computing

$$
\hat{x}(t) = E[x(t)|\mathcal{Y}^t].
\tag{1.2.4}
$$

Let us first remark that

$$
\bar{x}(t) = Ex(t)
$$

is the solution of

$$
\begin{aligned}
\frac{\mathrm{d}\bar{x}}{\mathrm{d}t} &= F(t)\bar{x} + f(t) \\
\bar{x}(0) &= x_0.
\end{aligned}
\tag{1.2.5}
$$

Define also the covariance matrix

$$
\Pi(t) = E[x(t) - \bar{x}(t)][x(t) - \bar{x}(t)]^*.
\tag{1.2.6}
$$

Writing

$$\tilde{x}(t) = x(t) = \bar{x}(t)$$

we have

$$d\tilde{x} = F(t)\tilde{x}dt + dw$$
$$\tilde{x}(0) = \xi - x_0 \tag{1.2.7}$$

and from Ito's formula (with some formal notation)

$$\frac{d\Pi}{dt} = E[F(t)\tilde{x}\tilde{x}^* + \tilde{x}\tilde{x}^*F^*(t) + dw \cdot dw^*] \tag{1.2.8}$$
$$= F(t)\Pi(t) + \Pi(t)F^*(t) + Q(t)$$
$$\Pi(0) = P_0.$$

1.2.2 The best linear estimate

Define

$$\tilde{y}(t) = y(t) - \bar{y}(t) = y(t) - H(t)\bar{x}(t) - h(t)$$

hence

$$d\tilde{y} = H(t)\tilde{x}(t)dt + db(t)$$
$$\tilde{y}(0) = 0. \tag{1.2.9}$$

A linear filter is defined by a kernel $S(t) \in L^2[0, T; L(R^m; R^n)]$ in the following manner:

$$\mathcal{F}_S = \bar{x}(t) + \int_0^T S(t)d\tilde{y}(t). \tag{1.2.10}$$

The best linear filter is defined by the condition

$$\min \mathcal{L}(S) = E[x(T) - \mathcal{F}_S]^*[x(T) - \mathcal{F}_S] \tag{1.2.11}$$

$$\Lambda(t, s) = E\tilde{x}(t)\tilde{x}(s)^* \qquad \Lambda \in L(R^n; R^n). \tag{1.2.12}$$

Note that, if $M \in L(R^n; R^n)$,

$$E\tilde{x}(s)^* M\tilde{x}(t) = \text{tr}\Lambda(s, t)M.$$

Therefore we can compute

$$\mathcal{L}(S) = E\left(\tilde{x}(T) - \int_0^T S(t)d\tilde{y}(t)\right)^* \left(\tilde{x}(T) - \int_0^T S(t)d\tilde{y}(t)\right)$$

$$= E\tilde{x}^*(T)\tilde{x}(T) + E\left(\int_0^T \tilde{x}(t)^* H^*(t)S^*(t)dt\right)\left(\int_0^T S(t)H(t)\tilde{x}(t)dt\right)$$

$$+ E\left(\int_0^T db^*(t)S^*(t)S(t)db(t)\right) - 2E\left(\int_0^T \tilde{x}^*(t)H^*(t)S^*(t)dt\right)\tilde{x}(T)$$

i.e.

$$\mathcal{L}(S) = \text{tr}\Lambda(T,T) + \int_0^T \int_0^T \text{tr}\Lambda(s,t)H^*(t)S^*(t)S(s)H(s)\,dsdt$$

$$+ \text{tr} \int_0^T R(t)S^*(t)S(t)dt - 2\int_0^T \text{tr}\Lambda(t,T)S(t)H(t)dt. \qquad (1.2.13)$$

We then have the following:

Proposition 1.2.1 *There exists a unique S minimizing the functional $\mathcal{L}(S)$.*

Proof

We provide the set of S with the scalar product

$$(S,\tilde{S}) = \text{tr} \int_0^T S^*(t)S(t)dt.$$

Note that if $e_1,\ldots,e_n$ represents an orthonormal basis of R^n, then

$$\text{tr}\Lambda(s,t)H^*(t)S^*(t)S(s)H(s) = \sum_{i=1}^n e_i^*\Lambda(s,t)H^*(t)S^*(t)S(s)H(s)e_i$$

$$= E\sum_{i=1}^n e_i^*\tilde{x}(s)\tilde{x}^*(t)H^*(t)S^*(t)S(s)H(s)e_i$$

$$= E\,\tilde{x}^*(t)H^*(t)S^*(t)S(s)H(s)\sum_{i=1}^n e_i^*\tilde{x}(s)e_i$$

$$= E\,\tilde{x}^*(t)H^*(t)S^*(t)S(s)H(s)x(s)$$

hence

$$\int_0^T \int_0^T \text{tr}\Lambda(s,t)H^*(t)S(t)S(s)H(s)\,dsdt \geq 0.$$

Next consider that $S(t) = \begin{bmatrix} S_1^*(t) \\ \vdots \\ S_n^*(t) \end{bmatrix}$, where $S_1^*,\ldots,S_n^*$ are n lines.

Then

$$\text{tr}\,R(t)S^*(t)S(t) = \text{tr}\,R(t)\sum_{j=1}^n S_j(t)S_j^*(t)$$

$$= \sum_{j=1}^n S_j^*(t)R(t)S_j(t)$$

$$\geq \alpha\sum_{j=1}^n S_j^*(t)S_j(t)$$

$$= \alpha\,\text{tr}\,S^*(t)S(t).$$

This proves that $\mathcal{L}(S)$ is a quadratic coercive functional on the Hilbert space $L^2(0,T; L(R^m; R^n))$, hence the desired result. ∎

Let us denote by $\hat{S}$ the optimal linear filter. We can check that the best linear filter is also the best filter.

Proposition 2.2.1 *The optimal filter $\hat{x}(T)$ is given by*

$$\hat{x}(T) = \mathcal{F}_{\hat{S}} = \bar{x}(t) + \int_0^T \hat{S}(t)d\tilde{y}(t).$$

Proof

Define the estimate error

$$\epsilon(T) = x(T) - \mathcal{F}_S.$$

Then from the definition (1.2.11) one deduces that

$$E\epsilon(T)^* \int_0^T S(t)d\tilde{y}(t) = 0, \quad \forall S(\cdot).$$

It follows that $\epsilon(T)$ is not correlated with any vector $\tilde{y}(t_1), \ldots, \tilde{y}(t_p), \forall t_1, \ldots, t_p \leq T$, hence is independent of $\mathcal{Y}^T$.

1.2.3   The continuous time Kalman filter

An approach based on the innovation process as in the discrete time case, requires deep representation theorems on martingales, which we shall consider in the general non-linear filtering problem (see Chapter 4). We shall prefer a more elementary method in the present situation.

Define

$$P(T) = E\epsilon(T)\epsilon^*(T) \tag{1.2.15}$$

which is the covariance error. One easily checks that, $\forall \lambda \in R^n$,

$$\lambda^* P(T)\lambda = \min_S E[\lambda^*(x(T) - \mathcal{F}_S)]^2 \tag{1.2.16}$$

and that for any λ, $\hat{S}$ realizes the mininum in (1.2.16). Let us introduce a control problem related to (1.2.16). Let us set

$$-\frac{d\beta}{dt} = F^*(t)\beta(t) - H^*(t)S^*(t)\lambda \tag{1.2.17}$$
$$\beta(T) = \lambda.$$

Then from (1.2.7) and (1.2.17) we deduce

$$d\beta^* \tilde{x} = -(\beta^* F\tilde{x} - \lambda^* SH\tilde{x})dt + \beta^*(F\tilde{x}dt + dw)$$

hence, integrating between 0 and T,

$$\lambda^* \tilde{x}(T) - \beta^*(0)(\xi - \bar{x}_0) = \int_0^T \lambda^* SH\tilde{x}dt + \int_0^T \beta^* dw$$

which implies

$$\lambda^* \left(\tilde{x}(T) - \int_0^T S d\tilde{y} \right) = \beta^*(0)(\xi - \bar{x}_0) + \int_0^T \beta^* dw - \int_0^T \lambda^* S db \qquad (1.2.18)$$

and also

$$E \left[\lambda^*(\tilde{x}(T) - \int_0^T S(t) d\tilde{y}) \right]^2 = \beta^*(0) P_0 \beta(0) + \int_0^T \beta^*(t) Q(t) \beta(t) dt$$

$$+ \int_0^T \lambda^* S(t) R(t) S^*(t) \lambda dt. \qquad (1.2.19)$$

Therefore we can consider the optimal control problem

$$-\frac{d\beta}{dt} = F^*(t)\beta - H^*(t)u$$
$$\beta(T) = \lambda \qquad (1.2.20)$$

$$J(u) = \beta^*(0) P_0 \beta(0) + \int_0^T \beta^*(t) Q(t) \beta(t) dt + \int_0^T u^*(t) R(t) u(t) dt.$$

This optimal control problem admits as solution

$$\hat{u}(t) = \hat{S}^*(t)\lambda. \qquad (1.2.21)$$

The set of necessary conditions of optimality corresponding to (1.2.20) is written as

$$\frac{d\hat{\alpha}}{dt} = F\hat{\alpha} - Q\hat{\beta} \qquad (1.2.22)$$

$$-\frac{d\hat{\beta}}{dt} = F^*\hat{\beta} + H^* R^{-1} H \hat{\alpha}$$

$$\hat{\alpha}(0) = -P_0 \hat{\beta}(0)$$

$$\hat{\beta}(T) = \lambda$$

and

$$\hat{u}(t) = -R(t)^{-1} H(t)\hat{\alpha}(t). \qquad (1.2.23)$$

From (1.2.16), (1.2.19), (1.2.21) we have

$$\lambda^* P(T)\lambda = \hat{\beta}^*(0) P_0 \hat{\beta}(0) + \int_0^T \hat{\beta}^*(t) Q(t) \hat{\beta}(t) dt + \int_0^T \hat{u}^*(t) R(t) \hat{u}(t) dt \qquad (1.2.24)$$

using

$$\frac{d}{dt} \hat{\beta}^* \hat{\alpha} = -(\hat{\beta}^* F + \hat{\alpha}^* H^* R^{-1} H)\hat{\alpha} + \hat{\beta}^*(F\hat{\alpha} - Q\hat{\beta}).$$

Integrating between 0 and T, and comparing with (1.2.24) yields

$$\lambda^* P(T)\lambda = -\lambda^* \hat{\alpha}(T). \qquad (1.2.25)$$

We deduce next an equation for $P(t)$, by the decoupling argument

$$\hat{\alpha}(t) = -P(t)\hat{\beta}(t)$$

which, used in (1.2.22), yields

$$-FP\hat{\beta} - Q\hat{\beta} = -\frac{\mathrm{d}P}{\mathrm{d}t}\hat{\beta} + P(F^*\hat{\beta} - H^*R^{-1}HP\hat{\beta})$$

hence

$$-\frac{\mathrm{d}P}{\mathrm{d}t} + FP + PF^* - PH^*R^{-1}HP + Q = 0 \qquad (1.2.26)$$

$$P(0) = P_0.$$

We can then proceed to obtain a recursive formula for the filter.

Introduce instead of (1.2.22) the system

$$\frac{\mathrm{d}\hat{\gamma}}{\mathrm{d}t} = F\hat{\gamma} - Q\hat{\delta} \qquad (1.2.27)$$

$$-\frac{\mathrm{d}\hat{\delta}}{\mathrm{d}t} = F^*\hat{\delta} + H^*R^{-1}(H\hat{\gamma} - g)$$

$$\hat{\gamma}(0) = -P_0\hat{\delta}(0)$$

$$\hat{\delta}(T) = 0$$

where $g(\cdot)$ is a given input.

The relations (1.2.27) correspond to the set of necessary and sufficient conditions of optimality for the control problem

$$-\frac{\mathrm{d}\delta}{\mathrm{d}t} = F^*\delta - H^*(R^{-1}g + u) \qquad (1.2.28)$$

$$\delta(T) = 0$$

$$J(u) = \delta_0^* P_0 \delta_0 + \int_0^T \delta^* Q\delta \,\mathrm{d}t + \int_0^T u^* Ru \,\mathrm{d}t.$$

The optimal control is given by

$$\hat{u}(t) = -R^{-1}(t)H(t)\hat{\gamma}(t). \qquad (1.2.29)$$

Now comparing (1.2.21) and (1.2.23) we can assert

$$\hat{S}^*(t)\lambda = -R^{-1}(t)H(t)\hat{\alpha}(t)$$

and by transposition

$$\lambda^*\hat{S}(t) = -\hat{\alpha}^*(t)H^*(t)R^{-1}(t). \qquad (1.2.30)$$

Hence using (1.2.27) we can write

$$\int_0^T \lambda^*(t)\hat{S}(t)g(t)\mathrm{d}t = -\int_0^T \hat{\alpha}^*(t)H^*(t)R^{-1}(t)g(t)\mathrm{d}t$$

$$= \int_0^T \hat{\alpha}^* \left(-\frac{\mathrm{d}\hat{\delta}}{\mathrm{d}t} - F^*\hat{\delta} - H^*R^{-1}H\hat{\gamma}\right) dt$$

$$= -\hat{\beta}^*(0)P_0\hat{\delta}(0) + \int_0^T -(\hat{\beta}^* Q\hat{\delta} + \hat{\alpha}^* H^*R^{-1}H\hat{\gamma})dt$$

$$= \hat{\beta}^*(0)\hat{\gamma}(0) + \int_0^T \hat{\beta}^* \left(\frac{\mathrm{d}\hat{\delta}}{\mathrm{d}t} - F\hat{\gamma}\right) dt + \int_0^T \left(\frac{\mathrm{d}\hat{\beta}^*}{\mathrm{d}t} + \hat{\beta}^* F\right) \hat{\gamma}\mathrm{d}t$$

$$= \lambda^*\hat{\gamma}(T).$$

Therefore we have proved that

$$\int_0^T \hat{S}(t)g(t)\mathrm{d}t = \hat{\gamma}(T).$$ (1.2.31)

Now consider the function

$$r(t) = \hat{\gamma}(t) + P(t)\hat{\delta}(t).$$ (1.2.32)

We first notice that

$$r(T) = \hat{\gamma}(T).$$ (1.2.33)

Then

$$\begin{aligned}
\frac{\mathrm{d}r}{\mathrm{d}t} &= \frac{\mathrm{d}\hat{\gamma}}{\mathrm{d}t} + \frac{\mathrm{d}P}{\mathrm{d}t}\hat{\delta} + P\frac{\mathrm{d}\hat{\delta}}{\mathrm{d}t} \\
&= F\hat{\gamma} - Q\hat{\delta} + (FP + PF^* - PH^*R^{-1}HP + Q)\hat{\delta} \\
&\quad - P(F^*\hat{\delta} + H^*R^{-1}H\hat{\gamma} - H^*R^{-1}g) \\
&= (F - PH^*R^{-1}H)r + PH^*R^{-1}g.
\end{aligned}$$

Therefore we have established that

$$r(T) = \int_0^T \hat{S}(t)g(t)\mathrm{d}t$$ (1.2.34)

where $r(T)$ is the solution at T of the differential equation

$$\begin{aligned}
\frac{\mathrm{d}r}{\mathrm{d}t} &= (F - PH^*R^{-1}H)r + PH^*R^{-1}g \\
r(0) &= 0.
\end{aligned}$$ (1.2.35)

Define also for completeness

$$e(t) = \bar{x}(t) + r(t);$$ (1.2.36)

then $e(t)$ is the solution of

$$\begin{aligned}
\frac{\mathrm{d}e}{\mathrm{d}t} &= Fe + f + PH^*R^{-1}(g + H\bar{x} - He) \\
e(0) &= x_0
\end{aligned}$$ (1.2.37)

and

$$e(T) = \bar{x}(T) + \int_0^T \hat{S}(t)g(t)\mathrm{d}t.$$

Comparing with formula (1.2.14) we are led to introducing the stochastic differential equation

$$\begin{aligned}
\mathrm{d}\hat{x} &= (F\hat{x} + f)\mathrm{d}t + PH^*R^{-1}(\mathrm{d}y - (H\hat{x} + h)\mathrm{d}t) \\
\hat{x}(0) &= x_0
\end{aligned}$$ (1.2.38)

and $\hat{x}(T)$ is the optimal filter.

We have then proved the following.

Theorem 1.2.1 *For the model (1.2.1), (1.2.2) the optimal filter $\hat{x}(t)$ (Kalman filter) is given recursively by the equation (1.2.38), where the function $P(t)$ represents the covariance error and is the solution of equation (1.2.26).* ∎

Equation (1.2.26) is a Riccati equation.

1.2.4   The innovation process

The innovation process is defined by

$$I(t) = y(t) - \int_0^T (H\hat{x} + h)\mathrm{d}s. \tag{1.2.39}$$

Then one has the following.

Proposition 1.2.3 *The innovation process is a $\mathcal{Y}^t$ Wiener process with covariance matrix $R(\cdot)$.*

Proof

We have from (1.2.39)

$$\mathrm{d}I = H\,\epsilon\,\mathrm{d}t + \mathrm{d}b$$

hence for $s > t$

$$I(s) - I(t) = \int_t^s H(\lambda)\epsilon(\lambda)\mathrm{d}\lambda + b(s) - b(t)$$

$$E[I(s) - I(t)|\mathcal{Y}^t] = \int_t^s H(\lambda)E[\epsilon(\lambda)|\mathcal{Y}^t]\mathrm{d}\lambda$$
$$= 0.$$

Next

$$\mathrm{d}(I(s) - I(t))(I(s) - I(t))^* = [H(s)\epsilon(s)\mathrm{d}s + \mathrm{d}b][I(s) - I(t)]^*$$
$$+ [I(s) - I(t)][\epsilon^*(s)H^*(s)\mathrm{d}s + \mathrm{d}b]$$
$$+ R(s)\mathrm{d}s$$

hence

$$E[(I(s) - I(t))(I(s) - I(t))^*|\mathcal{Y}^t] = \int_t^s R(\lambda)\mathrm{d}\lambda$$

which completes the proof of the desired property.   ∎

Remark 1.2.1 From the equation (1.2.38) of the Kalman filter we can also write, using the innovation process,

$$\mathrm{d}\hat{x} = (F\hat{x} + f)\mathrm{d}t + PH^*R^{-1}\mathrm{d}I$$

hence $\hat{x}$ is clearly adapted to the filtration $\mathcal{I}^t = \sigma(I(s), s \leq t)$. Therefore from (1.2.38), $\mathcal{Y}^t \subset \mathcal{I}^t$ and since the reverse is also true, we have

$$\mathcal{Y}^t = \mathcal{I}^t.$$

This is not true for general non-linear filtering problems (see Chapter 4).   ∎

2   Optimal stochastic control for linear dynamic systems with quadratic payoff

Introduction

We develop in this chapter the theory of optimal control for linear systems with quadratic payoff. The problem can be deterministic or stochastic in which cases the information is respectively complete or partial. In the case of partial information the observation process has the form which yields the Kalman filter (see 1.1.2).

The general philosophy of the chapter is the following. It is possible to derive the optimal control *directly*, which means in a way that does not rely on a general theory like the Poutryagin maximum principle, or dynamic programming. This direct approach is based on algebraic manipulations, the general idea of which goes back to the search for the minimum of a quadratic function. When minimizing a quadratic function, one can set the gradient equal to zero, which is a necessary condition of optimality, not particularly related to the fact that the function is quadratic. On the other hand, one can also operate by completing squares, and this is an algebraic manipulation which is only possible because of the quadratic character of the payoff. What is interesting, as we show in this chapter, is that this general idea can be used for far more complicated control problems, provided the cost functional is quadratic. See Faurre (1972) for early developments of this kind.

2.1   A brief review of the deterministic problem

2.1.1   Setting of the problem

The system is described by the dynamics

$$\frac{\mathrm{d}x}{\mathrm{d}t} = F(t)x + G(t)v + f(t)$$
$$x(0) = x_0$$

(2.1.1)

where the control $v(\cdot) \in L^2(0,T;R^k)$, $x(\cdot) \in H^1(0,T;R^n)$.

We want to minimize the payoff

$$J(v(\cdot)) = \int_0^T (x^*Mx + v^*Nv + 2mx + 2nv)\mathrm{d}t$$
$$+ x(T)^*M_T x(T) + 2m_T x(T).$$

(2.1.2)

where $Q(\cdot)$ is symmetric non negative, $N(\cdot)$ is symmetric uniformly positive definite. All functions of time F, G, f, M, N, m, n are bounded.

2.1.2   Solution of the problem

The solution of (2.1.1), (2.1.2) is described as follows.

We solve the 'backward' Riccati equation

$$\frac{\mathrm{d}\Pi}{\mathrm{d}t} + \Pi F + F^*\Pi - \Pi G N^{-1} G^*\Pi + M = 0$$
$$\Pi(T) = M_T.$$

(2.1.3)

We then solve the linear equation

$$\frac{\mathrm{d}r}{\mathrm{d}t} + (F^* - \Pi G N^{-1} G^*)r + \Pi(f - GN^{-1}n^*) + m^* = 0$$
$$r(T) = m_T^*.$$

(2.1.4)

We use the feedback

$$\hat{\mathcal{L}}(x;t) = -N^{-1}(t)[n^*(t) + G^*(t)(\Pi(t)x + r(t))]$$

(2.1.5)

in the equation (2.1.1) which yields the state equation

$$\frac{\mathrm{d}y}{\mathrm{d}t} = (F - GN^{-1}G^*\Pi)y + f - GN^{-1}(n^* + G^*r)$$
$$y(0) = x_0.$$

(2.1.6)

The optimal control is then given by

$$u(t) = \hat{\mathcal{L}}(y(t); t). \tag{2.1.7}$$

Theorem 2.1.1 *The optimal control for the problem (2.1.1), (2.1.2) is given by (2.1.7), i.e. via a feedback $\hat{\mathcal{L}}$ given by (2.1.5). The feedback $\hat{\mathcal{L}}$ depends on two parameters $P(\cdot)$, $r(\cdot)$, the solutions respectively of (2.1.3), (2.1.4).*

2.1.3   Proof of Theorem 2.1.1

We make the change of function

$$v(t) = \Lambda(t)x + \lambda(t) + \mu(t) \tag{2.1.8}$$

in which $\Lambda(\cdot)$, $\lambda(\cdot)$ are fixed parameters to be chosen later on, and $\mu(\cdot)$ represents the new control.

We are trying to reduce the situation to one in which the optimal new control is $\hat{\mu} = 0$.

Define $y(\cdot)$ to be the solution of

$$\begin{aligned}
\frac{dy}{dt} &= (F + G\Lambda)y + f + G\lambda \\
y(0) &= x_0
\end{aligned} \tag{2.1.9}$$

and

$$u(t) = \Lambda(t)y(t) + \lambda(t). \tag{2.1.10}$$

For the control $\mu(\cdot)$ the state becomes the solution of

$$\begin{aligned}
\frac{dx}{dt} &= (F + G\Lambda)x + f + G\lambda + G\mu \\
x(0) &= x_0
\end{aligned} \tag{2.1.11}$$

hence

$$x(t) = y(t) + \tilde{x}(t) \tag{2.1.12}$$

where

$$\begin{aligned}
\frac{d\tilde{x}}{dt} &= (F + G\Lambda)\tilde{x} + G\mu \\
\tilde{x}(0) &= 0.
\end{aligned} \tag{2.1.13}$$

Moreover from (2.1.8), (2.1.10) it follows that

$$v(t) = u(t) + \Lambda(t)\tilde{x}(t) + \mu(t). \tag{2.1.14}$$

We can express the payoff as

$$\begin{aligned}
J(v(\cdot)) = K(\mu(\cdot)) =\, &J(u(\cdot)) \\
&+ \int_0^T \Big(\tilde{x}^* M \tilde{x} + (\Lambda \tilde{x} + \mu)^* N (\Lambda \tilde{x} + \mu) \Big) dt + \tilde{x}(T)^* M_T \tilde{x}(T) \\
&+ 2 \int_0^T \Big(\tilde{x}^* M y + (\Lambda \tilde{x} + \mu)^* N u + m \tilde{x} + n(\Lambda \tilde{x} + \mu) \Big) dt \\
&+ 2\tilde{x}(T)^* M_T y(T) + 2m_T \tilde{x}(T).
\end{aligned} \tag{2.1.15}$$

If we pick Λ, λ so that the mixed term in (2.1.15) becomes identically 0, then it will follow that $u(\cdot)$ give n by (2.1.10) is indeed optimal.

Define

$$p(t) = \Pi(t)y(t) + r(t)$$

$$X = \int_0^T \left(\tilde{x}^* My + (\Lambda\tilde{x} + \mu)^* N(\Lambda y + \lambda) + m\tilde{x} + n(\Lambda\tilde{x} + \mu) \right) dt$$
$$+ \tilde{x}^*(T)p(T).$$

We have

$$\frac{d}{dt}\tilde{x}^*(t)p(t) = \tilde{x}^*(F^* + \Lambda^* G^*)p + \mu^* G^* p$$

$$+ \tilde{x}^* \left[\frac{d\Pi}{dt} y + \Pi(F + G\lambda)y + \Pi(f + G\lambda) + \frac{dr}{dt} \right]$$

hence

$$X = \int_0^T \left\{ \tilde{x}^* My + (\Lambda\tilde{x} + \mu)^* N(\Lambda y + \lambda) + m\tilde{x} + n(\Lambda\tilde{x} + \mu) \right.$$
$$+ \tilde{x}^*(F^* + \Lambda^* G^*)(\Pi y + r) + \mu^* G^*(\Pi y + r)$$
$$+ \left. \tilde{x}^* \left[\frac{d\Pi}{dt} y + \Pi(F + G\lambda)y + \Pi(f + G\lambda) + \frac{dr}{dt} \right] \right\} dt.$$

We set the coefficients of $\tilde{x}^*$ and μ^* to zero. This yields the relations

$$My + \Lambda^* N(\Lambda y + \lambda) + m^* + \Lambda^* n^*$$
$$+ (F^* + \Lambda^* G^*)(\Pi y + r) + \frac{d\Pi}{dt} y + \Pi(F + G\Lambda)y \qquad (2.1.16)$$
$$+ \Pi(f + G\lambda) + \frac{dr}{dt} = 0,$$

$$N(\Lambda y + \lambda) + n^* + G^*(\Pi y + r) = 0. \qquad (2.1.17)$$

Matching in (2.1.16), (2.1.17) the coefficient of y and the constant term, we get

$$M + \Lambda^* N\Lambda + (F^* + \Lambda^* G^*)\Pi + \frac{d\Pi}{dt} + \Pi(F + G\Lambda) = 0$$

$$\Lambda^* N\lambda + m^* + \Lambda^* n^* + (F^* + \Lambda^* G^*)r + \Pi(f + G\lambda) + \frac{dr}{dt} = 0$$

$$N\Lambda + G^*\Pi = 0$$

$$N\lambda + n^* + G^* r = 0$$

hence

$$\Lambda(t) = -N^{-1}(t)G^*(t)\Pi(t)$$
$$\lambda(t) = -N^{-1}(t)n^*(t) - N^{-1}(t)G^*(t)r(t) \qquad (2.1.18)$$

and, by the choice of Π and r, the matching can be completed. Therefore $X = 0$, and

$$J(v(\cdot)) = K(\mu(\cdot))$$

$$= J(u(\cdot)) + \int_0^T \left(\tilde{x}^* M\tilde{x} + (\Lambda\tilde{x} + \mu)^* N(\Lambda\tilde{x} + \mu) \right) dt + \tilde{x}(T)^* M_T \tilde{x}(T).$$

The optimal μ is clearly $\mu = 0$, which completes the proof of Theorem 2.1.1. ∎

2.2   Optimal stochastic control with complete observation

2.2.1   Setting of the problem

Let $(\Omega, \mathcal{A}, P)$ be a probability space on which is given

$w(\cdot)$ a Wiener process with values in R^n, with covariance matrix $Q(\cdot)$. (2.2.1)

Let $\mathcal{F}^t = \sigma(w(s), s \leq t)$. The process $w(\cdot)$ will be the unique source of noise in the model, and we assume full information, i.e. $\mathcal{F}^t$ represents the set of observable events at time t.

An admissible control is a process $v(\cdot)$ adapted to $\mathcal{F}^t$, which is square integrable.

Let $v(\cdot)$ be an admissible control; the corresponding state is the solution of

$$\mathrm{d}x = (F(t)x + G(t)v + f(t))\mathrm{d}t + \mathrm{d}w$$
$$x(0) = x_0.$$ (2.2.2)

Our objective is to minimize the payoff

$$J(v(\cdot)) = E\left[\int_0^T (x^*Mx + v^*Nv + 2mx + 2nv)\mathrm{d}t \right.$$
$$\left. + x(T)^*M_T x(T) + 2m_T x(T)\right].$$ (2.2.3)

2.2.2   Solution of the problem

Let us describe the solution of the problem. In fact the optimal control is constructed through the same feedback as in the deterministic case. Indeed consider $\hat{\mathcal{L}}$ defined by (2.1.5). Using this feedback in (2.2.2) we derive the equation

$$\mathrm{d}y = [(F - GN^{-1}G^*\Pi)y + f - GN^{-1}(n^* + G^*r)]\mathrm{d}t + \mathrm{d}w$$
$$y(0) = x_0$$ (2.2.4)

Set

$$u(t) = \hat{\mathcal{L}}(y(t); t)$$ (2.2.5)

then $u(\cdot)$ is an admissible control since $y(\cdot)$ defined by (2.2.4) is adapted to $\mathcal{F}^t$, and we have

Theorem 2.2.1 *The optimal control for the problem (2.2.2), (2.2.3) is given by (2.2.5).*

2.2.3   Proof of Theorem 2.2.1

The method is similar to that of the deterministic case. Let $\mu(\cdot) \in L^2_{\mathcal{F}}$ (square integrable process adapted to the family $\mathcal{F}^t$). Solve the equation

$$dx = [(F + G\Lambda)x + f + G\lambda + G\mu]dt + dw$$
$$x(0) = x_0 \tag{2.2.6}$$

then

$$v(t) = \Lambda(t)x(t) + \lambda(t) + \mu(t) \tag{2.2.7}$$

is admissible.

Conversely, given any $v(\cdot)$, we can define $\mu(\cdot)$ by (2.2.7), where of course the process $x(\cdot)$ is the solution of (2.2.2). It is thus possible to consider $\mu(\cdot)$ as the unknown control variable.

Ignoring for a while the values of Λ, λ derived in the deterministic case, we follow the same route as in the deterministic case. Define y as

$$dy = [(F + G\Lambda)y + f + G\lambda]dt + dw$$
$$y(0) = x_0 \tag{2.2.8}$$

and

$$u(t) = \Lambda(t)y(t) + \lambda(t). \tag{2.2.9}$$

Define next

$$\tilde{x}(t) = x(t) - y(t)$$

which is the solution of

$$\frac{d\tilde{x}}{dt} = (F + G\Lambda)\tilde{x} + G\mu$$
$$\tilde{x}(0) = 0. \tag{2.2.10}$$

Note that the Wiener process $w(\cdot)$ disappears in the equation for $\tilde{x}$. We also have

$$v(t) = u(t) + \Lambda(t)\tilde{x}(t) + \mu(t). \tag{2.2.11}$$

It is clear that (cf. (2.1.15))

$$J(v(\cdot)) = K(\mu(\cdot))$$
$$= J(u(\cdot)) + E\int_0^T \left[\tilde{x}^* M\tilde{x} + (\Lambda\tilde{x} + \mu)^* N(\Lambda\tilde{x} + \mu)\right]dt$$
$$+ E\tilde{x}(T)^* M_T\tilde{x}(T) + 2X \tag{2.2.12}$$

with

$$X = E\int_0^T \left[\tilde{x}^* My + (\Lambda\tilde{x} + \mu)^* N(\Lambda y + \lambda) + m\tilde{x} + n(\Lambda\tilde{x} + \mu)\right]dt$$
$$+ E[\tilde{x}(T)^* M_T y(T) + m_T\tilde{x}(T)].$$

Defining again

$$p(t) = \Pi(t)y(t) + r(t) \tag{2.2.13}$$

we have

$$d\tilde{x}^*(t)p(t) = \left\{ \tilde{x}^*(F^* + \Lambda^*G^*)p + \mu^*G^*p + \tilde{x}^* \left(\frac{d\Pi}{dt}y + \frac{dr}{dt} \right) \right.$$
$$\left. + \tilde{x}^*\Pi[(F + G\Lambda)y + f + G\lambda] \right\} dt + \tilde{x}^*\Pi dw.$$

Integrating between 0 and T, and taking the mathematical expectation, the stochastic integral vanishes, and we are led to the same algebraic conditions as in the deterministic case. This concludes the proof of Theorem 2.2.1. ∎

2.4   A more elaborate model with control on the diffusion term

We shall now consider the model

$$dx = (Fx + Gv + f)dt + \sum_{i=1}^{d}(B_ix + D_iv + g_i)dw_i + dw \tag{2.2.14}$$
$$x(0) = x_0$$

where w is a Wiener process with covariance matrix Q, and $w_1, \ldots, w_d$ are scalar standard Wiener processes. To simplify we assume that all these Wiener processes are mutually independent.

The cost function is still defined by the expression (2.2.3).

This problem has been considered by Wonham (1968) and Bismut (1977). Let us describe the solution of this problem. Consider the Riccati equation

$$\frac{d\Pi}{dt} + F^*\Pi + \Pi F + \sum_{i=1}^{d} B_i^*\Pi B_i + M$$
$$- \left(\Pi G + \sum_{i=1}^{d} B_i^*\Pi D_i \right) \left(N + \sum_{i=1}^{d} D_i^*\Pi D_i \right)^{-1} \left(G^*\Pi + \sum_{i=1}^{d} D_i^*\Pi B_i \right) = 0 \tag{2.2.15}$$
$$\Pi(T) = M_T$$

and the function $r(\cdot)$, solution of

$$\frac{dr}{dt} + \left[F^* - \left(\Pi G + \sum_i B_i^*\Pi D_i \right) \left(N + \sum_i D_i^*\Pi D_i \right)^{-1} G^* \right] r + m^* + \Pi f$$
$$+ \sum_i B_i^*\Pi g_i - \left(\Pi G + \sum_i B_i^*\Pi D_i \right) \left(N + \sum_i D_i^*\Pi D_i \right)^{-1} \left(n^* + \sum_i D_i^*\Pi g_i \right) = 0$$
$$r(T) = m_T^*. \tag{2.2.16}$$

Define the feedback

$$\hat{\mathcal{L}}(x;t) = -\left(N + \sum_{i=1}^{d} D_i^*\Pi D_i\right)^{-1}\left[\left(G^*\Pi + \sum_{i=1}^{d} D_i^*\Pi B_i\right)x \right.$$
$$\left. +n^* + G^*r + \sum_{i=1}^{d} D_i^*\Pi g_i\right]. \tag{2.2.17}$$

We can then state

Theorem 2.2.2 *For the problem (2.2.14), (2.2.3) the feedback (2.2.17) is optimal.*

By optimal feedback we mean that inserting $\hat{\mathcal{L}}$ into (2.2.14) in place of v, and noting $y(\cdot)$, the corresponding solution, the expression

$$u(t) = \hat{\mathcal{L}}(y(t);t)$$

defines an optimal control.

2.2.5   Proof of Theorem 2.2.2

We use the same methodology. Let us make the change of control process

$$v(t) = \Lambda(t)x(t) + \lambda(t) + \mu(t) \tag{2.2.18}$$

and the state equation becomes

$$dx = [(F + G\Lambda)x + f + G\lambda + G\mu]dt$$
$$+ \sum_{i=1}^{d}[(B_i + D_i\Lambda)x + D_i\lambda + D_i\mu + g_i]dw_i + dw \tag{2.2.19}$$
$$x(0) = x_0$$

Denote by $y(\cdot)$ the state corresponding to $\mu = 0$, i.e. the solution of

$$dy = [(F + G\Lambda)y + f + G\lambda]dt$$
$$+ \sum_{i=1}^{d}[(B_i + D_i\lambda)y + D_i\lambda + g_i]dw_i + dw \tag{2.2.20}$$
$$y(0) = x_0$$

and the corresponding optimal control is

$$u(t) = \Lambda(t)y(t) + \lambda(t). \tag{2.2.21}$$

We write

$$x = y + \tilde{x}$$

thus $\tilde{x}$ is the solution of

$$d\tilde{x} = [(F + G\Lambda)\tilde{x} + G\mu]dt + \sum_{i=1}^{d}[(B_i + D_i\Lambda)\tilde{x} + D_i\mu]dw_i$$
$$\tilde{x}(0) = 0 \tag{2.2.22}$$
$$v(t) = \Lambda(t)\tilde{x}(t) + \mu(t) + u(t).$$

We can write the same expression as (2.2.12). It remains to check whether it is possible to pick Λ, λ to achieve $X = 0$.

Defining again $p(\cdot)$ by (2.2.13) we now have

$$
\begin{aligned}
\mathrm{d}\tilde{x}^* p =& [\tilde{x}^*(F^* + \Lambda^* G^*)p + \mu^* G^* p]\mathrm{d}t \\
&+ \sum_{i=1}^{d}[\tilde{x}^*(B_i^* + \Lambda^* D_i^*)p + \mu^* D_i^* p]\mathrm{d}w_i \\
&+ \tilde{x}^*\left(\frac{\mathrm{d}\Pi}{\mathrm{d}t}y + \frac{\mathrm{d}r}{\mathrm{d}t}\right)\mathrm{d}t + \tilde{x}^*\Pi[(F + G\Lambda)y + f + G\lambda]\mathrm{d}t \\
&+ \sum_{i=1}^{d}\tilde{x}^*\Pi[(B_i + D_i\Lambda)y + D_i\lambda + g_i]\mathrm{d}w_i + \tilde{x}^*\Pi\mathrm{d}w \\
&+ \sum_{i=1}^{d}[\tilde{x}^*(B_i^* + \Lambda^* D_i^*) + \mu^* D_i^*]\Pi[(B_i + D_i\Lambda)y + D_i\lambda + g_i].
\end{aligned}
$$

Collecting results we can express

$$
\begin{aligned}
X = E\int_0^T \Big\{ & \tilde{x}^* My + (\Lambda\tilde{x} + \mu)^* N(\Lambda y + \lambda) + m\tilde{x} \\
& + n(\Lambda\tilde{x} + \mu) + \tilde{x}^*(F^* + \Lambda^* G^*)(\Pi y + r) + \mu^* G^*(\Pi y + r) \\
& + \tilde{x}^*\left(\frac{\mathrm{d}\Pi}{\mathrm{d}t}y + \frac{\mathrm{d}r}{\mathrm{d}t}\right) + \tilde{x}^*\Pi(F + G\Lambda)y + \tilde{x}^*\Pi(f + G\lambda) \qquad (2.2.23) \\
& + \sum_{i=1}^{d}[\tilde{x}^*(B_i^* + \Lambda^* D_i^*) + \mu^* D_i^*]\Pi[(B_i + D_i\Lambda)y + D_i\lambda + g_i] \Big\}\mathrm{d}t.
\end{aligned}
$$

To achieve $X = 0$ we equate to 0 the coefficients of $\tilde{x}^*$ and μ^*, which yields

$$
\begin{aligned}
& My + \Lambda^* N(\Lambda y + \lambda) + m^* + \Lambda^* n^* + (F^* + \Lambda^* G^*)(\Pi y + r) \\
& + \frac{\mathrm{d}\Pi}{\mathrm{d}t}y + \frac{\mathrm{d}r}{\mathrm{d}t} + \Pi(F + G\Lambda)y + \Pi(f + G\lambda) \\
& + \sum_{i=1}^{d}(B_i^* + \Lambda^* D_i^*)\Pi[(B_i + D_i\Lambda)y + D_i\lambda + g_i] = 0
\end{aligned}
$$
$$(2.2.24)$$

$$
\begin{aligned}
& N(\Lambda y + \lambda) + n^* + G^*(\Pi y + r) \\
& + \sum_{i=1}^{d} D_i^*\Pi[(B_i + D_i\Lambda)y + D_i\lambda + g_i] = 0.
\end{aligned}
$$

These conditions split again by setting to 0 the coefficients of $y(\cdot)$ and the terms independent of y, giving

$$
\begin{aligned}
& M + \Lambda^* N\Lambda + (F^* + \Lambda^* G^*)\Pi + \frac{\mathrm{d}\Pi}{\mathrm{d}t} \\
& + \Pi(F + G\Lambda) + \sum_{i=1}^{d}(B_i^* + \Lambda^* D_i^*)\Pi(B_i + D_i\Lambda) = 0
\end{aligned}
$$
$$(2.2.25)$$

$$\Lambda^* N \lambda + m^* + \Lambda^* n^* + (F^* + \Lambda^* G^*)r + \frac{dr}{dt}$$

$$+ \Pi(f + G\lambda) + \sum_{i=1}^{d}(B_i^* + \Lambda^* D_i^*)\Pi(D_i \lambda + g_i) = 0 \qquad (2.2.26)$$

$$N\Lambda + G^* \Pi + \sum_{i=1}^{d} D_i^* \Pi(B_i + D_i \Lambda) = 0 \qquad (2.2.27)$$

$$N\lambda + n^* + G^* r + \sum_{i=1}^{d} D_i^* \Pi(D_i \lambda + g_i) = 0. \qquad (2.2.28)$$

Therefore we deduce

$$\Lambda = -\left(N + \sum_{i=1}^{d} D_i^* \Pi D_i\right)^{-1}\left(G^* \Pi + \sum_{i=1}^{d} D_i^* \Pi B_i\right)$$

$$\lambda = -\left(N + \sum_{i=1}^{d} D_i^* \Pi D_i\right)^{-1}\left(n^* + G^* r + \sum_{i=1}^{d} D_i^* \Pi g_i\right). \qquad (2.2.29)$$

Inserting those expressions in (2.2.25), (2.2.26) the desired result follows. ∎

2.3 Optimal stochastic control with partial information: simplified approach

2.3.1 Description of the model

Let $(\Omega, \mathcal{A}, P)$ be a probability space on which are given

ξ, a random variable with gaussian probability

law, with mean x_0 and covariance matrix P_0, $w(\cdot)$, $b(\cdot)$ Wiener

processes with covariance matrices $Q(\cdot)$, $R(\cdot)$ respectively.   (2.3.1)

The matrix $R(\cdot)$ is uniformly positive definite.

The variable ξ and the processes $w(\cdot)$ $b(\cdot)$ are mutually independent.

Define the processes $\alpha(\cdot)$, $\beta(\cdot)$ by

$$
\begin{aligned}
\mathrm{d}\alpha &= [F(t)\alpha + f(t)]\mathrm{d}t + \mathrm{d}w \\
\alpha(0) &= \xi \\
\mathrm{d}\beta &= [H(t)\beta + h(t)]\mathrm{d}t + \mathrm{d}b \\
\beta(0) &= 0
\end{aligned}
\tag{2.3.2}
$$

The processes α, β correspond to the state and observation processes when there is no control (more precisely the control is 0 as will be seen).

Let $v(t) \in L^2((0,T) \times \Omega, \mathrm{d}t \otimes \mathrm{d}P; R^k)$ (any square integrable process, no adaptation whatsoever is required at this stage).

Define x_1, z_1 by

$$
\begin{aligned}
\frac{\mathrm{d}x_1}{\mathrm{d}t} &= Fx_1 + Gv \\
x_1(0) &= 0
\end{aligned}
\tag{2.3.3}
$$

$$
\begin{aligned}
\frac{\mathrm{d}z_1}{\mathrm{d}t} &= Hx_1 \\
z_1(0) &= 0.
\end{aligned}
\tag{2.3.4}
$$

The relations (2.3.3), (2.3.4) are solved pointwise in w (i.e. for each sample).

To any control $v(\cdot)$ (not necessarily admissible, see below) we define

$$
\begin{aligned}
x(t) &= \alpha(t) + x_1(t) \\
z(t) &= \beta(t) + z_1(t).
\end{aligned}
\tag{2.3.5}
$$

We shall say that $x(\cdot)$ is the state corresponding to the control $v(\cdot)$, and $z(\cdot)$ is the corresponding observation process.

We consider the family of σ-algebras

$$\mathcal{Z}_v^t = \sigma(z(s), \ s \le t). \tag{2.3.6}$$

As is apparent we have emphasized in the notation the fact that the observation process depends on the control.

A natural definition of admissibility is

$$v(\cdot) \text{ admissible } \Leftrightarrow v(t) \text{ is adapted to } \mathcal{Z}_v^t. \tag{2.3.7}$$

This definition expresses the fact that we want the control to be determined by observing the observation process. However, there is an immediate difficulty because the observation process depends on the control. This is why the definition is split in two parts, as explained above.

Let us introduce the family of σ-algebras

$$\mathcal{F}^t = \sigma(\beta(s), \ s \le t). \tag{2.3.7'}$$

By virtue of the second relation (2.3.5), we see at once that

$$\text{if } v(\cdot) \text{ is admissible then } \mathcal{F}^t \subset \mathcal{Z}_v^t. \tag{2.3.8}$$

Define the payoff

$$J(v(\cdot)) = E\left[\int_0^T (x^* M x + v^* N v + 2mx + 2nv)\mathrm{d}t \right.$$
$$\left. + x(T)^* M_T x(T) + 2m_T x(T) \right]. \tag{2.3.9}$$

Then we can state the problem: minimize the payoff (2.3.9) among the set of admissible controls.

2.3.2   A more restrictive definition of admissibility

We shall restrict the set of admissible controls, which will facilitate the obtaining of an optimal control. More precisely we shall consider

$$v(\cdot) \text{ admissible } \Leftrightarrow v(t) \text{ is adapted to } \mathcal{F}^t \text{ and to } \mathcal{Z}_v^t. \tag{2.3.10}$$

With this definition of admissibility, it is immediate from the second of the relations (2.3.5) that

$$\text{if } v(\cdot) \text{ is admissible, then } \mathcal{F}^t = \mathcal{Z}_v^t. \tag{2.3.11}$$

This property makes it easy to obtain the Kalman filter, i.e. find an equation for the evolution of

$$\hat{x}(t) = E[x(t)|\mathcal{Z}_v^t] = E[x(t)|\mathcal{F}^t]. \tag{2.3.12}$$

Firstly x has a stochastic differential

$$dx = (Fx + f + Gv)dt + dw$$
$$x(0) = \xi$$

$$(2.3.13)$$

and the observation z satisfies

$$dz = (Hx + h)dt + db$$
$$z(0) = 0.$$

$$(2.3.14)$$

We then have

Proposition 2.3.1 *If $v(\cdot)$ is admissible in the sense (2.3.10), then the Kalman filter $\hat{x}(t)$ is the solution of*

$$d\hat{x} = (F\hat{x} + f + Gv)dt + PH^*R^{-1}(dz - (H\hat{x} + h)dt)$$
$$\hat{x}(0) = x_0$$

$$(2.3.15)$$

where P is the solution of

$$-\frac{dP}{dt} + FP + PF^* - PH^*R^{-1}HP + Q = 0$$
$$P(0) = P_0.$$

$$(2.3.16)$$

Proof

From (2.3.5) we have

$$\hat{x}(t) = x_1(t) + E[\alpha(t)|\mathcal{F}^t]$$
$$= x_1(t) + \hat{\alpha}(t)$$

$$(2.3.17)$$

and $\hat{\alpha}$ is the Kalman filter for the system (2.3.2). It has been defined in § 1.2.3; it is then given by the equation

$$d\hat{\alpha} = (F\hat{\alpha} + f) = dt + PH^*R^{-1}(d\beta - (H\hat{\alpha} + h)dt)$$
$$\hat{\alpha}(0) = x_0$$

$$(2.3.18)$$

where P is the solution of (2.3.16). Using (2.3.17) we deduce

$$d\hat{x} = (F\hat{x} + f + Gv)dt + PH^*R^{-1}[d\beta - (H\hat{\alpha} + h)dt]$$

But from (2.3.5)

$$d\beta - (H\hat{\alpha} + h)dt = dz - Hx_1dt - (H\hat{\alpha} + h)dt$$
$$= dz - (H\hat{x} + h)dt$$

hence (2.3.15). ∎

Let us introduce an important class of admissible controls, those defined by a linear feedback on the filter. More precisely, consider the innovation process

$$I(t) = \beta(t) - \int_0^t (H\hat{\alpha} + h)ds$$
$$= z(t) - \int_0^t (H\hat{x} + h)ds.$$

We know (cf. Proposition 1.2.3) that $I(t)$ is a $\mathcal{F}^t$ Wiener process with covariance matrix $R(\cdot)$. Let $\Lambda(\cdot)$, $\lambda(\cdot)$ be the parameters of a feedback

$$\Lambda(t)x + \lambda(t);$$

we can consider the stochastic differential equation

$$\begin{aligned} d\eta &= [(F + G\Lambda)\eta + f + G\lambda]dt + PH^*R^{-1}dI \\ \eta(0) &= x_0. \end{aligned} \tag{2.3.19}$$

The process η is clearly adapted to $\mathcal{F}^t$. Define

$$v(t) = \Lambda(t)\eta(t) + \lambda(t) \tag{2.3.20}$$

then the process $v(\cdot)$ is square integrable and adapted to $\mathcal{F}^t$.

We can consider the state $x(\cdot)$ and the observation $z(\cdot)$ associated to the control $v(\cdot)$ (2.3.20). Let us set

$$\begin{aligned} \hat{x}(t) &= E[x(t)|\mathcal{F}^t] \\ &= \hat{\alpha}(t) + x_1(t). \end{aligned}$$

We deduce for $\hat{x}$ the equation

$$\begin{aligned} d\hat{x} &= (F\hat{x} + f + Gv)dt + PH^*R^{-1}(d\beta - (H\hat{\alpha} + h)dt) \\ &= (F\hat{x} + G\Lambda\eta + G\lambda + f)dt + PH^*R^{-1}dI \\ \hat{x}(0) &= x_0. \end{aligned}$$

Comparing with (2.3.19) we deduce that

$$\eta(t) = E[x(t)|\mathcal{F}^t] = \hat{x}(t). \tag{2.3.21}$$

Moreover the observation z satisfies

$$dz = d\beta + Hx_1 dt$$

hence

$$dI = d\beta - (H\hat{\alpha} + h)dt = dz - (H\eta + h)dt.$$

Therefore equation (2.3.19) can be rewritten as

$$\begin{aligned} d\eta &= [(F + G\Lambda)\eta + f + G\lambda]dt + PH^*R^{-1}(dz - (H\eta + h)dt) \\ \eta(0) &= x_0. \end{aligned} \tag{2.3.22}$$

Under this form the process z appears as the unique source of noise driving (2.3.22). From the linearity of the equation it is easy to show that

$$\eta(t) \text{ is adapted to } \mathcal{Z}_v^t \tag{2.3.23}$$

(use a representation formula for η and an approximation scheme for the stochastic integral involving z). It follows from (2.3.23) and (2.3.20) that $v(t)$ is adapted to $\mathcal{Z}_v^t$. Since it is already adapted to $\mathcal{F}^t$, it is admissible.

Naturally $\eta(t)$ is the corresponding Kalman filter, and the feedback (2.3.20) appears as a feedback on the Kalman filter.

2.3.3   Optimality

Consider the values of Λ and λ which have been used for the full observation case (cf (2.1.5)), i.e

$$\begin{aligned}
\Lambda(t) &= -N^{-1}(t)G^*(t)\Pi(t) \\
\lambda(t) &= -N^{-1}[n^*(t) + G^*(t)r(t)]
\end{aligned} \tag{2.3.24}$$

where $\Pi(t)$, $r(t)$ are defined by (2.1.3), (2.1.4) (the same as in the full observation case).

Our objective is to prove

Theorem 2.3.1 *For the problem whose state and observation are defined by (2.3.2), (2.3.3), (2.3.4), (2.3.5) and for which the payoff is defined by (2.3.9), the control obtained through the feedback (2.3.24) in the way explained in (2.3.19), (2.3.20) is optimal among the set of admissible controls in the sense (2.3.10).*

Proof

Let Λ, λ be unspecified for the moment. We shall denote by $u(\cdot)$ the corresponding control, $y(\cdot)$ the corresponding state, $\xi(\cdot)$ the corresponding observation, and $\hat{y}(\cdot)$ the corresponding Kalman filter.

The following relations hold:

$$\begin{aligned}
\mathrm{d}\hat{y} &= [(F + G\Lambda)\hat{y} + f + G\lambda]\mathrm{d}t \\
&\quad + PH^*R^{-1}(\mathrm{d}\zeta - (H\hat{y} + h)\mathrm{d}t), \quad \hat{y}(0) = x_0 \\
u(t) &= \Lambda(t)\hat{y}(t) + \lambda(t) \\
\mathrm{d}y &= (Fy + f + Gu)\mathrm{d}t + \mathrm{d}w, \quad y(0) = \xi \\
\mathrm{d}\zeta &= (Hy + h)\mathrm{d}t + \mathrm{d}b, \quad \zeta(0) = 0.
\end{aligned} \tag{2.3.25}$$

Let us check that, as in the full observation case, we can make a change of control function.

Let $\mu(\cdot)$ be adapted to $\mathcal{F}^t$. Consider the equation

$$\begin{aligned}
\mathrm{d}\eta &= [(F + G\Lambda)\eta + f + G\lambda + G\mu]\mathrm{d}t + PH^*R^{-1}\mathrm{d}I \\
\eta(0) &= x_0.
\end{aligned} \tag{2.3.26}$$

The process η defined in this way is adapted to $\mathcal{F}^t$. Set

$$v(t) = \Lambda(t)\eta(t) + \lambda(t) + \mu(t) \tag{2.3.27}$$

then $v(\cdot)$ is $\mathcal{F}^t$ adapted. Consider the state $x(\cdot)$ and the observation $z(\cdot)$ corresponding to $v(\cdot)$; we deduce as in the end of the preceding paragraph that

$$\eta(t) = E[x(t)|\mathcal{F}^t] \tag{2.3.28}$$

and

$$dI = dz - (H\eta + h)dt. \tag{2.3.29}$$

If we impose, in addition the condition that the process $\mu(\cdot)$ be adapted to $\mathcal{Z}_v^t$, then from (2.3.29) and the form of the equation (2.3.26) we deduce that $\eta(t)$ is adapted to $\mathcal{Z}_v^t$, hence from (2.3.27) $v(\cdot)$ is admissible, and

$$\eta(t) = \hat{x}(t) = E[x(t)|\mathcal{Z}_v^t]. \tag{2.3.30}$$

On the other hand given an admissible control $v(\cdot)$, the process $\mu(\cdot)$ defined by (2.3.27) in which η is substituted for $\hat{x}$ is also $\mathcal{F}^t$ and $\mathcal{Z}_v^t$ measurable. Therefore the transformation (2.3.27) is always possible, without loss of generality. We can thus write for any $\mu(\cdot)$

$$\begin{aligned}
d\hat{x} =& [(F + G\Lambda)\hat{x} + f + G\lambda + G\mu]dt \\
&+ PH^*R^{-1}[dz - (H\hat{x} + h)dt], \\
\hat{x} =& x_0 \\
v(t) =& \Lambda(t)\hat{x}(t) + \lambda(t) + \mu(t) \\
dx =& (Fx + f + Gv)dt + dw, \quad x(0) = \xi \\
dz =& (Hx + h)dt + db, \quad z(0) = 0.
\end{aligned} \tag{2.3.31}$$

Comparing (2.3.31) and (2.3.25) (which corresponds to (2.3.31) with $\mu = 0$), it is useful to notice that

$$\begin{aligned}
dz - (H\hat{x} + h)dt =& d\zeta - (H\hat{y} + h)dt \\
=& d\beta - (H\hat{\alpha} + h)dt.
\end{aligned} \tag{2.3.32}$$

Set

$$\tilde{x}(t) = \hat{x}(t) - \hat{y}(t); \tag{2.3.33}$$

then by virtue of (2.3.32), we derive from (2.3.25), (2.3.31)

$$\begin{aligned}
\frac{d\tilde{x}}{dt} =& (F + G\Lambda)\tilde{x} + G\mu \\
\tilde{x}(0) =& 0.
\end{aligned} \tag{2.3.34}$$

Moreover using

$$\begin{aligned}
\hat{x}(t) =& \hat{\alpha}(t) + x_1(t) \\
\hat{y}(t) =& \hat{\alpha}(t) + y_1(t)
\end{aligned}$$

and

$$\begin{aligned}
\frac{d}{dt}(x_1 - y_1) =& F(x_1 - y_1) + G(v - u) \\
=& F(x_1 - y_1) + G\Lambda(x_1 - y_1) + G\mu
\end{aligned}$$

we have

$$x_1(0) - y_1(0) = 0.$$

Therefore we have proved that

$$\hat{x}(t) - \hat{y}(t) = x_1(t) - y_1(t) = x(t) - y(t) = \tilde{x}(t).$$

We can then compute the value of the cost functional

$$
\begin{aligned}
J(v(\cdot)) = K(\mu(\cdot)) = E\Bigg\{ \int_0^T &[(y(t) + \tilde{x}(t))^* M(y(t) + \tilde{x}(t)) \\
&+ (\Lambda\hat{y} + \lambda + \Lambda\tilde{x} + \mu)^* N(\Lambda\hat{y} + \lambda + \Lambda\tilde{x} + \mu) \\
&+ 2m(y + \tilde{x}) + 2n(\Lambda\hat{y} + \lambda + \Lambda\tilde{x} + \mu)]\mathrm{d}t \\
&+ (y(T) + \tilde{x}(T))^* M_T(y(T) + \tilde{x}(T)) \\
&+ 2m_T(y(T) + \tilde{x}(T)) \Bigg\}.
\end{aligned}
$$

Hence

$$
\begin{aligned}
J(v(\cdot)) = J(u(\cdot)) + E\Bigg\{ \int_0^T &[\tilde{x}^* M\tilde{x} + (\Lambda\tilde{x} + \mu)^* N(\Lambda\tilde{x} + \mu)]\mathrm{d}t \\
&+ \tilde{x}(T)^* M_T\tilde{x}(T) \Bigg\} + 2X
\end{aligned}
\tag{2.3.35}
$$

where

$$
\begin{aligned}
X = E\Bigg\{ \int_0^T &[\tilde{x}^*(t)My(t) + (\Lambda\tilde{x} + \mu)^* N(\Lambda\hat{y} + \lambda) + m\tilde{x} + n(\Lambda\tilde{x} + \mu)]\mathrm{d}t \\
&+ \tilde{x}^*(T)M_Ty(T) + m_T\tilde{x}(T) \Bigg\}.
\end{aligned}
$$

Noticing that

$$
\begin{aligned}
E\tilde{x}^*(t)My(t) &= E\{\tilde{x}^*(t)ME[y(t)|\mathcal{F}^t]\} \\
&= E\tilde{x}^*(t)M\hat{y}(t)
\end{aligned}
$$

the expression of X becomes identical to that obtained in (2.2.12). Therefore choosing Λ and λ as in (2.3.24) we can guarantee $X = 0$, and thus prove the optimality of $u(\cdot)$.

∎

2.4 Complete solution of the optimal stochastic control problem with partial information

2.4.1 Approach using Girsanov transformation

2.4.1.1 The model

We have emphasized in the preceding paragraphs the intrinsic difficulty arising from the fact that the control depends on the observation, which itself is dependent on the control. The approach via a change of probability offers a way to overcome this difficulty.

The idea is as follows. On a convenient probability space $\Omega, \mathcal{A}, P$, let us assume the existence of

ξ, a random variable with gaussian probability law with mean x_0

and covariance matrix P_0, $w(\cdot)$, a Wiener process with covariance $Q(\cdot)$;

ξ is independent of $w(\cdot)$. $\qquad (2.4.1)$

$z(\cdot)$, a Wiener process with covariance $R(\cdot)$, independent of $\xi, w(\cdot)$.

We assume $R(t) \geq \rho I$, $\rho > 0$, $\forall t \in [0, T]$. We define

$$F^t = \sigma(\xi, w(s), z(s), s \leq t)$$

$$\mathcal{Z}^t = \sigma(z(s), s \leq t).$$

Define admissible controls as being square integrable processes $v(\cdot)$, adapted to this family $\mathcal{Z}^t$ (note that in this framework $\mathcal{Z}^t$ is fixed).

More precisely

$$\mathcal{U} = L^2_{\mathcal{Z}}(0, T; R^k) \qquad (2.4.2)$$

We recall that the notation $L^2_{\mathcal{Z}}$ means the space of square integrable processes, adapted to the family $\mathcal{Z}^t$. For any admissible control $v(\cdot)$, define the state $x(\cdot)$ by

$$\mathrm{d}x = (F(t)x + G(t)v + f(t))\mathrm{d}t + \mathrm{d}w$$
$$x(0) = \xi. \qquad (2.4.3)$$

The process $x(\cdot)$ is adapted to F^t, and of course depends on $v(\cdot)$.

Let us next define

$$\eta^v(t) = \exp\left[\int_0^t (x^*H^* + h^*)R^{-1}\mathrm{d}z - \frac{1}{2}\int_0^t (x^*H^* + h^*)R^{-1}(Hx + h)\mathrm{d}s\right] \quad (2.4.4)$$

$$b^v(t) = z(t) - \int_0^t (Hx(s) + h)\mathrm{d}s. \quad (2.4.5)$$

The index $v = v(\cdot)$ means that the processes η, b depend on $v(\cdot)$ (via of course the state $x(\cdot)$).

The process η^v satisfies

$$\mathrm{d}\eta^v = \eta^v(x^*H^* + h^*)R^{-1}\mathrm{d}z, \quad \eta^v(0) = 1. \quad (2.4.6)$$

At this stage we can guarantee only that

$$E\eta^v(t) \le 1, \quad (2.4.7)$$

(cf. Gikhman and Skorokhod 1972).

Therefore $\eta^v(T)$ does not represent the Radon–Nikodym derivative of a probability P^v such that

$$\frac{\mathrm{d}P^v}{\mathrm{d}P} = \eta^v(T). \quad (2.4.8)$$

It would be possible to introduce restrictions on the controls $v(\cdot)$ in order to achieve the integrability condition

$$E\eta^v(T) = 1. \quad (2.4.9)$$

But these are not relevant for the control problem we have in mind. By using adequate terminology, we shall not need explicitly the property that the *measure* P^v (which is at any rate a positive finite measure) be a probability.

Let us write

$$x(t) = \alpha(t) + x_1(t) \quad (2.4.10)$$

with the definitions (2.3.2), (2.3.3). We shall define the cost function

$$J(v(\cdot)) = E\int_0^T \eta^v(t)(x^*Mx + v^*Nv + 2mx + 2nv)\mathrm{d}t \quad (2.4.11)$$
$$+ E\eta^v(T)(x(T)^*M_Tx(T) + 2m_Tx(T))$$

possibly accepting the value $+\infty$. Let us check that there cannot be ambiguity in the right hand side of (2.4.11) (no term of the form $+\infty - \infty$). Firstly

$$|x_1(t)| \le C\left(\int_0^T |v|^2\mathrm{d}s\right)^{1/2}. \quad (2.4.12)$$

Therefore, by virtue of the invertibility of N, the quantity

$$E\int_0^T \eta^v(t)(x^*Mx + v^*Nv + 2mx_1 + 2nv)\mathrm{d}t$$
$$+ E\eta^v(T)(x(T)^*M_Tx(T) + 2m_Tx_1(T))$$

is not ambiguous (although it can take the value $+\infty$). On the other hand by the next lemma, one has

$$E\eta^v(t)|\alpha(t)|^2 < C.$$

Using the definition (2.4.10) and splitting the right hand side of (2.4.11) into two parts, the non-ambiguity has been proven. The following result has been used.

Lemma 2.4.1 *The following estimate holds:*

$$E\eta^v(t)|\alpha(t)|^4 \leq C \tag{2.4.13}$$

with a constant independent of the control.

Proof

Consider the following approximation of η^v,

$$\eta_\epsilon^v(t) = \exp\left[\int_0^t (x_\epsilon^* H^* + h^*)R^{-1}\mathrm{d}z - \frac{1}{2}\int_0^t (x_\epsilon^* H^* + h^*)R^{-1}(Hx_\epsilon + h)\mathrm{d}s\right]$$

with

$$x_\epsilon(t) = \frac{x(t)}{(1 + \epsilon|x(t)|^2)^{1/2}}.$$

There exists a subsequence, still denoted by ϵ, such that

$$\text{almost surely (a.s.)}\quad \eta_\epsilon^v(t) \to \eta^v(t)\quad \forall t,$$

Now

$$\mathrm{d}|\alpha(t)|^4 = 4|\alpha(t)|^2\alpha^*[(F\alpha + f)\mathrm{d}t + \mathrm{d}w]$$
$$+ 2[|\alpha(t)|^2\mathrm{tr}\,Q + 2\alpha^*Q\alpha]\mathrm{d}t$$

and

$$\mathrm{d}\eta_\epsilon^v(t)|\alpha(t)|^4 = 4\eta_\epsilon^v(t)|\alpha(t)|^2\alpha^*[(F\alpha + f)\mathrm{d}t + \mathrm{d}w]$$
$$+ 2\eta_\epsilon^v(t)[|\alpha(t)|^2\mathrm{tr}\,Q + 2\alpha^*Q\alpha]\mathrm{d}t$$
$$+ \eta_\epsilon^v(t)|\alpha(t)|^4(x_\epsilon^* H^* + h^*)R^{-1}\mathrm{d}z.$$

Integrating between 0 and t and taking the mathematical expectation, we deduce

$$E\eta_\epsilon^v(t)|\alpha(t)|^4 = E|\xi|^4 + 4E\int_0^t \eta_\epsilon^v(s)|\alpha(s)|^2\alpha^*(s)(F\alpha + f)\mathrm{d}s$$
$$+ 2E\int_0^t \eta_\epsilon^v(s)[|\alpha(s)|^2\mathrm{tr}\,Q + 2\alpha^*Q\alpha]\mathrm{d}s$$

Using $E\eta_\epsilon^v(s) = 1\quad \forall s$, we easily obtain

$$E\eta_\epsilon^v(t)|\alpha(t)|^4 \leq E|\xi|^4 + C\left(1 + E\int_0^t \eta_\epsilon^v(s)|\alpha(s)|^4\mathrm{d}s\right)$$

where C is a constant independent of ϵ and $v(\cdot)$.

By Gronwall's lemma, it follows that

$$E\eta_\epsilon^v(t)|\alpha(t)|^4 \leq C \tag{2.4.14}$$

Letting ϵ tend to 0 and using Fatou's lemma, the desired result is obtained. ∎

Our problem is now to minimize $J(v(\cdot))$ defined by (2.4.11) among all $v(\cdot)$ in U.

The justification of this approach is the following. When $\eta^v(t)$ is a martingale, P^v defined by (2.4.8) is a probability. We can then consider the probability system $(\Omega, \mathcal{A}, P^v, F^t)$. By Girsanov's theorem we can assert that for this system b^v becomes a Wiener process with covariance matrix $R(\cdot)$. Moreover $\xi, w(\cdot), b^v(\cdot)$ remain mutually independent, $\xi, w(\cdot)$ keeping the same probability laws.

For the system $(\Omega, \mathcal{A}, P^v, F^t)$ the processes $x(\cdot)$, $z(\cdot)$ satisfy

$$
\begin{aligned}
\mathrm{d}x &= (Fx + Gv + f)\mathrm{d}t + \mathrm{d}w \\
x(0) &= \xi \\
\mathrm{d}z &= (Hx + h)\mathrm{d}t + \mathrm{d}b^v \\
z(0) &= 0
\end{aligned}
\tag{2.4.15}
$$

Therefore we recover the model described in §2.3.1, for a convenient probability space; the observation process is given a *priori*, but the probability and the noise on the observation depend on the control. Denoting by E^v the expectation with respect to the probability P^v, the cost function $J(v(\cdot))$ can be written as

$$
\begin{aligned}
J(v(\cdot)) = E^v \bigg[&\int_0^T (x^* M x + v^* N v + 2mx + 2nv)\mathrm{d}t \\
&+ x(T)^* M_T x(T) + 2m_T x(T) \bigg].
\end{aligned}
\tag{2.4.16}
$$

This expression explains our previous terminology.

2.4.1.2 Preliminary results

Let us now turn to the problem as defined in (2.4.11).

Recall that $\alpha(\cdot)$ is the solution of

$$
\begin{aligned}
\mathrm{d}\alpha &= (F\alpha + f)\mathrm{d}t + \mathrm{d}w \\
\alpha(0) &= \xi.
\end{aligned}
\tag{2.4.17}
$$

Set

$$
\begin{aligned}
\frac{\mathrm{d}z_1}{\mathrm{d}t} &= H(t)x_1(t) \\
z_1(0) &= 0
\end{aligned}
\tag{2.4.18}
$$

we have

$$
z(t) = z_1(t) + \beta^v(t)
\tag{2.4.19}
$$

with $\beta^v(\cdot)$ satisfying

$$
\begin{aligned}
\mathrm{d}\beta^v &= (H(t)\alpha(t) + h)\mathrm{d}t + \mathrm{d}b^v, \\
\beta^v(0) &= 0
\end{aligned}
\tag{2.4.20}
$$

(note the dependence of β with respect to $v(\cdot)$, through $b_v(\cdot)$).

Define then

$$
\mathcal{F}_v^t = \sigma(\beta^v(s), \ s \le t);
\tag{2.4.21}
$$

clearly from (2.4.19) we can assert that

$$\mathcal{F}_v^t \subset \mathcal{Z}^t. \tag{2.4.22}$$

Again notice the difference between the situation of § 2.3.1 and the present one, $v(\cdot)$ affects $\mathcal{F}_v^t$ and not $\mathcal{Z}^t$. Moreover we have only an inclusion (2.4.22), and not an equality of the two σ-algebras.

Let us now introduce a convenient terminology in this framework. Let

$$L_{\eta^v}^2(0, T; R^n) = \left\{ \xi(\cdot) \text{ adapted to } F^t \,\Big|\, E \int_0^T \eta^v(t)|\xi(t)|^2 dt < \infty \right\} \tag{2.4.23}$$

and similar definitions replacing R^n by other euclidean spaces, or changing the power.

Lemma 4.1 has shown that

$$\alpha(\cdot) \in L_{\eta^v}^4(0, T; R^n)$$

and in fact $L_{\eta^v}^p(0, T; R^n)$, $\forall p \geq 2$, $p < \infty$. Moreover, the norm is uniformly bounded with respect to $v(\cdot)$. It is important to notice that we can without loss of generality assume that admissible controls satisfy

$$v(\cdot) \in L_{\eta^v}^2(0, T; R^k), \text{ i.e. } E \int_0^T \eta^v(t)|v(t)|^2 dt < \infty. \tag{2.4.24}$$

Indeed the control $v(\cdot) = 0$ is such that $J(0) < \infty$, by virtue of Lemma 2.4.1. Therefore we can without loss of generality restrict ourselves to those controls satisfying

$$J(v(\cdot)) \leq J(0).$$

From the discussion preceding Lemma 2.4.1, by virtue of the invertibility of N, this implies that we can without loss of generality assume that

$$E \int_0^T \eta^v(t)|v(t)|^2 dt \leq C$$

(we have used $E\eta^v(T) \int_0^T |v(t)|^2 dt \leq E \int_0^T \eta^v(t)|v(t)|^2 dt$, since $\eta^v(t)$ is a F^t upper martingale).

The following lemma is instrumental.

Lemma 2.4.2 *If $v(\cdot)$ satisfies (2.4.24) then $\eta^v(t)$ is a F^t martingale.*

Proof

Let (omitting to write the index $v(\cdot)$)

$$\eta_\epsilon(t) = \frac{\eta(t)}{1 + \epsilon\eta(t)};$$

we have

$$d\eta_\epsilon = (1 + \epsilon\eta)^{-2}\eta(x^*H^* + h^*)R^{-1}dz - \epsilon(1 + \epsilon\eta)^{-3}\eta^2(x^*H^* + h^*)R^{-1}(Hx + h)dt$$

hence

$$E\eta_\epsilon(t) = 1 - \epsilon E \int_0^t (1+\epsilon\eta)^{-3}\eta^2(x^*H^* + h^*)R^{-1}(Hx+h)\mathrm{d}s. \qquad (2.4.25)$$

Clearly

$$H_\epsilon = \epsilon(1+\epsilon\eta)^{-3}\eta^2(x^*H^* + h)R^{-1}(Hx+h) \quad \text{a.e. a.s.}$$

and

$$0 \le H_\epsilon \le \eta(x^*H^* + h^*)R^{-1}(Hx+h).$$

On the other hand

$$E\int_0^T \eta(x^*H^* + h^*)R^{-1}(Hx+h)\mathrm{d}t \le CE\int_0^T \eta(t)\left(\int_0^t v^2(s)\mathrm{d}s\right)^{1/2}\mathrm{d}t$$

$$+ CE\int_0^T \eta(t)|\alpha(t)|^2\mathrm{d}t$$

$$\le C\left[E\int_0^T \eta(t)\left(\int_0^t |v(s)|^2\mathrm{d}s\right)\mathrm{d}s\right]^{1/2} + C$$

$$\le C\left[E\int_0^T \eta(t)|v(t)|^2\mathrm{d}t\right]^{1/2} + C$$

$$\le C$$

for controls satisfying (2.4.24).

From Lebesgue's theorem we conclude from (2.4.25) that

$$E\int_0^t H_\epsilon\mathrm{d}s \to 0$$

and

$$E\eta(t) = 1.$$

This and the fact that $\eta(t)$ is an F^t upper martingale implies that it is an F^t martingale. ∎

Therefore for controls satisfying (2.4.24), the measure P^v is a probability, and the framework described in § 2.4.1.1 applies, and in particular $b^v(\cdot)$ is an F^t, P^v Wiener process with covariance matrix $R(\cdot)$.

Define now

$$\hat{x}(t) = E^v[x(t)|\mathcal{Z}^t] \qquad (2.4.26)$$

then from (2.4.10) one has

$$\hat{x}(t) = x_1(t) + E^v[\alpha(t)|\mathcal{Z}^t]. \qquad (2.4.27)$$

We shall prove the important

Lemma 2.4.3 *The following property holds:*

$$E^v[\alpha(t)|\mathcal{Z}^t] = E^v[\alpha(t)|\mathcal{F}_v^t]. \qquad (2.4.28)$$

Proof

Let us prove (2.4.28) for $t = T$. From the definition of P^v (see (2.4.8)), one has

$$E^v[\alpha(T)|\mathcal{Z}^T] = \frac{E[\eta^v(T)\alpha(T)|\mathcal{Z}^T]}{E[\eta^v(T)|\mathcal{Z}^T]}. \tag{2.4.29}$$

We can write

$$\eta^v(T) = \exp\left[\int_0^T x_1^* H^* R^{-1} \mathrm{d}z - \frac{1}{2}\int_0^T x_1^* H^* R^{-1} H x_1 \mathrm{d}t\right.$$

$$+ \int_0^T (\alpha^* H^* + h^*) R^{-1} \mathrm{d}z - \frac{1}{2}\int_0^T (\alpha^* H^* + h^*) R^{-1}(H\alpha + h)\mathrm{d}t$$

$$\left.- \int_0^T x_1^* H^* R^{-1}(H\alpha + h)\mathrm{d}t\right]$$

and since the two first integrals are $\mathcal{Z}^T$ measurable, we can assert that from (2.4.29)

$$E^v[\alpha(T)|\mathcal{Z}^T] = \frac{E[\alpha(T)\Lambda^v(T)|\mathcal{Z}^T]}{E[\Lambda^v(T)|\mathcal{Z}^T]} \tag{2.4.30}$$

where we have set

$$\Lambda^v(T) = \exp\left[\int_0^T (\alpha^* H^* + h^*) R^{-1} \mathrm{d}z - \frac{1}{2}\int_0^T (\alpha^* H^* + h^*) R^{-1}(H\alpha + h)\mathrm{d}t\right.$$

$$\left.- \int_0^T x_1^* H^* R^{-1}(H\alpha + h)\mathrm{d}t\right]$$

$$= \exp\left[\int_0^T (\alpha^* H^* + h^*) R^{-1} \mathrm{d}\beta_v - \frac{1}{2}\int_0^T (\alpha^* H^* + h^*) R^{-1}(H\alpha + h)\mathrm{d}t\right], \tag{2.4.31}$$

noting that α is independent of z (for the probability P) and that β_v is $\mathcal{Z}^T$ measurable. Assume for the moment that $R^{-1}H$, $R^{-1}h$ are differentiable in t. Then

$$\Lambda^v(T) = \exp\left[(\alpha^*(T)H^*(T) + h^*(T))R^{-1}(T)\beta_v(T) - \right.$$

$$\left.- \int_0^T \beta_v^*(R^{-1}H\mathrm{d}\alpha + (R^{-1}H)'\alpha\mathrm{d}t + (R^{-1}h)'\mathrm{d}t)\right]$$

and $E[\Lambda^v(T)|\mathcal{Z}^T]$ is obtained by freezing the value of β_v as a parameter and taking the expectation with respect to α (independent from $\mathcal{Z}^T$). It is clear from the formulas obtained that

$$E[\Lambda^v(T)|\mathcal{Z}^T] \quad \text{is } \mathcal{F}_v^T \text{ measurable.}$$

The same can be said about the numerator of the right hand side of (2.4.30). Therefore

$$E^v[\alpha(T)|\mathcal{Z}^T] \quad \text{is } \mathcal{F}_v^T \text{ measurable.} \tag{2.4.32}$$

and because of the inclusion (2.4.22) and (2.4.32)

$$E^v\left\{E^v[\alpha(T)|\mathcal{Z}^T]|\mathcal{F}_v^T\right\} = E^v[\alpha(T)|\mathcal{Z}^T]$$

$$= E^v[\alpha(T)|\mathcal{F}_v^T]$$

$$= \hat{\alpha}(T).$$

which is the desired result. ∎

The Kalman filter theory implies that $\hat{\alpha}$ is the solution of

$$d\hat{\alpha} = (F\hat{\alpha} + f)dt + PH^*R^{-1}(d\beta_v - (H\hat{\alpha} + h)dt)$$
$$\hat{\alpha}(0) = x_0.$$

$$(2.4.33)$$

2.4.1.3   Optimal control

It follows from (2.4.33) and (2.4.27) that $\hat{x}(t)$ is the solution of

$$d\hat{x} = [(F - PH^*R^{-1}H)\hat{x} + f - PH^*R^{-1}h + Gv]dt$$
$$\qquad + PH^*R^{-1}dz$$
$$\hat{x}(0) = x_0.$$

The error $\epsilon = x - \hat{x}$ satisfies

$$d\epsilon = (F - PH^*R^{-1}H)\epsilon dt + dw - PH^*R^{-1}db_v$$
$$\epsilon(0) = \xi - x_0.$$

$$(2.4.35)$$

Therefore for the probability P^v, ϵ is a gaussian process with mean 0 and covariance $P(\cdot)$.

It follows that

$$J(v(\cdot)) = E^v\left[\int_0^T (\hat{x}^*M\hat{x} + v^*Nv + 2m\hat{x} + 2nv)dt + \hat{x}(T)^*M_T\hat{x}(T) + 2m_T\hat{x}(T)\right]$$
$$+ \int_0^T \operatorname{tr} MP dt + \operatorname{tr} M_T P(T).$$

$$(2.4.36)$$

Now we consider the innovation

$$I_v(t) = z(t) - \int_0^t (H(s)\hat{x}(s) + h)ds$$
$$= b_v(t) - \int_0^t (H(s)\hat{\alpha}(s) + h)ds.$$

From the filtering theory, we know that $I_v(t)$ is (for P^v) a Wiener process with covariance $R(\cdot)$. It is an $\mathcal{F}_v^t$ Wiener, and also a $\mathcal{Z}^t$ Wiener since

$$E^v[I_v(t) - I_v(s)|\mathcal{Z}^s] = 0 \quad \forall t \geq s$$

$$E^v[(I_v(t) - I_v(s))(I_v(t) - I_v(s))^*|\mathcal{Z}^s] = \int_s^t R(\lambda)d\lambda.$$

Therefore the problem can be formulated as

$$d\hat{x} = (F\hat{x} + f + Gv)dt + PH^*R^{-1}dI_v$$
$$\hat{x}(0) = x_0$$

$$(2.4.37)$$

with the payoff

$$J_1(v(\cdot)) = E^v\left[\int_0^T (\hat{x}^*M\hat{x} + v^*Nv + 2m\hat{x} + 2nv)\mathrm{d}t\right.$$
$$\left. + \hat{x}(T)^*M_T\hat{x}(T) + 2m_T\hat{x}(T)\right]. \qquad (2.4.38)$$

This problem is a stochastic control problem with full observation; $z(\cdot)$ is the unique source of noise and the control is adapted to the family $\mathcal{Z}^t$.

The only thing which differs from the presentation of § 2.2.2 is the fact that P^v varies with $v(\cdot)$. But let us set

$$v(t) = -N^{-1}(t)[n^*(t) + G^*(t)(\Pi(t)\hat{x}(t) + r(t))] + \mu(t) \qquad (2.4.39)$$

and let $\eta(\cdot)$ be the solution of

$$\mathrm{d}\eta = [(F - GN^{-1}G^*\Pi)\eta + f - GN^{-1}(n^* + G^*r)]\mathrm{d}t$$
$$+ PH^*R^{-1}\mathrm{d}I_v \qquad (2.4.40)$$
$$\eta(0) = x_0$$

(note that $\eta(\cdot)$ depends on $v(\cdot)$ because of I_v, but its probability law does not). Let also

$$\hat{x} - \eta = \tilde{x}$$
$$\chi = -N^{-1}[n^* + G^*(\Pi\eta + r)] \qquad (2.4.41)$$

hence

$$v = \chi - N^{-1}G^*\Pi\tilde{x} + \mu.$$

A computation shows that

$$J_1(v(\cdot)) = E^v\left[\int_0^T (\eta^*M\eta + \chi^*N\chi + 2m\eta + 2n\chi)\mathrm{d}t\right.$$
$$\left. + \eta(T)^*M_T\eta(T) + 2m_T\eta(T)\right]$$
$$+ E^v\left\{\int_0^T [\tilde{x}^*M\tilde{x} + (-\tilde{x}^*\Pi GN^{-1} + \mu^*)N(-N^{-1}G^*\Pi\tilde{x} + \mu)]\mathrm{d}t\right.$$
$$+ [-\tilde{x}(T)^*\Pi(T)G(T)N^{-1}(T) + \mu^*(T)]$$
$$\left. M_T[-N^{-1}(T)G^*(T)\Pi(T)\tilde{x}(T) + \mu(T)]\right\}$$

since the double product term vanishes, by the choice of Π and r.

It follows that

$$J_1(v(\cdot)) \geq E^v\left[\int_0^T (\eta^*M\eta + \chi^*N\chi + 2m\eta + 2n\chi)\mathrm{d}t\right.$$
$$\left. + \eta(T)^*M_T\eta(T) + 2m_T\eta(T)\right]. \qquad (2.4.42)$$

But one checks that

$$d[\eta(t)^*\Pi(t)\eta(t) + 2r(t)^*\eta(t)] = [-(\eta^*M\eta + 2m\eta + \chi^*N\chi + 2n x) + 2f^*r$$
$$- (r^*G + n)N^{-1}(G^*r + n^*) + \text{tr } R^{-1}HP\Pi PH^*]dt$$
$$+ 2(\eta^*\Pi + r^*)PH^*R^{-1}dI_v$$

hence (2.4.42) implies

$$J_1(v(\cdot)) \geq \int_0^T [2f^*r - (r^*G + n)N^{-1}(G^*r + n^*) \tag{2.4.43}$$
$$+ \text{tr } R^{-1}HP\Pi H^*]dt.$$

We shall now define a particular control denoted by $u(\cdot)$. Let $\hat{y}(\cdot)$ be the process solution of

$$d\hat{y} = [(F - PH^*R^{-1}H)\hat{y} + f - PH^*R^{-1}h - GN^{-1}n^* - GN^{-1}G^*(\Pi\hat{y} + r)]dt$$
$$+ PH^*R^{-1}dz$$
$$\hat{y}(0) = x_0.$$
$$\tag{2.4.44}$$

Then define $u(\cdot)$ by

$$u(t) = -N^{-1}(t)[n^*(t) + G^*(t)(\Pi(t)\hat{y}(t) + r(t))]. \tag{2.4.45}$$

Clearly $\hat{y}(\cdot)$ is adapted to $\mathcal{Z}^t$, hence $u(\cdot) \in \mathcal{U}$. Define $y(\cdot)$ to be the state corresponding to the control $u(\cdot)$, and $\eta^u(\cdot)$ associated to $u(\cdot)$, by (cf. (2.4.6))

$$d\eta^u = \eta^u(y^*H^* + h^*)R^{-1}dz, \quad \eta^u(0) = 1. \tag{2.4.46}$$

Now follows

Lemma 2.4.4 *We have*

$$E\eta^u(t)|\hat{y}(t)|^2 \leq C. \tag{2.4.47}$$

Proof

Define $y_\epsilon(t)$ by

$$y_\epsilon(t) = \frac{y(t)}{(1 + \epsilon|y(t)|^2)^{1/2}}$$

and let $\eta^u\epsilon$ be defined by (2.4.46) with y replaced by y_ϵ. Note that

$$d|\hat{y}(t)|^2 = 2\hat{y}^*\{[(F - PH^*R^{-1}H)\hat{y} - PH^*R^{-1}h + f - GN^{-1}n^*$$
$$- GN^{-1}G^*(\Pi\hat{y} + r)]dt + PH^*R^{-1}dz\} + PH^*R^{-1}HPdt$$

and

$$d\eta_\epsilon^u(t)|\hat{y}(t)|^2 = [\eta_\epsilon^u|\hat{y}(t)|^2(y_\epsilon^*H^* + h^*)R^{-1} + 2\eta_\epsilon^u\hat{y}^*PH^*R^{-1}]dz$$
$$+ \eta_\epsilon^u\{2\hat{y}^*[(F - PH^*R^{-1}H)\hat{y} - PH^*R^{-1}h + f$$
$$- GN^{-1}n^* - GN^{-1}G^*(\Pi\hat{y} + r)] + PH^*R^{-1}HP$$
$$+ 2(y_\epsilon^*H^* + h^*)R^{-1}HP\hat{y}\}dt.$$

Since $E|\hat{y}(t)|^4 < \infty$ and $E(\eta_\epsilon^u)^2 < \infty$, we can integrate and take the mathematical expectation. It yields

$$E\eta_\epsilon^u(t)|\hat{y}(t)|^2 = |x_0|^2 + E\int_0^t \eta_\epsilon^u\{2\hat{y}^*[(F - PH^*R^{-1}H)\hat{y} - PH^*R^{-1}h + f$$
$$- GN^{-1}n^* - GN^{-1}G^*(\Pi\hat{y} + r)]$$
$$+ PH^*R^{-1}HP + 2(y_\epsilon^*H^* + h^*)R^{-1}HP\hat{y}\}ds.$$

We can easily majorize the right hand side by

$$C + E\int_0^t \eta_\epsilon^u(s)|\hat{y}(s)|^2 ds.$$

By Gronwall's inequality we deduce

$$E\eta_\epsilon^u(t)|\hat{y}(t)|^2 \leq C$$

and from Fatou's lemma, (2.4.47) obtains. ∎

From Lemma 2.4.4 and formula (2.4.45) we can assert that $u(\cdot)$ satisfies (2.4.24). Therefore $\eta^u(t)$ is a $\mathcal{F}^t$ martingale and P^u is a probability. Considering (2.4.34) with $v(\cdot) = u(\cdot)$, and the equation (2.4.44), we obtain that

$$\hat{y}(t) = E^u[y(t)|\mathcal{Z}^t]. \tag{2.4.48}$$

Considering the innovation

$$I_u(t) = z(t) - \int_0^t (H\hat{y} + h)ds$$

we find that $\eta(\cdot)$, the solution of (2.4.40) with $v(\cdot) = u(\cdot)$, coincides with $\hat{y}$.

Hence

$$\tilde{x} = \hat{y} - \eta = 0$$

also

$$\chi = u, \quad \mu = 0.$$

Therefore

$$J_1(u(\cdot)) = E^u\left[\int_0^T (\hat{y}^*M\hat{y} + u^*Nu + 2m\hat{y} + 2nu)dt\right.$$
$$\left. + y(T)^*M_Ty(T) + 2m_Ty(T)\right]$$
$$= \int_0^T [2f^*r - (r^*G + n)N^{-1}(G^*r + n^*) + \text{tr } R^{-1}HP\Pi H^*]dt.$$

Comparing this formula with (2.4.43) we have

$$J_1(v(\cdot)) \geq J_1(u(\cdot)) \quad \forall v(\cdot)$$

hence we can state

Theorem 2.4.1 *The control $u(\cdot)$ defined by (2.4.45) is optimal for the problem of minimizing (2.4.11) on the space $\mathcal{U}$.*

2.4.2   General solution

2.4.2.1   Return to the model of Section 2.3

We now return to the model of Section 2.3, described in § 2.3.1. We have seen in § 2.3.2 that, provided one accepts restricting the class of admissible controls to that described by (2.3.10), then the optimal control can be found.

We have seen in § 2.4.1 that the approach through the Girsanov transformation allows us to consider admissible controls just as processes adapted to the observation. But this is at the price of a slight extension of the definition of the problem.

The question remains whether the class of admissible controls described by (2.3.10) is necessary in the framework of § 2.3.2 or not.

We shall see in this paragraph that, using techniques inspired by those of § 2.4.1, one can overcome the restriction imposed by the class (2.3.10).

We thus consider the situation of § 2.3.1. An admissible control satisfies (2.3.7).

2.4.2.2   Preliminaries

Let us define the process

$$\rho^v(t) = \exp\left[-\int_0^t (x^*H^* + h^*)R^{-1}\mathrm{d}b - \frac{1}{2}\int_0^t (x^*H^* + h^*)R^{-1}(Hx + h)\mathrm{d}s\right] \quad (2.4.49)$$

and let

$$\mathcal{F}^t = \sigma(\xi, w(s), b(s), s \leq t);$$

then $\rho^v(t)$ is an F^t upper martingale, and thus

$$E\rho^v(t) \leq 1.$$

We first have

Lemma 2.4.5 *One has the property*

$$E\rho^v(t)|\alpha(t)|^m \leq C_m, \quad \forall\, m\; 1 \leq m < \infty, \;\; t \in [0, T]. \quad (2.4.51)$$

Proof

Let us for simplicity take $m = 2$. Replace x by

$$x_\epsilon(t) = \frac{x(t)}{(1 + \epsilon|x(t)|^2)^{1/2}}$$

and let $\rho_\epsilon^v(t)$ be the process defined by (2.4.49) with x_ϵ replaced for x. Then

$$\mathrm{d}\rho_\epsilon^v = -\rho_\epsilon^v(x_\epsilon^*H^* + h^*)R^{-1}\mathrm{d}b.$$

Furthermore

$$d|\alpha(t)|^2 = 2\alpha^*[(F\alpha + f)dt + dw] + \operatorname{tr} Qdt$$
$$d\rho_\epsilon^v|\alpha(t)|^2 = [2\alpha^*(F\alpha + f) + \operatorname{tr} Q]\rho_\epsilon^v dt$$
$$+ 2\rho_\epsilon^v \alpha^* dw - |\alpha(t)|^2 \rho_\epsilon^v (x_\epsilon^* H^* + h^*) R^{-1} db$$

hence

$$E\rho_\epsilon^v(t)|\alpha(t)|^2 = |x_0|^2 + E \int_0^t \rho_\epsilon^v(s)[2\alpha^*(F\alpha + f) + \operatorname{tr} Q]ds$$

from which it easily follows that

$$E\rho_\epsilon^v(t)|\alpha(t)|^2 \le C$$

and, from Fatou's lemma, the desired result is obtained.   ∎

Lemma 2.4.6 *Assume that* $v(\cdot)$ *satisfies*

$$E \int_0^T \rho^v(t)|v(t)|^2 dt < \infty \tag{2.4.52}$$

then $\rho^v(t)$ *is a* P, $\mathcal{F}^t$ *martingale.*

Proof
It is similar to that of Lemma 2.4.2.   ∎

Let us prove the following important result, similar to that of Lemma 2.4.3.
Lemma 2.4.7 *One has the property*

$$E[\alpha(t)|\mathcal{Z}_v^t] = E[\alpha(t)|\mathcal{F}^t] \tag{2.4.53}$$

Proof.
The definition of $\mathcal{F}^t$ has been given in (2.3.7), that of $\mathcal{Z}_v^t$ in (2.3.6).

Let us set

$$\hat{\alpha}(t) = E[\alpha(t)|\mathcal{F}^t].$$

By virtue of (2.3.8), $\hat{\alpha}(t)$ is $\mathcal{Z}_v^t$ measurable.

Let us check that it is sufficient to prove (2.4.53) for controls which are bounded by a deterministic constant. Indeed, suppose (2.4.53) is true in this case; consider $\psi(z(t_1), z(t_2), \ldots, z(t_k))$ where ψ is a continuous bounded function on R^{mk}, and $z(\cdot)$ is the observation corresponding to a control $v(\cdot)$, with $t_1, \ldots, t_k$ fixed values of time less than or equal to t.

Define

$$v_\epsilon(t) = \frac{v(t)}{(1 + \epsilon|v(t)|^2)^{1/2}}$$

$x_{1\epsilon}(t) =$ solution of (2.3.3) corresponding to $v(\cdot)$ being replaced by $v_\epsilon(\cdot)$

$x_\epsilon(t) = x_{1\epsilon}(t) + \alpha(t)$

$z_{1\epsilon}(t) =$ solution of (2.3.4) with $x_{1\epsilon}(\cdot)$ replacing $x_1(\cdot)$

$z_\epsilon(t) = z_{1\epsilon}(t) + \beta(t)$

$\mathcal{Z}_{v_\epsilon}^t = \sigma(z_\epsilon(s), \ s \le t).$

By the assumption, we have

$$E[\alpha(t)|\mathcal{Z}_{v_\epsilon}^t] = E[\alpha(t)|\mathcal{F}^t] = \hat{\alpha}(t)$$

therefore, in particular

$$E\alpha(t)\psi(z_\epsilon(t_1), \ldots, z_\epsilon(t_k)) = E\hat{\alpha}(t)\psi(z_\epsilon(t_1), \ldots, z_\epsilon(t_k)).$$

Letting ϵ tend to 0, we deduce

$$E\alpha(t)\psi(z(t_1), \ldots, z(t_k)) = E\hat{\alpha}(t)\psi(z(t_1), \ldots, z(t_k)).$$

Since, in this equality, ψ is an arbitrary continuous bounded function, and $t_1, \ldots, t_k$ are arbitrary values less than or equal to t, we also have

$$E\alpha(t)\xi_t = E\hat{\alpha}(t)\xi_t$$

for any ξ_t where ξ_t is a bounded $\mathcal{Z}_v^t$ random variable. Since $\hat{\alpha}(t)$ is $\mathcal{Z}_v^t$ measurable, (2.4.53) necessarily holds. We can thus assume that

$$|v(t)| \leq C \tag{2.4.54}$$

in which C is a deterministic constant.

But then, in particular (2.4.52) holds, and $\rho^v(t)$ is a $P, \mathcal{F}^t$ martingale. Define now

$$
\begin{aligned}
\eta^v(t) &= \frac{1}{\rho^v(t)} \\
&= \exp\left[\int_0^t (x^*H^* + h^*)R^{-1}\mathrm{d}b + \frac{1}{2}\int_0^t (x^*H^* + h^*)R^{-1}(Hx + h)\mathrm{d}s\right] \\
&= \exp\left[\int_0^t (x^*H^* + h^*)R^{-1}\mathrm{d}z - \frac{1}{2}\int_0^t (x^*H^* + h^*)R^{-1}(Hx + h)\mathrm{d}s\right] \\
\mathrm{d}P^v &= \rho^v(T)\mathrm{d}P.
\end{aligned}
$$

then P^v is a probability, and

$$\mathrm{d}P = \eta^v(T)\mathrm{d}P^v.$$

Moreover, by Girsanov's theorem, for the system $(\Omega, \mathcal{A}, P^v, \mathcal{F}^t)$ the processes $w(\cdot), z(\cdot)$ are $\mathcal{F}^t$ Wiener processes, which are independent and have $Q(\cdot), R(\cdot)$ respectively as covariance matrices. It follows also that $w(\cdot)$ and $z(\cdot)$ are independent of ξ. As a consequence of the independence of $z(\cdot)$, and $w(\cdot)$, ξ, the process $\alpha(\cdot)$ is independent of $\mathcal{Z}_v^T$.

We now have a situation which is in a sense converse to that of § 2.4.1, the role of P (in § 2.4.1) being played by P^v here, and the role of P^v (in § 2.4.1) being played by P here. By analogy to what has been done in Lemma 2.4.3, we have

$$E[\alpha(T)|\mathcal{Z}_v^T] = \frac{E^v[\eta^v(T)\alpha(T)|\mathcal{Z}_v^T]}{E^v[\eta^v(T)|\mathcal{Z}_v^T]}.$$

A calculation similar to that leading to (2.4.30), yields

$$E[\alpha(T)|\mathcal{Z}_v^T] = \frac{E^v[\alpha(T)\Lambda(T)|\mathcal{Z}_v^T]}{E^v[\Lambda(T)|\mathcal{Z}_v^T]}$$

where

$$\Lambda(T) = \exp\left[\int_0^T (\alpha^* H^* + h^*)R^{-1}d\beta - \frac{1}{2}\int_0^T (\alpha^* H^* + h^*)R^{-1}(H\alpha + h)dt\right].$$
(2.4.55)

Reasoning as in Lemma 2.4.3, we conclude that $E[\alpha(T)|\mathcal{Z}_v^T]$ is $\mathcal{F}^T$ measurable, and thus the desired result holds.   ■

2.4.2.3   Optimality

Consider the innovation

$$I(t) = \beta(t) - \int_0^t (H\hat{\alpha}(s) + h)ds$$

which is an $\mathcal{F}^t$ Wiener process, with covariance $R(\cdot)$, and also a $\mathcal{Z}_v^t$ Wiener process, since

$$I(t) = z(t) - \int_0^t (H\hat{\alpha}(s) + h)ds$$

with

$$\hat{x}(t) = x_1(t) + \hat{\alpha}(t) = E[x(t)|\mathcal{Z}_v^t].$$

We have, from Lemma 2.4.7,

$$d\hat{x} = (F\hat{x} + f + Gv)dt + PH^*R^{-1}dI$$
$$\hat{x}(0) = x_0$$
(2.4.56)

and

$$J(v(\cdot)) = E\left[\int_0^T (\hat{x}^* M\hat{x} + v^* Nv + 2m\hat{x} + 2nv)dt\right.$$
$$\left. + \hat{x}(T)^* M_T \hat{x}(T) + 2m_T \hat{x}(T)\right] + \int_0^T \text{tr } MPdt + \text{tr } M_T P(T).$$
(2.4.57)

Consider next the process

$$d\hat{y} = [(F - GN^{-1}G^*\Pi)\hat{y} + f - GN^{-1}(n^* + G^*r)]dt$$
$$+ PH^*R^{-1}dI$$
$$\hat{y}(0) = x_0$$
(2.4.58)

and the control

$$u(t) = -N^{-1}G^{-1}\Pi\hat{y} - GN^{-1}(n^* + G^*r).$$
(2.4.59)

Since $I(t)$ is a $\mathcal{F}^t$ and $\mathcal{Z}_v^t$ Wiener process, $u(\cdot)$ is admissible.

We can proceed as in the proof of Theorem 2.4.1, to obtain

Theorem 2.4.2 *The control $u(\cdot)$ defined by (2.4.59) is optimal for the problem (2.3.9) among all admissible controls (i.e. satisfying (2.3.7)).*

Comments

In this chapter the optimal control which has been determined is always obtained through a *linear feedback*.

When the problem is deterministic the feedback operates on the state which is observable since there are no disturbances.

In the stochastic case, we can make several remarks. Consider first the full observation case. The state is observable and, as we have seen in both problems studied in Section 2.2, the optimal control is again expressed as a linear feedback on the state. However, for the model whose dynamics are given by (2.2.2) the optimal feedback is the same as in the deterministic case. This is not true for the model whose dynamics are given by (2.2.14). In (2.2.2) the noise is purely external, which means that it cannot be modified by the state or the control. In (2.2.14) on the other hand the noise is dependent on the state and the control. We see that in (2.2.2) the optimal feedback does not depend on the intensity of the noise, so long as it is external. This intensity may be unknown. This justifies the use of the feedback, instead of the open loop control, even in deterministic models. Indeed the feedback control is *robust* with respect to external disturbances and these may be unknown. However, when there is interaction between the noise and the state or the control, the nature of this interaction should be known and influences the optimal feedback.

In the partial observation case we can only consider (in the framework of this chapter) external noises. We see that the optimal control is obtained as a linear feedback on the Kalman filter. This is intuitive; the state, being unobservable, is replaced by its best estimate, the Kalman filter. Note that the coefficients of the feedback are the same as those of the deterministic case. It is remarkable that this 'heuristic' decision rule is optimal in the linear quadratic gaussian model considered here. This is referred to as the *separation principle*. This principle expresses the fact that, to design the optimal control, one first estimates the state of the system and then applies the optimal feedback of the deterministic case. In other words the operations of estimation and control can be put in sequence and thus separated. Strictly speaking when expressing the separation principle one can slightly generalize by allowing the optimal feedback to be different from the deterministic case, provided again one applies it on the Kalman filter. This generalization is very useful, since it permits us to extend the principle to some nonlinear situations (Wonham's (1968) separation principle; for a more up to date presentation see Bensoussan and Lions (1978)). When the optimal feedback is that of the deterministic case, as for the model of Section 2.3, we say that the *certainty equivalence principle* applies. This terminology was introduced in the economic literature in the 1950s, hence before being rediscovered in the control literature (see Holt *et al.* 1960).

Let me finally make a remark concerning the restriction on admissible controls introduced by (2.3.10). For a long time the removal of this restriction was somewhat open (see Fleming and Rishel 1975) and it is in teresting to notice that even though the control problem is linear quadratic it is not easy to remove (2.3.10). In fact, however, when one refers to the theory of a non linear filtering, this restriction can be easily removed. This is acceptable since the introduction of a control may transform the model into non linear one. Nevertheless we can overcome the use of non linear filtering theory at the price of some manipulations, which in fact derive what is essential in the present case. For earlier treatments, see Liptser and Shiryaev (1977), Messulam (1983).

3   Optimal control of linear stochastic systems with an exponential-of-integral performance index

Introduction

In Chapter 2 (see comments) we have seen the concepts of certainty equivalence and the separation principle. Again the main idea is that an optimal control for a stochastic control problem with partial observation can be obtained by a feedback rule (deterministic of course) applied on the Kalman filter (the best estimate of the state). As we shall see in later chapters, this situation is by no means general. The 'sufficient statistics', to which one applies a feedback rule, are in general infinite dimensional (as will be seen it is the conditional probability distribution), and thus does not reduce to a single moment, the conditional mean. Some natural questions arise at this stage. Can we find examples in which the sufficient statistics are finite dimensional, for instance several moments (conditional mean and conditional variance)? Note that if this holds, then the dimension of the sufficient statistics, although *finite*, is larger than that of the state. In this chapter we will study a different situation. We shall meet a situation where the optimal control is given by feedback on sufficient statistics, which are finite dimensional, with a dimension which is the same as that of the state, but does not coincide with the conditional mean, the best estimate of the state. It is clearly a situation where the separation principle does not hold, but from a computational viewpoint offers the same simplicity as that when the separation principle does hold.

The situation that we consider here is naturally very specific again, and corresponds to linear dynamics, with linear observation, and a payoff which is the exponential of the standard quadratic functional. It turns out that this type of cost is well suited for modelling risk aversion or attraction, and thus is important for applications. The problem has been studied and solved in the discrete time case by Whittle (1982, 1983). In the continuous time case, the existence of finite dimensional sufficient statistics was an open problem for some time and studied in particular cases by Jacobson (1973) who introduced the problem, Speyer, Deyst and Jacobson (1974), Speyer (1976), Kumar and Van Schuppen (1981). The full problem has been solved by Bensoussan and Van Schuppen (1985) which we follow here with some improvement in the presentation.

3.1 The full observation case

3.1.1 Setting of the problem

The dynamic system is described by the model of § 3.2.1, i.e.

$$dx = (F(t)x + G(t)v + f(t))dt + dw$$
$$x(0) = x_0$$

(3.1.1)

where F, G, f are bounded in time.

$$w(\cdot) \text{ is a Wiener process with values in } R^n, \text{ correstricted on}$$
$$\text{a probability space } (\Omega, \mathcal{A}, P), \text{ with covariance matrix } Q(\cdot).$$

(3.1.2)

Let $\mathcal{F}^t = \sigma(w(s), s \leq t)$.

An *admissible* control is a process adapted to the family $\mathcal{F}^t$, which satisfies

$$\int_0^T |v(t)|^2 dt < \infty, \text{ a.s}, \ v(t) \text{ is } \mathcal{F}^t \text{ measurable.}$$

(3.1.3)

The cost function to be minimized is now given by

$$J(v(\cdot)) = E\left\{\theta \exp \theta \left[\int_0^T (x^* M x + v^* N v + 2mx + 2nv)dt \right. \right.$$
$$\left. \left. + x(T)^* M_T x(T) + 2m_T x(T) \right] \right\}$$

(3.1.4)

in which

$$M, N, m, n \text{ are bounded functions,}$$
$$M, M_T \text{ are symmetric non negative definite,}$$
$$N \text{ is symmetric positive definite (uniformly in time).}$$

(3.1.5)

$$\theta \text{ is a real number.}$$

(3.1.6)

Note that $J(v(\cdot))$ can take the value $+\infty$.

3.1.2 Interpretation of the number θ

Let us set

$$X = \int_0^T (x^* M x + v^* N v + 2mx + 2nv)dt$$
$$+ x(T)^* M_T x(T) + 2m_T x(T).$$

The cost function (3.1.4) is thus

$$J(v(\cdot)) = E\theta \exp \theta X. \qquad (3.1.7)$$

We can interpret X as a random cost and the expression (3.1.7) can be interpreted as the *utility* of the random cost X. Such a utility can be compared to the classical expectation EX, used in Chapter 2. To some extent, θ expresses risk preference or risk aversion. Indeed we have

$$\theta \exp \theta X = \theta + \theta^2 X + \theta^3 X^2 \int_0^1 \int_0^1 x \exp(y\theta X) \mathrm{d}x\mathrm{d}y.$$

Therefore

$$J(v(\cdot)) = \theta + \theta^2 EX + \theta^3 EX^2 \int_0^1 \int_0^1 x \exp(y\theta X) \mathrm{d}x\mathrm{d}y.$$

If $\theta > 0$, the second order moment introduces a penalty to the cost. There is risk aversion. On the other hand, if $\theta < 0$, the second order moment has a favorable effect: there is risk preference. When θ is small, the important term to minimize becomes EX; we recover the standard mathematical expectation.

3.1.3   Preliminaries

Let us consider for any admissible control $v(\cdot)$ the stochastic process

$$\begin{aligned}
h_t &= \int_0^t (x^* M x + v^* N v + 2mx + 2nv)\mathrm{d}s \\
&\quad + x_t^* S_t x_t + 2s_t x_t + \rho_t
\end{aligned} \qquad (3.1.8)$$

in which S_t, s_t are deterministic functions†, which will be chosen later on, except that we require

$$S_T = M_T, \quad s_T = m_T, \quad \rho_T = 0 \qquad (3.1.9)$$

and $\dot{S}_t, \dot{s}_t, \dot{\rho}_t$ are L^2 functions; S_t symmetric. The process h_t has the stochastic differential

$$\begin{aligned}
\mathrm{d}h_t =&[x_t^*(\dot{S}_t + S_t F_t + F_t^* S_t + M_t)x_t \\
&+ v_t^* N_t v_t + 2v_t^* G_t^* S_t x_t + 2(\dot{s}_t + s_t F_t + f_t^* S_t + m_t)x_t \\
&+ 2(n_t + s_t G_t)v_t + \dot{\rho}_t + \mathrm{tr}\, S_t Q_t + 2s_t f_t]\mathrm{d}t \\
&+ 2(x_t^* S_t + s_t)\mathrm{d}w_t
\end{aligned} \qquad (3.1.10)$$

Consequently by Ito's formula

$$\begin{aligned}
\mathrm{d}\theta \exp \theta h_t =&\theta^2 \exp \theta h_t \{x_t^*[\dot{S}_t + S_t F_t + F_t^* S_t \\
&- S_t(G_t N_t^{-1} G_t^* - 2\theta Q_t)S_t + M_t]x_t \\
&+ [v_t + N_t^{-1} G_t^* S_t x_t + N_t^{-1}(G_t^* s_t^* + n_t^*)]^* N_t[v_t + N_t^{-1} G_t^* S_t x_t \\
&+ N_t^{-1}(G_t^* s_t^* + n_t^*)] + 2[\dot{s}_t + s_t(F_t + 2\theta Q_t S_t) + f_t^* S_t + m_t \\
&- (s_t G_t + n_t)N_t^{-1} G_t^* S_t]x_t + \dot{\rho}_t + \mathrm{tr}\, S_t Q_t \\
&+ 2\theta s_t Q_t s_t^* + 2s_t f_t - (s_t G_t + n_t)N_t^{-1}(G_t^* s_t^* + n_t^*)\}\mathrm{d}t \\
&+ 2\theta^2 \exp \theta h_t(x_t^* S_t + s_t)\mathrm{d}w_t.
\end{aligned} \qquad (3.1.11)$$

---

† To save notation we write x_t instead of $x(t)$, etc.

Suppose now that we can find S_t, s_t, ρ_t satisfying the relations

$$\dot{S}_t + S_t F_t + F_t^* S_t - S_t (G_t N_t^{-1} G_t^* - 2\theta Q_t) S_t + M_t = 0 \qquad (3.1.12)$$

$$\dot{s}_t + s_t (F_t + 2\theta Q_t S_t - G_t N_t^{-1} G_t^* S_t) + f_t^* S_t + m_t - n_t N_t^{-1} G_t^* S_t = 0 \qquad (3.1.13)$$

$$\dot{\rho}_t + \text{tr } S_t Q_t + 2\theta s_t Q_t s_t^* + 2s_t f_t - (s_t G_t + n_t) N_t^{-1} (G_t^* s_t^* + n_t^*) = 0 \qquad (3.1.14)$$

then (3.1.11) becomes

$$
\begin{aligned}
\mathrm{d}\theta \exp \theta h_t =& \theta^2 \exp \theta h_t [v_t + N_t^{-1} G_t^* S_t x_t + N_t^{-1} (G_t^* s_t^* + n_t^*)]^* \\
& \times N_t [v_t + N_t^{-1} G_t^* S_t x_t + N_t^{-1} (G_t^* s_t^* + n_t^*)] \mathrm{d}t \\
& + 2\theta^2 \exp \theta h_t (x_t^* S_t + s_t) \mathrm{d}w_t.
\end{aligned}
$$

hence also

$$
\begin{aligned}
\theta \exp \theta h_t =& \theta \exp \theta h_0 \exp\left\{ \theta \int_0^t [v_t + N_t^{-1} G_t^* S_t x_t + N_t^{-1} (G_t^* s_t^* + n_t^*)]^* \right. \\
& \left. \times N_t [v_t + N_t^{-1} G_t^* S_t x_t + N_t^{-1} (G_t^* s_t^* + n_t^*)] \mathrm{d}t \right\} \\
& \times \exp\left\{ 2\theta \int_0^t (x_t^* S_t + s_t) \mathrm{d}w_t - 2\theta^2 \int_0^t (x_t^* S_t + s_t) Q_t (S_t x_t + s_t^*) \mathrm{d}t \right\}.
\end{aligned}
$$
$$(3.1.15)$$

3.1.4   Optimal control

In what follows, we make precise the assumptions. We first assume that
the solution of (3.1.12) with final condition (3.1.9) exists on $[0, T]$ and
is unique. Moreover $S(t)$ is symmetric and non negative. $\qquad (3.1.16)$

Note that (3.1.15) is automatically satisfied whenever $\theta \leq 0$. We shall address the
following class of admissible controls:

$$v(\cdot) \in L_{\mathcal{F}}^2 (0, T; R^k) \qquad (3.1.17)$$

and

$$E \exp\left[2\theta \int_0^T (x_t^* S_t + s_t) \mathrm{d}w_t - 2\theta^2 \int_0^T (x_t^* S_t + s_t) Q_t (S_t x_t + s_t^*) \mathrm{d}t \right] = 1.$$

Define now a particular control denoted by $u(t)$ such that

$$u(t) = -N_t^{-1} (G_t^* S_t y_t + G_t^* s_t^* + n_t^*) \qquad (3.1.18)$$

with $y(t)$ defined by

$$
\begin{aligned}
\mathrm{d}y =& [(F_t - G_t N_t^{-1} G_t^* S_t) y_t - G_t N_t^{-1} (G_t^* s_t^* + n_t^*) + f_t] \mathrm{d}t + \mathrm{d}w \\
y(0) =& x_0.
\end{aligned}
$$
$$(3.1.19)$$

It is clear that $u(\cdot) \in L^2_{\mathcal{F}}(0, T; R^k)$. We then state

Theorem 3.1.1 *We assume (3.1.2), (3.1.5), (3.1.6), (3.1.16). Then the control $u(\cdot)$ defined by (3.1.18), (3.1.19) belongs to the class (3.1.17) and is optimal in this class for the payoff (3.1.4).*

Proof

We must prove that

$$E \exp\left[2\theta \int_0^T (y_t^* S_t + s_t) dw_t - 2\theta^2 \int_0^T (y_t^* S_t + s_t) Q_t (S_t y_t + s_t^*) dt\right] = 1. \quad (3.1.20)$$

This will be done in the following lemma.

Let $\hat{h}_t$ be the process h_t corresponding to the choice of the control $v_t = u_t$.

From (3.1.15) we deduce

$$\theta \exp \theta \hat{h}_T = \theta \exp \theta h_0 \exp\left[2\theta \int_0^T (y_t^* S_t + s_t) dw_t\right.$$
$$\left. -2\theta^2 \int_0^t (y_t^* S_t + s_t) Q_t (S_t y_t + s_t^*) dt\right]$$

and from (3.1.20) it follows that

$$E\theta \exp \theta \hat{h}_T = \theta \exp \theta h_0$$

i.e.

$$J(u(\cdot)) = \theta \exp \theta h_0. \quad (3.1.21)$$

On the other hand, for a control $v(\cdot)$ belonging to the class (3.1.17), we have from (3.1.15)

$$\theta \exp \theta h_T \geq \theta \exp \theta h_0 \exp\left[\int_0^T 2\theta(x_t^* S_t + s_t) dw_t\right.$$
$$\left. -2\theta^2 \int_0^T (x_t^* S_t + s_t) Q_t (S_t x_t + s_t^*) dt\right]$$

and from (3.1.17)

$$E\theta \exp \theta h_T \geq \theta \exp \theta h_0$$

i.e.

$$J(v(\cdot)) \geq \theta \exp \theta h_0.$$

Comparing with (3.1.21), the desired optimality follows. ∎

`It remains to prove

Lemma 3.1.1 *The property (3.1.20) holds.*

Proof

Let

$$\phi_t = 2\theta(S_t y_t + s_t^*).$$

Then from (3.1.19) ϕ_t is a gaussian random process with expectation and covariance matrix denoted respectively by m_t and Π_t. They are bounded in t. Let δ be such that

$$I - 2\delta\Pi_t \geq \beta I, \quad \beta > 0, \ \forall t \in [0, T]$$

then we have

$$E \exp \delta |\phi_t|^2 = \frac{\exp\left[\delta|m_t|^2 + 2\delta^2 m_t^* \Pi_t^{1/2}(I - 2\delta\Pi_t)^{-1}\Pi_t^{1/2}m_t\right]}{\sqrt{|I - 2\delta\Pi_t|}}$$

$$\leq C.$$

This criterion is sufficient to ensure (3.1.20) (see for instance Gikhman and Skorokhod 1972), hence the desired result. ∎

3.2   The partial observation case

3.2.1   Setting of the problem

We shall now consider the same problem as in Section 3.1, in the case of partial observation. To state the problem, we shall use the technique developed in § 2.4.1, based on the use of the Girsanov transformation. It has the advantage that the observation is introduced *a priori*, independently of the control. We briefly repeat the general set up of § 2.4.1.1. Let $(\Omega, \mathcal{A}, P)$ be a convenient probability space, on which are defined

ξ, a random variable with gaussian probability law, with mean x_0
and covariance matrix P_0, $w(\cdot)$ a Wiener process with covariance
matrix $Q(\cdot)$, $z(\cdot)$ a Wiener process with covariance matrix $R(\cdot)$.
$\qquad$ (3.2.1)
The processes $w(\cdot)$, $z(\cdot)$ and the variable ξ are mutually independent.

We assume that

$$R(t) \geq \rho I, \quad \rho > 0, \quad \forall t \in [0, T].$$

We define

$$\mathcal{F}^t = \sigma(\xi, w(s), z(s), s \leq t)$$
$$\mathcal{Z}^t = \sigma(z(s), s \leq t).$$

An admissible control is for the time being a process belonging to the space

$$\mathcal{U} = L_Z^2(0, T; R^k). \qquad (3.2.2)$$

For any admissible control $v(\cdot)$, define the state $x(\cdot)$ by

$$dx = (F(t)x + G(t)v + f(t))dt + dw$$
$$x(0) = \xi. \qquad (3.2.3)$$

The process $x(\cdot)$ is adapted to $\mathcal{F}^t$.

Define next the processes (depending on $v(\cdot)$ as is the case for $x(\cdot)$)

$$\eta^v(t) = \exp\left[\int_0^t (x^*H^* + h^*)R^{-1}dz - \frac{1}{2}\int_0^t (x^*H^* + h^*)R^{-1}(Hx + h)ds\right] \quad (3.2.4)$$

$$b^v(t) = z(t) - \int_0^t (Hx(s) + h)ds. \qquad (3.2.5)$$

The process η^v satisfies

$$d\eta^v = \eta^v(x^* H^* + h^*)R^{-1}dz$$
$$\eta^v(0) = 1. \qquad (3.2.6)$$

and we know that

$$E\eta^v(t) \leq 1. \qquad (3.2.7)$$

Define the positive finite measure P^v by the Radon–Nikodym derivative

$$\frac{dP^v}{dP} = \eta^v(T) \qquad (3.2.8)$$

(note that at this stage P^v is not necessary a probability).

We write again

$$x(t) = x_1(t) + \alpha(t) \qquad (3.2.9)$$

where

$$\frac{dx_1}{dt} = Fx_1 + Gv,$$
$$x_1(0) = 0 \qquad (3.2.10)$$

$$d\alpha = (F\alpha + f)dt + dw$$
$$\alpha(0) = \xi. \qquad (3.2.11)$$

The cost function is defined by

$$J(v(\cdot)) = \theta E\eta^v(T)\exp\left\{\theta\left[\int_0^T (x^* Mx + v^* Nv + 2mx + 2nv)dt \right. \right.$$
$$\left. \left. + x(T)^* M_T x(T) + 2m_T x(T)\right]\right\}. \qquad (3.2.12)$$

All coefficients $F, G, f, M, N, m, n, M_T, m_T$ are the same as in the full observation case.

Again we take the expectation of positive random variables, which exists and is possibly equal to $+\infty$.

The value $-\infty$ is not possible. Indeed this requires $\theta < 0$. In this case

$$E\eta^v(T)\exp\left\{\theta\left[\int_0^T (x^* Mx + v^* Nv + 2mx + 2nv)dt \right. \right.$$
$$\left. \left. + x(T)^* M_T x(T) + 2m_T x(T)\right]\right\}$$

$$\leq E\eta^v(T)\exp\left\{\theta\left[\int_0^T (v^* Nv + 2mx_1 + 2nv)dt + 2m_T x_1(T)\right]\right\} \qquad (3.2.13)$$

$$\times \exp\left\{\theta\left[\int_0^T 2m\alpha dt + 2m_T\alpha(T)\right]\right\}.$$

It is easy to check from (3.2.10) and the invertibility of the matrix N that

$$\int_0^T (v^* Nv + 2mx_1 + 2nv)dt + 2m_T x_1(T) \geq -C$$

where C is a constant.

To the left hand side of (3.2.13) is finite, it is thus sufficient to prove that

$$E\eta^v(T)\exp\left\{\theta\left[\int_0^T 2m\alpha dt + 2m_T\alpha(T)\right]\right\} < C_1$$

where C_1 is a constant. In fact, all these constants, $C, C_1, \ldots$ will depend on $v(\cdot)$.

By the Cauchy–Schwartz inequality

$$E\eta^v\exp\left\{\theta\left[\int_0^T 2m\alpha dt + 2m_T\alpha(T)\right]\right\}$$

$$\leq \left[E\eta^v(T)\exp\left(\theta\int_0^T 4m\alpha dt\right)\right]^{1/2}\left\{E\eta^v(T)\exp\left[4\theta m_T\alpha(T)\right]\right\}^{1/2}.$$

From Jensen's inequality

$$E\eta^v(T)\exp\left(\theta\int_0^T 4m\alpha dt\right) \leq \frac{1}{T}E\eta^v(T)\int_0^T \exp\left(4\theta Tm\alpha\right)dt$$

$$\leq \frac{1}{T}\int_0^T E\eta^v(t)\exp\left(4\theta Tm\alpha\right)dt$$

since $\eta^v(t)$ is an upper $\mathcal{F}^t$ martingale.

The desired result will then follow from the next lemma.

Lemma 3.2.1 *Let $k(t) \in (R^n)^*$ be such that $|k(t)| \leq k_0$, then*

$$E\eta^v(t)\exp\left[k(t)\alpha(t)\right] \leq C, \quad \forall t \in [0, T] \qquad (3.2.14)$$

where C does not depend on $v(\cdot)$.

Proof

We fix $t = T$, to simplify notation. The same proof applies to any $t < T$.

Therefore we want to prove that

$$E\eta^v(T)\exp\left[k\alpha(T)\right] \leq C. \qquad (3.2.15)$$

Define $\nu(t)$ such that

$$\dot{\nu} + \nu F = 0, \quad \nu(T) = k,$$

and consider $\eta^v(T)\exp[\nu(t)\alpha(t)]$. By Ito's formula

$$d\eta^v(t)\exp[\nu(t)\alpha(t)] = \eta^v(t)\exp[\nu(t)\alpha(t)][(x^*H^* + h^*)R^{-1}dz$$
$$+ \nu dw + (\nu f + \frac{1}{2}\nu Q\nu^*)dt]$$

hence

$$\frac{d\eta^v(t)\exp[\nu(t)\alpha(t)]}{1 + \epsilon\eta^v(t)\exp[\nu(t)\alpha(t)]}$$

$$= (1 + \epsilon\eta\exp\nu\alpha)^{-2}\eta\exp\nu\alpha[(x^*H^* + h^*)R^{-1}dz + \nu dw + (\nu f + \frac{1}{2}\nu Q\nu^*)dt]$$

$$- \epsilon[1 + \epsilon\eta\exp\nu\alpha]^{-3}[(x^*H^* + h^*)R^{-1}(Hx + h) + \nu Q\nu^*]\eta^2\exp 2\nu\alpha dt.$$

Integrating between 0 and t and taking the mathematical expectation, it follows that

$$E\frac{\eta^v(t)\exp[\nu(t)\alpha(t)]}{1+\epsilon\eta^v(t)\exp[\nu(t)\alpha(t)]} \leq E\frac{\exp[\nu(0)\alpha(0)]}{1+\epsilon\exp[\nu(0)\alpha(0)]}$$

$$+ E\int_0^t \frac{(\nu f+\frac{1}{2}\nu Q\nu^*)\eta^v(t)\exp[\nu(t)\alpha(t)]}{\{1+\epsilon\eta^v(t)\exp[\nu(t)\alpha(t)]\}^2}\,dt.$$

From which we deduce

$$E\frac{\eta^v(t)\exp[\nu(t)\alpha(t)]}{1+\epsilon\eta^v(t)\exp[\nu(t)\alpha(t)]} \leq C.$$

Letting ϵ tend to 0, the desired result (3.2.14) obtains.   ■

3.2.2   Preliminary calculations

Let us define for any $v(\cdot)$ in $\mathcal{U}$ (see (3.2.2)), the random variable

$$X_v = \theta\eta^v(T)\exp\left\{\theta\left[\int_0^T (x^*Mx+v^*Nv+2mx+2nv)dt\right.\right.$$
$$\left.\left. + x(T)^*M_Tx(T)+2m_Tx(T)\right]\right\} \tag{3.2.16}$$

hence from (3.2.12)

$$J(v(\cdot)) = EX_v$$

possibly equal to $+\infty$.

We shall make the following assumptions

$$(R^{-1}H)', \quad (R^{-1}h)' \in L^\infty(0,T). \tag{3.2.17}$$

The Riccati equation

$$\dot{\Sigma} + \Sigma F + F^*\Sigma - \Sigma Q\Sigma + H^*R^{-1}H - 2\theta M = 0$$
$$\Sigma(T) = -2\theta M_T \tag{3.2.18}$$

has a continuous solution on $[0,T]$.

The assumption (3.2.17) is purely technical. Note also that (3.2.18) holds whenever $\theta < 0$.

Let us next define $\sigma(t) \in (R^n)^*$ and $\beta(t) \in R$ by the formulae

$$\dot{\sigma} + \sigma(F-Q\Sigma) - z^*[R^{-1}H(F-Q\Sigma)+(R^{-1}H)']$$
$$+ 2\theta m - (f^*+v^*G^*)\Sigma - h^*R^{-1}H = 0 \tag{3.2.19}$$
$$\sigma(T) = 2\theta m_T + z^*(T)R^{-1}(T)H(T).$$

$$\beta(t) = -\int_t^T [2\theta(v^*Nv+2nv)+z^*R^{-1}HQH^*R^{-1}z - h^*R^{-1}h - 2(h^*R^{-1})'z$$
$$+ \sigma Q\sigma^* - 2z^*R^{-1}H(Q\sigma^*+f+Gv)+2\sigma(f+Gv) - \operatorname{tr}\Sigma Q]ds$$
$$- 2h^*(T)R^{-1}(T)z(T). \tag{3.2.20}$$

The functions $\sigma(\cdot)$, $\beta(\cdot)$ depend parametrically on $z(\cdot)$, in particular through $v(\cdot)$, and thus are stochastic processes. But they are independent from $\xi, w(\cdot)$.

We can then state

Lemma 3.2.2 *We assume (3.2.17), (3.2.18). Then we can express X_v as*

$$X_v = \theta \exp\left\{-\frac{1}{2}[\xi^*\Sigma(0)\xi - 2\sigma(0)\xi + \beta(0)]\right\}$$
$$\times \exp\left[\int_0^T -(z^*R^{-1}H + x^*\Sigma - \sigma)\mathrm{d}w - \frac{1}{2}|H^*R^{-1}z + \Sigma x - \sigma^*|^2\mathrm{d}t\right]. \tag{3.2.21}$$

Proof

We write $\eta(T)$ instead of $\eta^v(T)$ to simplify notation. Since

$$\mathrm{d}(x^*H^* + h^*)R^{-1}z = (x^*H^* + h^*)R^{-1}\mathrm{d}z + (x^*F^* + v^*G^* + f^*)H^*R^{-1}z\mathrm{d}t$$
$$+ \mathrm{d}w^*H^*R^{-1}z + [x^*(H^*R^{-1})' + (h^*R^{-1})']z\mathrm{d}t$$

we have

$$\eta(T) = \exp\Bigg\{(x^*(T)H^*(T) + h^*(T))R^{-1}(T)z(T)$$
$$- \int_0^T [(x^*F^* + v^*G^* + f^*)H^*R^{-1}z$$
$$+ (x^*(H^*R^{-1})' + (h^*R^{-1})')z - \frac{1}{2}(x^*H^* + h^*)R^{-1}(Hx + h)]\mathrm{d}t$$
$$- \int_0^T z^*R^{-1}H\mathrm{d}w\Bigg\}.$$

Therefore, from formula (3.2.16) it follows that

$$X_v = \theta \exp\Bigg\{\int_0^T [\theta(x^*Mx + v^*Nv + 2mx + 2nv)$$
$$- (x^*F^* + v^*G^* + f^*)H^*R^{-1}z$$
$$- (x^*(H^*R^{-1})' + (h^*R^{-1})')z - \frac{1}{2}(x^*H^* + h^*)R^{-1}(Hx + h)]\mathrm{d}t$$
$$- \int_0^T z^*R^{-1}H\mathrm{d}w\Bigg\}$$
$$\times \exp\{\theta[x^*(T)M_Tx(T) + 2m_Tx(T)]$$
$$+ (x^*(T)H^*(T) + h^*(T))R^{-1}(T)z(T)\}. \tag{3.2.22}$$

Let us define the function

$$p(x,t) = \exp\left\{-\frac{1}{2}[x^*\Sigma(t)x - 2\sigma(t)x - \beta(t)]\right\} \tag{3.2.23}$$

in which x is a parameter. Note that

$$p(x(T),T) = \exp\{\theta[x^*(T)M_Tx(T) + 2m_Tx(T)]$$
$$+ (x^*(T)H^*(T) + h^*(T))R^{-1}(T)z(T)\}. \tag{3.2.24}$$

Let also $\lambda(t)$ be defined by

$$
\begin{aligned}
\lambda(t) = \exp\Bigg\{ & \int_0^t [\theta(x^*Mx + v^*Nv + 2mx + 2nv) - (x^*F^* + v^*G^* + f^*)H^*R^{-1}z \\
& - (x^*(H^*R^{-1})' + (h^*R^{-1})')z - \frac{1}{2}(x^*H^* + h^*)R^{-1}(Hx + h)]\mathrm{d}s \\
& - \int_0^t z^*R^{-1}H\mathrm{d}w \Bigg\}
\end{aligned}
$$

(3.2.25)

then from (3.2.22) we get

$$
X_v = \theta\lambda(T)p(x(T),T). \tag{3.2.26}
$$

We note that $p(x,t)$ depends parametrically on $z(\cdot)$. Since $z(\cdot)$ and $x(\cdot)$ are independent we can express $p(x(T),T)$ by computing the Ito differential $\mathrm{d}p(x(t),t)$ while considering $z(\cdot)$ frozen. This can be done using Ito's formula, provided $p(x,t)$ is continuously differentiable in time (it is already C^2 in x). This is not exactly the case, looking at the values of $\dot{\sigma}$ and $\dot{\beta}$ derived from (3.2.19), (3.2.20). It is on the other hand possible when the functions of time $F, Q, (R^{-1}H)'(R^{-1}h)', m, f, G$ are continuous and for controls $v(\cdot)$ which are continuous processes. But, considering the equality (3.2.21) to be proven, it is easy to check that whenever it is valid for slightly more regular coefficients, and continuous controls, it remains valid when the above coefficients are just L^∞ and the control $v(\cdot)$ is in U. Therefore we may assume the desired regularity. From Ito's formula, we have

$$
\begin{aligned}
\mathrm{d}p(x(t),t) = p\Bigg\{ & \Big[-\frac{1}{2}(x_t^*\dot{\Sigma}_t x_t - 2\dot{\sigma}_t x_t + \dot{\beta}_t) - (x_t^*\Sigma_t - \sigma_t)(F_f x_t + f_t + G_t v_t) \\
& + \frac{1}{2}(x_t^*\Sigma_t - \sigma_t)Q_t(\Sigma_t x_t - \sigma_t^*) - \frac{1}{2}\mathrm{tr}\,\Sigma_t Q_t\Big]\mathrm{d}t \\
& - (x_t^*\Sigma_t - \sigma_t)\mathrm{d}w_t\Bigg\}.
\end{aligned}
$$

A calculation allows us to express $\mathrm{d}\lambda(t)p(x(t),t)$.

By the choice of Σ, σ, β, it readily follows that

$$
\mathrm{d}\lambda(t)p(x(t),t) = -\lambda(t)p(x(t),t)[z_t^*R_t^{-1}H_t + x_t^*\Sigma_t - \sigma_t]\mathrm{d}w
$$

and thus

$$
\begin{aligned}
\lambda(t)p(x(t),t) = p(\xi,0)\exp\Bigg[& \int_0^t -(z^*R^{-1}H + x^*\Sigma - \sigma)\mathrm{d}w \\
& - \frac{1}{2}\int_0^t |H^*R^{-1}z + \Sigma x - \sigma^*|^2\mathrm{d}s\Bigg].
\end{aligned}
$$

Then from (3.2.26) and the definition of $p(x,t)$, we get immediately (3.2.21). ∎

The next step is to prove

Lemma 3.2.3 *We assume (3.2.17), (3.2.18) and*

$$
I + P_0^{1/2}\Sigma(0)P_0^{1/2} > 0 \tag{3.2.27}
$$

then one has

$$E[X_v|\mathcal{Z}^T] = \frac{\theta}{|I + P_0^{1/2}\Sigma(0)P_0^{1/2}|^{1/2}} \exp\{-\frac{1}{2}[x_0^*\Sigma(0)x_0 - 2\sigma(0)x_0 + \beta(0)$$
$$- (x_0^*\Sigma(0) - \sigma(0))P_0^{1/2}(I + P_0^{1/2}\Sigma(0)P_0^{1/2})^{-1}$$
$$\times P_0^{1/2}(\Sigma(0)x_0 - \sigma(0)^*)]\}.$$

(3.2.28)

Proof

Since $X_v > 0$ (up to the multiplicative constant θ), we can consider its expectation (or conditional expectation), with a possible infinite value.

The main point consists in noticing that

$$E\left\{\exp\left[\int_0^T -(z^*R^{-1}H + x^*\Sigma - \sigma)\mathrm{d}w - \frac{1}{2}|H^*R^{-1}z + \Sigma x - \sigma^*|^2\mathrm{d}t\right]\,\Big|\,\mathcal{Z}^T, \xi\right\}$$
$$= 1 \quad a.s.$$

(3.2.29)

Indeed since $z(\cdot), \xi, w(\cdot)$ are mutually independent, we compute (3.2.29) by freezing $z(\cdot), \xi$ as parameters and taking the expectation with respect to $w(\cdot)$. The argument $-(z^*R^{-1}H + x^*\Sigma - \sigma)$ can be written as the sum of $-(z^*R^{-1}H + x_1^*\Sigma - \sigma)$ which is deterministic and $-\alpha^*\Sigma$ which is gaussian. The result is then a consequence of the general property used in Lemma 3.1.1.

But from the definition of X_v, we deduce at once that

$$E[X_v|\mathcal{Z}^T, \xi] = \theta \exp\left\{-\frac{1}{2}[\xi^*\Sigma(0)\xi - 2\sigma(0)\xi + \beta(0)]\right\}.$$

We next take the expectation of the right hand side with respect to the random variable ξ, while freezing the value of $z(\cdot)$, since ξ and $z(\cdot)$ are independent. An explicit calculation is possible since ξ is gaussian, with mean x_0 and covariance P_0. The assumption (3.2.27) is used to guarantee that the expectation with respect to ξ is finite. Its value corresponds to the right hand side of (3.2.28). ∎

We shall now give another expression for the conditional expectation (3.2.28).

We first introduce a new Riccati equation as follows:

$$\dot{P} - FP - PF^* + P(H^*R^{-1}H - 2\theta M)P - Q = 0$$
$$P(0) = P_0$$

(3.2.30)

and we shall assume

$$P^{-1}(t) + \Sigma(t) \geq \delta I, \quad \delta > 0, \quad \forall t \in [0, T].$$

(3.2.31)

We next introduce the processes

$$\mathrm{d}r = [(F - PH^*R^{-1}H + 2\theta PM)r + Gv + f + 2\theta Pm^*$$
$$- PH^*R^{-1}h]\mathrm{d}t + PH^*R^{-1}\mathrm{d}z$$

(3.2.32)

$$r(0) = x_0$$

and

$$s_t = x_0^* P_0^{-1} x_0 + \log[(2r)^n |P_0|]$$
$$+ \int_0^t [\text{tr} \, (P^{-1}Q + 2F) - r^* P^{-1} Q P^{-1} r + 2r^* P^{-1}(Gv + f) \qquad (3.2.33)$$
$$- 2\theta(v^* Nv + 2nv) + 2z^*(R^{-1}h)' + h^* R^{-1} h] ds$$

The justification of these formulas will become apparent in the following.

Note that $r(t), s(t)$ are forward processes, whereas $\sigma(t), \beta(t)$ arising in the definition of $p(x,t)$ (see (3.2.23)) are backward processes.

The main idea is described as follows. Note that (3.2.28) can be read as

$$E[X_v | \mathcal{Z}^T] = \theta \int p(x, 0) \Pi(x) dx \qquad (3.2.34)$$

where

$$\Pi(x) = \frac{\exp[-\frac{1}{2}(x - x_0)^* P_0^{-1}(x - x_0)]}{(2n)^{n/2} \sqrt{|P_0|}}$$

represents the density of the initial random variable ξ. We shall try to identify a process $q(x,t)$ (which is forward) such that

$$\int p(x, t) q(x, t) dx = \int p(x, 0) q(x, 0) dx, \quad \forall t \in [0, T] \qquad (3.2.35)$$

and

$$q(x, 0) = \Pi(x).$$

To check that this is possible, we first notice that $p(x,t)$ is solution of a parabolic equation. An easy calculation shows that p is the solution of

$$-\frac{\partial p}{\partial t} = \frac{1}{2} \text{tr} \, D^2 p Q + Dp(Fx + Gv + f - QH^* R^{-1} z)$$
$$+ \frac{p}{2} \{ -x^*(H^* R^{-1} H - 2\theta M)x + 2\theta(v^* Nv + 2nv)$$
$$+ 4\theta mx - 2z^*[R^{-1}H(Fx + Gv + f) + (R^{-1}H)'x + (R^{-1}h)'] \qquad (3.2.36)$$
$$- 2h^* R^{-1} Hx + z^* R^{-1} HQH^* R^{-1} z - h^* R^{-1} h \}$$

$$p(x, T) = \exp \left[\theta(x^* M_T x + 2m_T x) + (x^* H^*(T) + h^*(T)) R^{-1}(T) z(T) \right]$$

Therefore to achieve (3.2.35), one can introduce the formal parabolic equation adjoint to (3.2.36), namely

$$\frac{\partial q}{\partial t} = \frac{1}{2} \text{tr} \, D^2 q Q - Dq(Fx + Gv + f - QH^* R^{-1} z)$$
$$+ \frac{q}{2} \{ -x^*(H^* R^{-1} H - 2\theta M)x + 2\theta(v^* Nv + 2nv)$$
$$+ 4\theta mx - 2z^*[R^{-1}H(Fx + Gv + f) + (R^{-1}H)'x + (R^{-1}h)'] \qquad (3.2.37)$$
$$- 2h^* R^{-1} Hx + z^* R^{-1} HQH^* R^{-1} z - 2\text{tr} \, F - h^* R^{-1} h \}$$

$$q(x, 0) = \Pi(x).$$

A simple calculation shows that the process

$$q(x,t) = \exp\left\{-\frac{1}{2}[x^*P^{-1}(t)x - 2x^*P^{-1}(t)(r(t)\right.$$

$$\left. - P(t)H^*(t)R^{-1}(t)z(t)) + s_t]\right\}$$

(3.2.38)

is the solution of (3.2.37).

By virtue of the assumption (3.2.31), we can, after multiplying (3.2.37) by p, perform the desired integrations by parts in x, and thus (3.2.35) obtains. ∎

We can then prove the following

Lemma 3.2.4 *We assume (3.2.17), (3.2.18), (3.2.27), (3.2.31); then*

$$E[X_v|\mathcal{Z}^T] = \theta \exp\left\{\theta[r_T^*(I - 2\theta M_T P(T))^{-1}(M_T r_T + 2m_T^*)\right.$$

$$+ 2\theta^2 m_T(P^{-1}(T) - 2\theta M_T)^{-1}m_T^*]$$

$$+ \int_0^T (r^*H^* + h^*)R^{-1}dz$$

$$- \frac{1}{2}\int_0^T (r^*H^* + h^*)R^{-1}(Hr + r)dt$$

$$+ \theta \int_0^T (r^*Mr + 2mr + v^*Nv + 2nv)dt$$

$$\left. + \theta \int_0^T \text{tr } PM dt\right\}|I - 2\theta M_T P(T)|^{-1/2}.$$

(3.2.39)

Proof

Using (3.2.34) and (3.2.35) we deduce

$$E[X_v|\mathcal{Z}^T] = \theta \int p(x,T)q(x,T)dx.$$

Using (3.2.23), (3.2.19), (3.2.20) and (3.2.38) it follows that

$$E[X_v|Z^T] = \theta \int \exp\left\{-\frac{1}{2}[x^*(P^{-1}(T) - 2\theta M_T)x - 2x^*(P^{-1}(T)r(T) + 2\theta m_T^*)\right.$$

$$\left. + s_T - 2h^*(T)R^{-1}(T)z(T)]\right\}dx.$$

(3.2.40)

The next step is to prove that

$$s_t = r_t P^{-1}(t)r_t + \log[(2n)^n|P(t)|] + 2h^*(t)R^{-1}(t)z(t)$$

$$- 2\int_0^t (r^*H^* + h^*)R^{-1}dz + \int_0^t (r^*H^* + h^*)R^{-1}(Hr + h)ds$$

$$- 2\theta \int_0^t (r^*Mr + 2mr + v^*Nv + 2nv)dt - 2\theta \int_0^t \text{tr } PM dt.$$

(3.2.41)

This follows from a simple calculation, noting in particular that since the equation (3.2.30) can be written

$$\dot{P} = (F + PF^*P^{-1} - PH^*R^{-1}H + 2\theta PM + QP^{-1})P$$

then one has the property

$$\frac{d}{dt} \log |P(t)| = \text{tr}(F + PF^*P^{-1} - PH^*R^{-1}H + 2\theta PM + QP^{-1})$$
$$= \text{tr}\,(2F - PH^*R^{-1}H + QP^{-1} + 2\theta PM).$$

Computing the integral at the right hand side of (3.2.40) and collecting results, the desired result obtains. ∎

3.2.3   Solution of the LEG stochastic control problem

By LEG we mean linear exponential gaussian. Under the assumptions of Lemma 3.2.4, the original problem (3.2.2), (3.2.3), (3.2.12) has been reduced to a problem with full observation, similar to the one treated in Section 3.1. Namely, the state of the system is governed by the dynamics

$$dr = [(F - PH^*R^{-1}H + 2\theta PM)r + Gv + f + 2\theta Pm^* - PH^*R^{-1}h]dt$$
$$+ PH^*R^{-1}dz \tag{3.2.42}$$

$$r(t) = x_0$$

and the payoff is defined by

$$J(v(\cdot)) = \theta E \exp\Bigg\{ \theta[r_T^*(I - 2\theta M_T P(T))^{-1}(M_T r_T + 2m_T^*)]$$
$$+ \int_0^T (r^*H^* + h^*)R^{-1}dz - \frac{1}{2}\int_0^T (r^*H^* + h^*)R^{-1}(Hr + h)dt$$
$$+ \theta \int_0^T (r^*Mr + 2mr + v^*Nv + 2nv)dt \Bigg\} \tag{3.2.43}$$

(with a possible value $+\infty$). The control $v(\cdot)$ belongs to U. Note that the functional (3.2.43) is not exactly (3.2.12). To recover (3.2.12) we have to multiply (3.2.43) by the constant

$$\frac{\exp[2\theta^2 m_T(P^{-1}(T) - 2\theta M_T)^{-1}m_T^* + \theta \int_0^T \text{tr}\, MPdt]}{|I - 2\theta M_T P(T)|^{1/2}}$$

We then follow the method used in the full observation case (see § 3.1.3).

We introduce for any control $v(\cdot)$, the process

$$\phi_t = \int_0^t (r^*H^* + h^*)R^{-1}dz - \frac{1}{2}\int_0^t (r^*H^* + h^*)R^{-1}(Hr + h)ds$$
$$+ \theta \int_0^t (r^*Mr + 2mr + v^*Nv + 2nv)dt \tag{3.2.44}$$
$$+ \theta r_t^*S(t)r_t + 2\theta s_t r_t + \theta \rho_t$$

where $S(\cdot)$, $s(\cdot)$, $\rho(\cdot)$ are defined later.

At this stage we require

$$S(T) = \frac{1}{2}[(I - 2\theta M_T P(T))^{-1} M_T + M_T (I - 2\theta P(T) M_T)^{-1}]$$
$$s(T) = m_T (I - 2\theta P(T) M_T)^{-1}$$
$$\rho(T) = 0.$$

(3.2.45)

With these definitions, we deduce from (3.2.43) that

$$J(v(\cdot)) = \theta E \exp \phi_{\rho T}.$$

(3.2.46)

We then compute

$$\begin{aligned}
\mathrm{d}\exp\phi_t = {} & \\
& \exp\phi_t \Big\{ \theta r_t^*[\dot{S}_t + S_t(F + 2\theta PM) \\
& + (F^* + 2\theta MP)S_t - S_t(GN^{-1}G^* - 2\theta PH^*R^{-1}HP)S_t + M]r_t \\
& + 2\theta[\dot{s}_t + s_t(F + 2\theta PH^*R^{-1}HPS + 2\theta PM - GN^{-1}G^*S) + m_t \\
& + (f^* + 2\theta mP - nN^{-1}G^*)S]r_t \\
& + \theta[\dot{\rho}_t + 2\theta s_t PH^*R^{-1}HPs_t^* - (s_tG_t + n_t)N_t^{-1}(G_t^*s_t^* + n_t^*) \\
& + 2s_t(f_t + 2\theta P_t m_t^*) + \mathrm{tr}\, PSPH^*R^{-1}H] \\
& + \theta[v_t^* + (r_t^*S_t + s_t)G_tN_t^{-1} + n_tN_t^{-1}]N_t[v_t + N_t^{-1}G_t^*(S_tr_t + s_t^*) + N_t^{-1}] \Big\}\mathrm{d}t \\
& + \exp\phi_t \{[r_t^*(I + 2\theta S_t P_t) + 2\theta s_t P_t]H^* + h^*\}\, R^{-1}\mathrm{d}z.
\end{aligned}$$

We then fix the (deterministic) functions $S(t), s(t), \rho(t)$ as follows:

$$\begin{aligned}
& \dot{S}_t + S_t(F + 2\theta PM) + (F^* + 2\theta MP)S_t \\
& \quad - S_t(GN^{-1}G^* - 2\theta PH^*R^{-1}HP)S_t + M = 0
\end{aligned}$$

(3.2.47)

$$\begin{aligned}
& \dot{s}_t + s_t(F + 2\theta PH^*R^{-1}HPS + 2\theta PM - GN^{-1}G^*S) \\
& + m_t + (f^* + 2\theta mP - nN^{-1}G^*)S_t = 0
\end{aligned}$$

(3.2.48)

$$\begin{aligned}
& \dot{\rho}_t + 2\theta PH^*R^{-1}HPs_t - (s_tG_t + n_t)N_t^{-1}(G_t^*s_t^* + n_t^*) \\
& + 2s_t(f_t + 2\theta P_t m_t^*) + \mathrm{tr}\, PSPH^*R^{-1}H = 0
\end{aligned}$$

(3.2.49)

provided, naturally, that the solution of the Riccati equation (3.2.47) (with initial condition (3.2.45)) exists on the integral $(0, T)$. With this choice, one obtains

$$\begin{aligned}
\mathrm{d}\exp\phi_t = {} & \exp\phi_t \left([r_t^*(I + 2\theta S_t P_t) + 2\theta s_t P_t]H^* + h^*\right) R^{-1}\mathrm{d}z \\
& + \theta \exp\phi_t[v_t^* + (r_t^*S_t + s_t)G_tN_t^{-1} + n_tN_t^{-1}]N_t \\
& \times [v_t + N_t^{-1}G_t^*(S_tr_t + s_t^*) + N_t^{-1}n_t^*]\mathrm{d}t
\end{aligned}$$

hence

$$
\begin{aligned}
\theta \exp \phi_T = &\theta \exp \phi_0 \exp \Bigg(\int_0^T \{[r_t^*(I + 2\theta S_t P_t) + 2\theta s_t P_t]H^* + h^*\} R^{-1} dz \\
&- \frac{1}{2} \int_0^T \{[r_t^*(I + 2\theta S_t P_t) + 2\theta s_t P_t]H^* + h^*\} R^{-1} \\
&\times \{H[(I + 2\theta P_t S_t)r_t + 2\theta P_t s_t^*] + h\} dt \Bigg) \\
&\times \exp \theta \int_0^T [v_t^* + (r_t^* S_t + s_t)G_t N_t^{-1} + n_t N_t^{-1}] \\
&\times N_t[v_t + N_t^{-1}G_t^*(S_t r_t + s_t^*) + N_t^{-1} n_t^*] dt.
\end{aligned}
\tag{3.2.50}
$$

In a way similar to what has been done in the full observation case (cf. § 3.1.4), we are led to consider the following class of admissible controls

$$
v(\cdot) \in U,
\tag{3.2.51}
$$

and

$$
\begin{aligned}
E \exp \Bigg(&\int_0^T \{[r_t^*(I + 2\theta S_t P_t) + 2\theta s_t P_t]H^* + h^*\} R^{-1} dz \\
&- \frac{1}{2} \int_0^T \{[r_t^*(I + 2\theta S_t P_t) + 2\theta s_t P_t]H^* + h^*\} R^{-1} \\
&\times H[(I + 2\theta P_t S_t)r_t + 2\theta P_t s_t^*] + h\} dt \Bigg) = 1.
\end{aligned}
$$

We define a particular control denoted by $u(t)$ such that

$$
u_t = -N_t^{-1}G_t^*(S_t \hat{r}_t + s_t^*) - N_t^{-1}n_t^*
\tag{3.2.52}
$$

with the process $\hat{r}_t$ defined by

$$
\begin{aligned}
d\hat{r} = &[(F - PH^*R^{-1}H + 2\theta PM - GN^{-1}G^*S)\hat{r} - PH^*R^{-1}h \\
&- GN^{-1}(G^*s^* + n^*) + f + 2\theta Pm^*]dt + PH^*R^{-1}dz
\end{aligned}
\tag{3.2.53}
$$
$$
\hat{r}(0) = x_0.
$$

We then can state

Theorem 3.2.1 *We assume (3.2.1), (3.2.17). We also assume that the Riccati equations (3.2.18), (3.2.30), (3.2.47) have a global solution on $[0, T]$ such that (3.2.31) holds true. Consider the class of admissible controls defined by (3.2.51) and the payoff (3.2.43) (or (3.2.12), which is the same up to a positive multiplicative constant). Then the control defined by formulas (3.2.52), (3.2.53) is optimal.*

Proof

From formulas (3.2.50) and (3.2.46) we deduce immediately that, whenever $v(\cdot)$ belongs to the class of admissible controls,

$$
J(v(\cdot)) \geq \theta \exp \phi_0.
\tag{3.2.54}
$$

On the other hand, for the control $u(\cdot)$, the corresponding process ϕ, denoted by $\hat{\phi}$, satisfies

$$
\begin{aligned}
\theta \exp \hat{\phi}_T = \theta \exp \phi_0 \exp\bigg(& \int_0^T \left\{ [\hat{r}_t^*(I + 2\theta S_t P_t) + 2\theta s_t P_t] H^* + h^* \right\} R^{-1} \mathrm{d}z \\
& - \frac{1}{2} \int_0^T \hat{r}_t \left\{ [r_t^*(I + 2\theta S_t P_t) + 2\theta s_t P_t] H^* + h^* \right\} R^{-1} \quad (3.2.55) \\
& \times \left\{ H[(I + 2\theta P_t S_t)\hat{r}_t + 2\theta P_t s_t^*] + h \right\} \mathrm{d}t \bigg).
\end{aligned}
$$

But $\hat{r}(t)$ is a gaussian process and, for reasons already explained in Lemma 3.1.1, the expectation of the second exponential on the right hand side of (3.2.55) is 1. Hence

$$
E\theta \exp \hat{\phi}_T = \theta \exp \phi_0.
$$

This proves at the same time that $u(\cdot)$ is admissible, and that it is optimal. ∎

3.3   Additional remarks to the partial information case

3.3.1   The case $\theta = 0$

Take $\theta = 0$, then (3.2.30) becomes

$$\dot{P} - FP - PF^* + PH^*R^{-1}HP - Q = 0$$
$$P(0) = P_0 \tag{3.3.1}$$

and (3.2.47), (3.2.48), (3.2.49) reduce to

$$\dot{S} + SF + F^*S - SGN^{-1}G^*S + M = 0$$
$$S(T) = M_T \tag{3.3.2}$$

$$\dot{s} + s(F - GN^{-1}G^*S) + m + (f^* - nN^{-1}G^*)S = 0$$
$$s(T) = m_T \tag{3.3.3}$$

$$\dot{\rho} - (sG + n)N^{-1}(G^*s^* + n^*) + \operatorname{tr} PSPH^*R^{-1}H + 2sf = 0$$
$$\rho(T) = 0. \tag{3.3.4}$$

The process (3.2.5) reduces to

$$d\hat{r} = [(F - PH^*R^{-1}H - GN^{-1}G^*S)\hat{r} - GN^{-1}(G^*s^* + n^*)$$
$$- PH^*R^{-1}h + f]dt + PH^*R^{-1}dz \tag{3.3.5}$$
$$\hat{r}(0) = x_0.$$

The above solution is the one obtained in Section 2.3, for the quadratic payoff. This is consistent with the remark already made in § 3.1.2, namely that when θ is small the problem amounts to minimizing the quadratic payoff.

3.3.2   The full information case

We try now to recover the full information case, i.e. the case studied in Section 3.1. Assume that $R = \epsilon\bar{R}$, then (see (3.2.30))

$$P \sim \sqrt{\epsilon}\bar{P}$$

with

$$\bar{P}H^*\bar{R}^{-1}H\bar{P} = Q$$

and S, s, ρ reduce to (see (3.2.47), (3.2.48), (3.2.49))

$$\dot{S} + SF + F^*S - S(GN^{-1}G^* - 2\theta Q)S + M = 0$$
$$S(T) = M_T, \tag{3.3.6}$$

$$\dot{s} + s(F + 2\theta QS - GN^{-1}G^*S) + m + (f^* - nN^{-1}G^*)S = 0$$
$$s(T) = m_T. \tag{3.3.7}$$

$$\dot{\rho} + 2\theta sQs^* - (sG + n)N^{-1}(G^*s^* + n^*) + \text{tr } SQ + 2sf = 0$$
$$\rho(T) = 0. \tag{3.3.8}$$

We recover the formulas already obtained in (3.1.12), (3.1.13), (3.1.14).

3.3.3   Sufficient statistics

We have obtained a linear feedback optimal control for the full problem, although the separation principle does not hold. Indeed the sufficient statistics (3.2.53) take account of the payoff, and the feedback rule takes account of the covariance of the disturbances.

4   Non linear filtering theory

Introduction

In this chapter we present the theory of non linear filtering. Given a signal governed by a diffusion equation

$$dx_t = g(x_t, t)dt + \sigma(x_t, t)dw_t, \quad x(0) = \xi$$

and an observation process described by

$$dz = h(x_t, t)dt + db_t, \quad z(0) = 0$$

the problem at hand is to compute the conditional probability

$$\Pi(t)(\psi) = E\left[\psi(x_t)|z(s), s \leq t\right]$$

where ψ is any test function.

The main core of the theory consists in characterizing $\Pi(t)(\psi)$ as the solution of an evolution equation. We study this equation and prove existence and uniqueness in some adequate space of measures. Moreover the conditional probability can be written as a ratio

$$\Pi(t)(\psi) = \frac{p(t)(\psi)}{p(t)(1)}$$

where $p(t)(\psi)$ is called the unnormalized conditional probability. It is also characterized as the solution of an evolution equation, which is in principle simpler than the other one, since it is linear.

The complete study of these two equations, known as the Kushner–Shatonovitch equation and the Zakai equation, is the objective of Sections 4.1–4.3.

In Section 4.4 we explicitly solve them in the linear case, and recover the Kalman filter.

In Section 4.5 we extend the results to the case when the signal noise and the observation noise are correlated. In Section 4.6 we look for densities. Namely, $p(t)(\psi)$ that will appear as

$$p(t)(\psi) = \int \psi(x)p(x, t)dx.$$

The process $p(x, t)$ will be the solution of a stochastic PDE (partial differential equation). In fact, when there is no correlation and h is sufficiently smooth, it is possible by a transformation called the robust form, to reduce this equation to an ordinary PDE with random coefficients. We develop this approach and use it to give representation formulas for $p(x, t)$. These formulas can be viewed as the analogue of the probabilistic interpretation of the solution of parabolic equations.

In Section 4.4 we study the stochastic PDEs directly. In fact we introduce forward and backward stochastic PDEs which are linked by a duality relation, and extend the classical situation with parabolic PDEs.

4.1　　　Non linear filtering equation

4.1.1　　　Assumptions: notation

Let $(\Omega, \mathcal{A}, P)$ be a probability space equipped with a filtration $\mathcal{F}^t$ on which are given two $\mathcal{F}^t$ Wiener processes $w(t)$ and $b(t)$ in R^n and R^m respectively. We assume that

$$
\begin{aligned}
&w(\cdot) \text{ has covariance matrix } Q(\cdot), \\
&b(\cdot) \text{ has covariance matrix } R(\cdot); \\
&R(\cdot) \text{ is uniformly positive definite,} \\
&w(\cdot) \text{ and } b(\cdot) \text{ are mutually independent.}
\end{aligned}
\tag{4.1.1}
$$

Let also

ξ be a random variable, $\mathcal{F}^0$ measurable, independent of $w(\cdot)$ and $b(\cdot)$, with values in R^n. The probability measure of ξ is denoted by π_0.

$$
\tag{4.1.2}
$$

Next let

$$
\begin{aligned}
g(x,t) &: \ R^n \times (0, \infty) \to R^n \\
\sigma(x,t) &: \ R^n \times (0, \infty) \to \mathcal{L}(R^n; R^n)
\end{aligned}
\tag{4.1.3}
$$

be Borel functions such that

$$
\begin{aligned}
&|g(x_1, t) - g(x_2, t)| \leq k|x_1 - x_2| \\
&\|\sigma(x_1, t) - \sigma(x_2, t)\| \leq k|x_1 - x_2| \\
&g(0, t), \ \sigma(0, t) \text{ are bounded}
\end{aligned}
$$

and take

$$
h(x; t): \ R^n \times (0, \infty) \to R^m, \ \text{Borel}
\tag{4.1.4}
$$

where

$$
|h(x, t)| \leq k(1 + |x|).
$$

We also assume that

$$
E|\xi|^2 < \infty \quad \text{i.e.} \quad \int |x|^2 \mathrm{d}\Pi_0(x) < \infty.
\tag{4.1.5}
$$

We can define the processes

$$
\begin{aligned}
\mathrm{d}x &= g(x_t, t)\mathrm{d}t + \sigma(x_t, t)\mathrm{d}w_t \\
x(0) &= \xi
\end{aligned}
\tag{4.1.6}
$$

$$dz = h(x_t, t)dt + db_t$$
$$z(0) = 0. \tag{4.1.7}$$

From the classical theory of stochastic differential equations, in the sense of Ito, there exists one and only one process x_t solution of (4.1.6) and

$$x(\cdot) \in L^2(\Omega, \mathcal{A}; P; C(0, T; R^n)), \quad \forall T \tag{4.1.8}$$

Having defined $x(\cdot)$, the process $z(\cdot)$ is explicitly defined by (4.1.7).

We denote

$$\mathcal{Z}^t = \sigma(z(s), s \leq t) \subset \mathcal{F}^t.$$

Let ψ be a Borel bounded function on R^n†. The problem of non linear filtering consists in computing

$$\Pi(t)(\psi) = E\left[\psi(x(t))|\mathcal{Z}^t\right], \quad \forall t. \tag{4.1.9}$$

Clearly one has

$$\Pi(0)(\psi) = \Pi_0(\psi) = \int \psi(x)d\Pi_0(x). \tag{4.1.10}$$

The process $x(t)$ is the signal process and $z(t)$ the observation process.

4.1.2 Change of probability measure

It will be convenient to modify the probability on Ω in order to transform the process $z(\cdot)$ into a Wiener process. For that we introduce the process $\rho(t)$ defined by

$$d\rho = -\rho h^*(x_t, t)R_t^{-1}db_t$$
$$\rho(0) = 1 \tag{4.1.11}$$

where h^* denotes the line vector h; explicitly

$$\rho(t) = \exp\left(-\int_0^t h^*(x_s, s)R_s^{-1}db_s - \frac{1}{2}\int_0^t h^*(x_s)R_s^{-1}h(x_s)ds\right). \tag{4.1.12}$$

Let us check

Lemma 4.1.1 *One has*

$$E\rho(t) = 1. \tag{4.1.13}$$

Proof

Let us first check that

$$E\rho(t)|x(t)|^2 < C. \tag{4.1.14}$$

Indeed by Ito's formula

$$d|x(t)|^2 = 2x^*(gdt + \sigma dw_t) + \operatorname{tr} \sigma Q\sigma^* dt \tag{4.1.15}$$
$$d\rho_t|x(t)|^2 = -\rho_t|x_t|^2 h^* db + 2\rho_t x_t^* \sigma dw_t$$
$$+ \rho_t(2x_t^* g + \operatorname{tr} \sigma Q\sigma^*)dt \tag{4.1.16}$$

† We shall also need to consider unbounded functions. This will be done later.

$$d\frac{\rho_t|x_t|^2}{1+\epsilon\rho_t|x_t|^2} = \frac{1}{1+\epsilon\rho_t|x_t|^2)^2}\left(-\rho_t|x_t|^2h^*db + 2\rho_t x_t^*\sigma dw_t\right)$$

$$+\left[\frac{\rho_t}{(1+\epsilon\rho_t|x_t|^2)^2}\left(2x_t^*g + \operatorname{tr}\sigma Q\sigma^*\right)\right.$$

$$\left.-\frac{\epsilon}{(1+\epsilon\rho_t|x_t|^2)^3}\left(\rho_t^2|x_t|^4 h^*Qh + 4\rho_t^2 x_t^*\sigma R\sigma^* x_t\right)\right]dt.$$

Integrating between 0 and t, we can take the mathematical expectation, which yields after some easy simplifications

$$\frac{d}{dt}E\frac{\rho_t|x_t|^2}{1+\epsilon\rho_t|x_t|^2} \le E\frac{\rho_t\left(2x_t^*g + \operatorname{tr}\sigma Q\sigma^*\right)}{1+\epsilon\rho_t|x_t|^2}$$

$$\le K\left(E\frac{\rho_t|x_t|^2}{1+\epsilon\rho_t|x_t|^2} + 1\right)$$

where we have used the fact that $E\rho_t \le 1$.

It follows that

$$E\frac{\rho_t|x_t|^2}{1+\epsilon\rho_t|x_t|^2} \le C \quad \forall t \in [0,T]$$

and, by Fatou's lemma, the result (4.1.14) obtains.

Let us next use the fact that

$$d\frac{\rho_t}{1+\epsilon\rho_t} = -\frac{\rho_t h^*db}{(1+\epsilon\rho_t)^2} - \frac{\epsilon\rho_t^2 h^*(x_t)R_t h(x_t)dt}{(1+\epsilon\rho_t)^3} \tag{4.1.17}$$

hence also

$$E\frac{\rho_t}{1+\epsilon\rho_t} = 1 - E\int_0^t \frac{\epsilon\rho_s^2 h^*(x_s)R_s h(x_s)ds}{(1+\epsilon\rho_s)^3}.$$

But

$$\frac{\epsilon\rho_s^2 h^*(x_s)R_s h(x_s)}{(1+\epsilon\rho_s)^3} \to 0 \quad \text{a.s. a.e. as } \epsilon \to 0$$

and remains bounded by $C\rho_s(|x_s|^2+1)$ where C is a constant. By Lebesgue's theorem

$$E\int_0^t \frac{\epsilon\rho_s^2 h^*(x_s)R_s h(x_s)ds}{(1+\epsilon\rho_s)^3} \to 0 \text{ as } \epsilon \to 0.$$

On the other hand, since $E\rho_t < 1$,

$$E\frac{\rho_t}{1+\epsilon\rho_t} \to E\rho_t$$

hence necessarily (4.1.13) holds.   ∎

By virtue of Lemma 4.1.1, we may define a new probability $\tilde{P}$, by setting

$$\left.\frac{d\tilde{P}}{dP}\right|_{\mathcal{F}^t} = \rho(t). \tag{4.1.18}$$

For the system $(\Omega, \mathcal{A}, \tilde{P}, \mathcal{F}^t)$ the processes $w(\cdot)$ and $z(\cdot)$ are $\mathcal{F}^t$ Wiener processes with $Q(\cdot)$ and $R(\cdot)$ respectively as covariance matrices. They also are independent of ξ.

Let us set

$$\eta(t) = \frac{1}{\rho(t)} = \exp\left(\int_0^t h^*(x_s)R_s^{-1}dz_s - \frac{1}{2}\int_0^t h^*(x_s)R_s^{-1}h(x_s)ds\right)$$

then

$$\tilde{E}\eta(t) = E\rho(t)\eta(t) = 1$$

and

$$\frac{dP}{d\tilde{P}}\bigg|_{\mathcal{F}^t} = \eta(t). \tag{4.1.19}$$

We can then express the conditional expectation $\Pi(t)(\psi)$ by the following formula called the Kallianpur–Striebel formula (1968).

Lemma 4.1.2 *The following formula holds:*

$$\Pi(t)(\psi) = \frac{\tilde{E}[\psi(x(t))\eta(t)|\mathcal{Z}^t]}{\tilde{E}[\eta(t)|\mathcal{Z}^t]}. \tag{4.1.20}$$

Proof

Let ξ_t be $\mathcal{Z}^t$ measurable and bounded. By definition

$$E\Pi(t)(\psi)\xi_t = E\psi(x_t)\xi_t$$

hence from (4.1.18)

$$\tilde{E}\Pi(t)(\psi)\xi_t\eta_t = \tilde{E}\psi(x_t)\xi_t\eta_t$$

and

$$\tilde{E}\left\{\Pi(t)(\psi)\tilde{E}[\eta_t|\mathcal{Z}^t]\xi_t\right\} = \tilde{E}\left\{\tilde{E}[\psi(x_t)\eta_t|\mathcal{Z}^t]\xi_t\right\}$$

from which (4.1.20) follows immediately. ∎

It is very useful for computational purposes, which will be emphasized later on, to introduce the so-called *unnormalized conditional probability*

$$p(t)(\psi) = \tilde{E}[\psi(x(t))\eta(t)|\mathcal{Z}^t]. \tag{4.1.21}$$

Noting that

$$\Pi(t)(\psi) = \frac{p(t)(\psi)}{p(t)(1)} \tag{4.1.22}$$

one understands easily the meaning of the word unnormalized. Let us make precise the functional spaces on which the linear operator $p(\cdot)$ operates. Let us use the definition

$$\tilde{L}_z^r(0,T) = \left\{\theta(t;\omega) \in L^r((0,T)\times\Omega; dt\otimes d\tilde{P}),\right.$$
$$\left. \text{for almost all } t, \ \theta_t \in L^r(\Omega,\mathcal{Z}^t,\tilde{P})\right\} \tag{4.1.23}$$

$$L_z^r(0,T) = \left\{\theta(t;\omega) \in L^r((0,T)\times\Omega; dt\otimes dP),\right.$$
$$\left. \text{for almost all } t, \ \theta_t \in L^r(\Omega,\mathcal{Z}^t,P)\right\}, \ r \geq 1.$$

We can then state

Lemma 4.1.3 *Let $\psi(x,t)$ be a Borel function such that*

$$|\psi(x,t)| \le K(1 + |x|^2) \qquad (4.1.24)$$

then setting for any t (with values in $L^1(\Omega, \mathcal{Z}^t, \tilde{P})$)

$$p(t)(\psi(t)) = \tilde{E}[\psi(x_t, t)\eta_t | \mathcal{Z}^t], \qquad (4.1.25)$$

$\psi(t)$ *coincides, up to a modification of Lebesgue measure 0 in time, with an element of $\tilde{L}^1_Z(0,T)$.*

Proof

The process $\psi(x_t, t)\eta_t$ belongs to $L^1((0,T) \times \Omega; \mathrm{d}t \otimes \mathrm{d}\tilde{P})$. Indeed,

$$\tilde{E} \int_0^T |\psi(x_t, t)| \eta_t \mathrm{d}t = E \int_0^T |\psi(x_t, t)| \mathrm{d}t$$

$$\le KE \int_0^T (|x_t|^2 + 1) \mathrm{d}t < \infty.$$

Now first consider $\psi \ge 0$ and set

$$X_t^k = \frac{k\psi(x_t, t)}{k + \psi(x_t, t)} \eta_t \wedge k.$$

We have

$$X_t^k \uparrow \psi(x_t, t)\eta_t \text{ as } k \uparrow \infty$$

The process X_t^k belongs (among other things) to $L^2((0,T) \times \Omega; \mathrm{d}t \otimes \mathrm{d}\tilde{P})$. Consider its projection $\hat{X}_t^k$ on $L^2_Z(0,T)$. It is well known that

$$\hat{X}_t^k = \tilde{E}\left[X_t^k | \mathcal{Z}^t\right] \text{ for almost all } t \text{ in } L^2(\Omega, \mathcal{Z}^t, \tilde{P}). \qquad (4.1.26)$$

We have

$$\tilde{E} \int_0^T |\hat{X}_t^k| \mathrm{d}t = \tilde{E} \int_0^T \hat{X}_t^k \operatorname{sign} \hat{X}_t^k \mathrm{d}t$$

$$= \tilde{E} \int_0^T X_t^k \operatorname{sign} \hat{X}_t^k \mathrm{d}t \qquad (4.1.27)$$

$$\le \tilde{E} \int_0^T \psi(x_t, t)\eta_t \, \mathrm{d}t.$$

But

$$\forall t, \ \tilde{E}\left[X_t^k | \mathcal{Z}^t\right] \uparrow p(t)(\psi(t)) \text{ a.s.}$$

On the other hand, from (4.1.27) and Fatou's lemma, $\hat{X}_t^k \uparrow$ a.e., a.s. to an element of $\tilde{L}^1_Z(0,T)$. Therefore $p(t)(\psi(t))$ is, up to a modification of Lebesgue measure 0 on $(0,T)$, an element of $\tilde{L}^1_Z(0,T)$. Taking respectively the positive and negative parts of ψ, the desired result obtains. ∎

4.1.3   A density result

Let us consider the space $L^1(\Omega, \mathcal{Z}^T, \tilde{P})$ extended to complex valued random variables; T is arbitrarily fixed. We recall that for $\tilde{P}$, $z(\cdot)$ is a Wiener process with values in R^m, with covariance matrix $R(\cdot)$.

Consider next $\beta(\cdot)$ to be a deterministic function, with values in R^m, Borel bounded, and the process

$$\theta(t) = \exp\left(i \int_0^t \beta^* R^{-1} \mathrm{d}z + \frac{1}{2} \int_0^t \beta^* R^{-1} \mathrm{d}s \right). \tag{4.1.28}$$

The use of this type of function in the context of non linear filtering is due to Krylov and Rozovskii (1978).

Then we have the following result.

Lemma 4.1.4 *Let $\xi_T \in L^1(\Omega, \mathcal{Z}^T, \tilde{P})$ such that*

$$\tilde{E}\xi_T\theta_T = 0 \tag{4.1.29}$$

for any choice of $\beta(\cdot)$ in (4.1.28).

Then necessarily

$$\xi_T = 0. \tag{4.1.30}$$

Proof

The condition (4.1.29) is clearly equivalent to

$$\tilde{E}\left[\xi_T \exp\left(i \int_0^T \beta^* R^{-1} \mathrm{d}z \right) \right] = 0 \quad \forall \beta(\cdot). \tag{4.1.31}$$

Now pick $t_1 < t_2 \cdots < t_p$ in $(0, T)$ and note that

$$\sum_{h=1}^p \lambda_h^* z(t_h) = \sum_{h=1}^p \mu_h^*(z(t_h) - z(t_{h-1}))$$

$$= \int_0^T \beta^* R^{-1} \mathrm{d}z$$

where $\lambda_1, \ldots, \lambda_p$ are given, $t_0 = 0$,

$$\mu_p = \lambda_p, \quad \mu_{p-1} = \lambda_p + \lambda_{p-1}, \cdots, \quad \mu_1 = \lambda_p + \cdots + \lambda_1$$

$$\beta(t) = \begin{cases} R(t)\mu_h & \text{for } t \in (t_{h-1}, t_h), h = 1, \ldots, p \\ 0 & \text{for } t \in (t_p, T) \end{cases}$$

then clearly (4.1.31) implies

$$\tilde{E}\left[\xi_T \exp\left(i \sum_{h=1}^p \lambda_h^* z(t_h) \right) \right] = 0$$

and also

$$\tilde{E}\left[\xi_T \sum_{k=1}^{K} c_k \exp\left(i \sum_{h=1}^{p} \lambda_{h,k}^* z(t_h)\right)\right] = 0 \qquad (4.1.32)$$

$\forall K, \forall c_1, \ldots, c_k$ complex, $\lambda_{h,k}$ real.

Let $\psi(x_1, \ldots, x_p)$ be a continuous bounded complex valued function defined on $(R^m)^p$. By Weierstrass' theorem, there exists a uniformly bounded sequence of functions of the form

$$P^\alpha(x_1, \ldots, x_p) = \sum_{k=1}^{K_\alpha} c_k^\alpha \exp\left(i \sum_{h=1}^{p} \lambda_{h,k}^{\alpha*} x_h\right)$$

such that

$$P^\alpha(x_1, \ldots, x_p) \rightarrow \psi(x_1, \ldots, x_p) \ \forall x_1, \ldots, x_p.$$

Therefore (4.1.32) implies immediately

$$\tilde{E}\xi_T\psi(z(t_1), \ldots, z(t_p)) = 0.$$

By a further approximation argument, we can take ψ to be any Borel bounded function. Since moreover the points $t_1, \ldots, t_p$ are arbitrary, we necessarily have (4.1.30). ■

We can also infer from Lemma 4.1.4

Corollary 4.1.1 *Let* $\xi_T \in L^1(\Omega, \mathcal{Z}^T, P)$ *such that*

$$E\xi_T\theta_T = 0$$

for any choice of $\beta(\cdot)$. *Then necessarily* $\xi_T = 0$.

Proof

Indeed, by the assumption, one has

$$\tilde{E}\xi_T\theta_T\eta_T = 0$$

hence, according to Lemma 4.1.4,

$$\xi_T\tilde{E}\left[\eta_T|\mathcal{Z}^T\right] = 0.$$

But $\tilde{E}\left[\eta_T|\mathcal{Z}^T\right] > 0$ a.s. for $\tilde{P}$, since $\eta_T > 0$. Hence $\xi_T = 0$ a.s. for $\tilde{P}$. ■

4.1.4   Zakai equation

We shall assume in this paragraph that

$$E|\xi|^3 < \infty, \text{ i.e. } \int |x|^3 \mathrm{d}\Pi_0(x) < \infty. \qquad (4.1.33)$$

(A stronger assumption than (4.1.5).)

It is easy to deduce

Lemma 4.1.5 *The state process $x(\cdot)$ satisfies*

$$E|x(t)|^3 < C. \tag{4.1.34}$$

Proof

From Ito's formula one has

$$d|x_t|^3 = 3|x_t|\Bigg(x_t^* g(x_t) + \operatorname{tr}\sigma(x_t)Q_t\sigma^*(x_t)$$
$$+ \frac{1}{2}\frac{x_t^*}{|x_t|}\sigma(x_t)Q_t\sigma^*(x_t)\frac{x_t}{|x_t|}\Bigg)dt + 3|x_t|\sigma(x_t)dw_t$$

from which one easily deduces (4.1.34). ∎

We shall use some notation. Let us set

$$a(x,t) = \frac{1}{2}\sigma(x,t)Q(t)\sigma^*(x,t) \tag{4.1.35}$$

and let $A(t)$ be the second order differential operator

$$A(t) = -g^*D - \operatorname{tr} aD^2$$
$$= -\sum_i g_i\frac{\partial}{\partial x_i} - \sum_{i,j} a_{i,j}\frac{\partial^2}{\partial x_i\partial x_j}. \tag{4.1.36}$$

Our objective is to prove

Theorem 4.1.1. *We assume (4.1.1), (4.1.2), (4.1.3), (4.1.4) and (4.1.33). Let $\psi(x,t) \in C_b^{2,1}(R^n \times [0,\infty))$, then one has the relation*

$$p(t)(\psi(t)) = \Pi_0(\psi(0)) + \int_0^t p(s)\left(\frac{\partial\psi}{\partial s} - A(s)\psi(s)\right)ds$$
$$+ \int_0^t p(s)(h^*(s)\psi(s))R_s^{-1}dz, \quad \text{a.s. } \forall t. \tag{4.1.37}$$

Proof

We shall decompose the proof into three steps.

(a) Approximation

Let

$$\eta^\epsilon = \frac{\eta}{1 + \epsilon\eta}$$

then successively

$$d\eta^\epsilon = (1 + \epsilon\eta)^{-2}\eta h^* R^{-1}dz$$
$$- 2\epsilon(1 + \epsilon\eta)^{-3}\eta^2 h^* R^{-1}hdt \tag{4.1.38}$$

$$d\eta^\epsilon\psi = \left[\eta^\epsilon\left(\frac{\partial\psi}{\partial t} - A(t)\psi\right) - 2\epsilon\psi(1 + \epsilon\eta)^{-3}\eta^2 h^* R^{-1}h\right]dt$$
$$+ \eta^\epsilon D\psi^*\sigma dw + \psi(1 + \epsilon\eta)^{-2}\eta h^* R^{-1}dz \tag{4.1.39}$$

and using the process θ defined in (4.1.28) it follows that

$$
\begin{aligned}
d\theta\eta^\epsilon\psi = &\left[\theta\eta^\epsilon\left(\frac{\partial\psi}{\partial t} - A(t)\psi\right) - 2\epsilon\theta\psi(1 + \epsilon\eta)^{-3}\eta^\epsilon h^* R^{-1}h \right. \\
&\left. + i\theta\psi\eta(1 + \epsilon\eta)^{-2}h^* R^{-1}\beta\right]dt + \theta\eta^\epsilon D\psi^*\sigma dw \\
&- \theta\psi(1 + \epsilon\eta)^{-2}\eta h^* R^{-1}dz.
\end{aligned}
\tag{4.1.40}
$$

Integrating between 0 and t and taking the mathematical expectation yields

$$
\begin{aligned}
\tilde{E}\theta_t\eta_i^\epsilon\psi(x_t) = &\frac{\Pi_0(\psi(0))}{1 + \epsilon} + \tilde{E}\int_0^t\left[\theta\eta(1 + \epsilon\eta)^{-1}\left(\frac{\partial\psi}{\partial x} - A(s)\psi\right)\right. \\
&- 2\epsilon\theta\psi(1 + \epsilon\eta)^{-3}\eta^2 h^* R^{-1}h \\
&\left. + i\theta\psi\eta(1 + \epsilon\eta)^{-2}h^* R^{-1}\beta\right]ds.
\end{aligned}
\tag{4.1.41}
$$

We use successively the following observations

$$
\begin{aligned}
\tilde{E}\theta_t\eta_i^\epsilon\psi(x_t) = &\tilde{E}\theta_t\tilde{E}^{Z^t}\left(\psi(x_t)\eta_t(1 + \epsilon\eta_t)^{-1}\right) \\
&+ \tilde{E}\int_0^t\left[\theta\eta(1 + \epsilon\eta)^{-1}\left(\frac{\partial\psi}{\partial s} - A(s)\psi\right) - 2\epsilon\theta\psi(1 + \epsilon\eta)^{-3}\eta^2 h^* R^{-1}h\right]ds \\
= &\tilde{E}\theta_t\int_0^t\tilde{E}^{Z^*}\left[\eta(1 + \epsilon\eta)^{-1}\left(\frac{\partial\psi}{\partial s} - A(s)\psi\right) - 2\epsilon\psi\eta^2(1 + \epsilon\eta)^{-3}h^* R^{-1}h\right]ds,
\end{aligned}
$$

$$
i\tilde{E}\int_0^t\theta\psi\eta(1 + \epsilon\eta)^{-2}h^* R^{-1}\beta ds = \tilde{E}\theta_t\int_0^t\tilde{E}^{Z^*}\left[\psi\eta(1 + \epsilon\eta)^{-2}h^*\right]R^{-1}dz.
$$

From (4.1.41) and Lemma 4.1.4 we easily deduce the relation

$$
\begin{aligned}
\tilde{E}^{Z^t}\left[\psi(x_t)\eta_t(1 + \epsilon\eta_t)^{-1}\right] = &\frac{\Pi_0(\psi(0))}{1 + \epsilon} + \int_0^t\tilde{E}^{Z^t}\left[\eta(1 + \epsilon\eta)^{-1}\left(\frac{\partial\psi}{\partial s} - A(s)\psi\right)\right. \\
&\left. - 2\epsilon\psi\eta^2(1 + \epsilon\eta)^{-3}h^* R^{-1}h\right]ds \\
&+ \int_0^t\tilde{E}^{Z^*}\left[\psi\eta(1 + \epsilon\eta)^{-2}h^*\right]R^{-1}dz \quad \forall t, \text{ a.s.}
\end{aligned}
\tag{4.1.42}
$$

Then let ϵ tend to 0. Clearly

$$
\frac{\eta_t}{1 + \epsilon\eta_t} \uparrow \eta_t \quad \text{a.s.},
$$

$$
\tilde{E}^{Z^t}\left[\psi(x_t)\eta_t(1 + \epsilon\eta_t)^{-1}\right] \to p(t)(\psi(t)) \quad \text{a.s.}
$$

$$
\text{a.e. } \tilde{E}^{Z^*}\left[\eta_s(1 + \epsilon\eta_s)^{-1}\left(\frac{\partial\psi}{\partial s} - A(s)\psi\right)\right] \to p(s)\left(\frac{\partial\psi}{\partial s} - A(s)\psi\right) \quad \text{a.s.}
\tag{4.1.43}
$$

and remains bounded by $\tilde{E}^{Z^*}\left[\eta_s|\partial\psi/\partial s - A(s)\psi|\right]$ which belongs to $\tilde{L}^1_Z(0, T)$, by virtue of Lemma 4.1.3, therefore

$$
\int_0^t\tilde{E}^{Z^*}\left[\eta(1 + \epsilon\eta)^{-1}\left(\frac{\partial\psi}{\partial s} - A(s)\psi\right)\right]ds \to \int_0^t p(s)\left(\frac{\partial\psi}{\partial s} - A(s)\psi\right)ds \text{ a.s.} \tag{4.1.44}
$$

Furthermore

$$\left| \epsilon \int_0^t \tilde{E}^{Z^\bullet} \left[\psi \eta^2 (1 + \epsilon \eta)^{-3} h^* R^{-1} h \right] \mathrm{d}s \right| \le C \int_0^t \tilde{E}^{Z^\bullet} \eta_s (1 + |x_s|^2) \mathrm{d}s$$

and tends a.s. to 0.

It remains to study the limit of the stochastic integral

$$\int_0^t \tilde{E}^{Z^\bullet} \left[\psi \eta (1 + \epsilon \eta)^{-2} h^* \right] R^{-1} \mathrm{d}z.$$

The difficulty is that we can only guarantee that

$$\tilde{E}^{Z^\bullet} \left[\psi \eta (1 + \epsilon \eta)^{-2} h^* \right] \to \tilde{E}^{Z^\bullet} \left[\psi \eta h^* \right] \quad \text{a.e., a.s.}$$

and in $\tilde{L}_Z^1(0, T)$. Although the left hand side is a bounded process, the limit process is not. The second part of the proof will be devoted to additional information.

(b) An a priori bound

We shall prove that

$$\text{a.s.} \quad \tilde{E}^{Z^\bullet} \left[\psi(x_s) \eta_s h(x_s) \right] \in L^\infty(0, T; R^m), \tag{4.1.45}$$

for a convenient modification.†

Therefore the stochastic integral $\int_0^t p(s)(\psi(x_s)h^*(x_s)) R_s^{-1} \mathrm{d}z$ is well defined for any t. To prove (4.1.45) it is sufficient to prove that for a convenient modification

$$\text{a.s.} \quad \tilde{E}^{Z^\bullet} [\eta_s (1 + |x_s|^2)^{1/2}] \in L^\infty(0, T). \tag{4.1.46}$$

We apply (4.1.42) with $\psi(x, t) = \chi = (1 + |x|^2)^{1/2}$. The only drawback is that χ is unbounded. But one can apply (4.1.42) to

$$\chi_\lambda(x) = \left(\frac{1 + |x|^2}{1 + \lambda |x|^2} \right)^{1/2}$$

and let λ tend to 0. Note in particular that

$$\left| \tilde{E}^{Z^\bullet} \left[\chi_\lambda(x_s) \eta_s (1 + \epsilon \eta_s)^{-2} h(x_s) \right] \right| \le \frac{C}{\epsilon} \tilde{E}^{Z^\bullet} [1 + |x_s|^2] = \frac{C}{\epsilon} \tilde{E} (1 + |x_s|^2) \le \frac{C}{\epsilon}$$

which permits us to conclude that

$$\tilde{E}^{Z^\bullet} \left[\chi_\lambda(x_s) \eta_s (1 + \epsilon \eta_s)^{-2} h(x_s) \right] \to \tilde{E}^{Z^\bullet} \left[\chi(x_s) \eta_s (1 + \epsilon \eta_s)^{-2} h(x_s) \right]$$
in $\tilde{L}_Z^2(0, T)$, as $\lambda \to 0$.

Hence (4.1.42) is valid for $\psi(x, t) = \chi = (1 + |x|^2)^{1/2}$. We notice that

$$\frac{\partial \chi}{\partial t} - A(t)\chi = -A(t)\chi = (1 + |x|^2)^{-1/2} \Big[-g^*(x, t)x - \mathrm{tr}\, a(x, t) \\ + x^* a x (1 + |x|^2)^{-1} \Big]. \tag{4.1.47}$$

---

† $\tilde{\xi}_t$ is a modification of ξ_t if $\forall t$ $\tilde{\xi}_t = \xi_t$ a.s. Note that $\tilde{\xi}_t$ and ξ_t are representations of the same element in $\tilde{L}_Z^1(0, T)$.

To simplify the writing we shall continue to use the notation $\chi(x)$ for $(1 + |x|^2)^{1/2}$. Note that from (4.1.47)

$$|A(t)\chi| \leq k\chi. \tag{4.1.48}$$

From (4.1.42) we deduce

$$\tilde{E}\left\{\tilde{E}^{Z^t}[\chi(x_t)\eta_t(1 + \epsilon\eta_t)^{-1}] - \frac{\Pi_0(\chi)}{1 + \epsilon}\right.$$
$$\left. - \int_0^t \tilde{E}^{Z^\lambda}\left[-\eta(1 + \epsilon\eta)^{-1}A\chi - 2\epsilon\chi\eta^2(1 + \epsilon\eta)^{-3}h^*R^{-1}h\right]d\lambda \,\middle|\, Z^s\right\}$$
$$= \tilde{E}^{Z^s}[\chi(x_s)\eta_s(1 + \epsilon\eta_s)^{-1}] - \frac{\Pi_0(\chi)}{1 + \epsilon}$$
$$- \int_0^s \tilde{E}^{Z^\lambda}\left[-\eta(1 + \epsilon\eta)^{-1}A\chi - 2\epsilon\chi\eta^2(1 + \epsilon\eta)^{-3}h^*R^{-1}h\right]d\lambda, \quad t > s.$$

It is possible to take the limit in ϵ in the above relation. Note in particular that

$$\epsilon\chi\eta^2(1 + \epsilon\eta)^{-3}h^*R^{-1}h \leq C\eta(1 + |x|^2)^{3/2}$$

and thus we can make use of the Lebesgue theorem, by virtue of the estimate of Lemma 4.1.5. We conclude that

$$\tilde{E}^{Z^t}[\chi(x_t)\eta_t] - \Pi_0(\chi) + \int_0^t \tilde{E}^{Z^s}(\eta A\chi)ds = M_t$$

is a Z^t, $\tilde{P}$ martingale. Clearly

$$\tilde{E}M_t = \tilde{E}\chi(x_t)\eta_t - \Pi_0(\chi) + \int_0^t \tilde{E}\eta A\chi ds = 0.$$

From Meyer (1966, pp. 128–129) we know that M_t has a modification which is continuous to the right. We can thus assume that M_t is continuous to the right. But a right continuous martingale has bounded trajectories on compact intervals. Therefore (4.1.46) obtains.

(c) Taking the limit in (4.1.42)

Let us show that

$$\int_0^T \left|\tilde{E}^{Z^s}\left[\frac{\epsilon\eta^2(2 + \epsilon\eta)}{(1 + \epsilon\eta)^2}(1 + |x_s|^2)^{1/2}\right]\right|^2 ds \to 0 \tag{4.1.49}$$

a.s., at least for a convenient subsequence.

Indeed

$$\tilde{E}\int_0^T \tilde{E}^{Z^s}\left[\epsilon\eta^2\frac{(2 + \epsilon\eta)}{(1 + \epsilon\eta)^2}(1 + |x_s|^2)^{1/2}\right]ds$$
$$= \tilde{E}\int_0^T \epsilon\eta^2\frac{(2 + \epsilon\eta)}{(1 + \epsilon\eta)^2}(1 + |x_s|^2)^{1/2}ds \to 0$$

using Lebesgue's theorem. Therefore for a convenient subsequence we have

$$\tilde{E}^{Z^s}\left[\epsilon\eta^2\frac{(2 + \epsilon\eta)}{(1 + \epsilon\eta)^2}(1 + |x_s|^2)^{1/2}\right] \to 0 \quad \text{a.s., a.e.}$$

which implies

$$\text{a.s. for } \tilde{P}, \quad \tilde{E}^{Z^*}\left[\eta^2 \frac{(2+\epsilon\eta)}{(1+\epsilon\eta)^2}(1+|x_s|^2)^{1/2}\right] \to 0 \quad \text{a.e. } s.$$

hence also

$$\text{a.s. for } \tilde{P}, \quad \left|\tilde{E}^{Z^*}\left[\eta^2 \frac{(2+\epsilon\eta)}{(1+\epsilon\eta)^2}(1+|x_s|^2)^{1/2}\right]\right|^2 \to 0 \quad \text{for almost all } s.$$

This function is also bounded above by $4|\tilde{E}^{Z^*}\left[\eta(1+|x_s|^2)^{1/2}\right]|^2$ which belongs to $L^\infty(0,T)$, therefore

$$\text{a.s. for } \tilde{P}, \quad \int_0^T \left|\tilde{E}^{Z^*}\left[\eta^2 \frac{(2+\epsilon\eta)}{(1+\epsilon\eta)^2}(1+|x_s|^2)^{1/2}\right]\right|^2 ds \to 0$$

which implies the property (4.1.49).

For the above subsequence the property (4.1.49) implies

$$\sup_{0\le t\le T}\left|\int_0^t \tilde{E}^{Z^*}\left[\psi\eta((1+\epsilon\eta)^{-2}-1)h^*\right]R^{-1}dz\right| \to 0$$

in probability (classical property of stochastic integrals, see for instance Bensoussan and Lions (1978, p. 34).

It is therefore possible to take the limit in (4.1.42), obtaining the desired relation (4.1.37). The proof has been completed. ∎

Remark 4.1.1 For any t, (4.1.3) is an equality of random variables belonging to $L^1(\Omega, \mathcal{Z}^t, \tilde{P})$. The stochastic integral is well defined since $p(s)(\psi(s)h(s)) \in L^\infty(0,T;R^m)$ a.s. but the fact that $\int_0^t p(s)(h^*(s)\psi(s))R_s^{-1}dz \in L^1(\Omega, \mathcal{Z}^t, \tilde{P})$ is a consequence of the relation (4.1.37) itself. Note that its mean is 0, again by virtue of relation (4.1.37). ∎

It is important (for uniqueness) to derive an integrated form (4.1.37), which is not an immediate consequence of (4.1.37), since we cannot easily manipulate the stochastic integral. This is the objective of the next theorem.

Theorem 4.1.2 *We assume (4.1.1), (4.1.2), (4.1.3), (4.1.4), (4.1.5). Let $\theta(t)$ be the process defined in (4.1.28); then one has*

$$\tilde{E}\theta(t)p(t)(\psi(t)) = \Pi_0(\psi(0)) + \tilde{E}\int_0^t \theta(s)p(s)$$
$$\left(\frac{\partial\psi}{\partial s} - A(s)\psi(s) + i\psi(s)h^*(s)R_s^{-1}\beta(s)\right)ds, \forall\beta(\cdot). \tag{4.1.50}$$

Proof
Consider the relation (4.1.41). We can let ϵ tend to 0.
Since

$$\left|\epsilon\theta\psi(1+\epsilon\eta)^{-3}\eta^2 h^*R^{-1}h\right| \le C\eta_t(1+|x_t|^2)$$

and

$$\tilde{E}\eta_t[1+|x_t|^2] = E[1+|x_t|^2] < \infty$$

we can take the limit in (4.1.41) and obtain the relation (4.1.50). Hence the desired result is proved. ∎

Remark 4.1.2 The assumption (4.1.33) is not necessary to get the integrated form (4.1.50). ∎

Remark 4.1.3 The two forms (4.1.37), (4.1.50) can be derived one from one another, when it is possible to manipulate $\int_0^t p(s)(h^*(s)\psi(s))R_s^{-1}\mathrm{d}z$ by the rules of stochastic calculus, which is not *a priori* true under our assumptions. ∎

4.2　　Uniqueness theorem

4.2.1　　Preliminaries

We shall need some information about the system of partial differential equations

$$\frac{\partial v}{\partial t} - A(t)v + \mathrm{i}vh^* R^{-1}\beta = 0$$
$$v(x,T) = \phi(x) \tag{4.2.1}$$

where $h(x,t)$ satisfies (4.1.4), $A(t)$ is the second order partial differential operator (4.1.36), $\beta(t) \in L^\infty(0,T; R^m)$, and ϕ is complex valued smooth, bounded. Note that if $v = v_1 + \mathrm{i}v_2$, $\phi = \phi_1 + \mathrm{i}\phi_2$, the equations (4.2.1) become

$$\frac{\partial v_1}{\partial t} - A(t)v_1 - v_2 h^* R^{-1}\beta = 0, \quad v_1(x,T) = \phi_1(x)$$
$$\frac{\partial v_2}{\partial t} - A(t)v_2 + v_1 h^* R^{-1}\beta = 0, \quad v_2(x,T) = \phi_2(x). \tag{4.2.2}$$

Since we have not made at this stage any ellipticity assumption on the operator $A(t)$, we cannot hope for regularity of the solution v_1, v_2.

Let us begin with a particular case. We assume

$$g(x,t) = F(t)x + f(t) \tag{4.2.3}$$
$$\sigma(x,t) = \sigma(t) \tag{4.2.4}$$
$$h(x,t) = H(t)x + h(t) \tag{4.2.5}$$
$$\phi(x) = \exp \mathrm{i}\lambda^* x \tag{4.2.6}$$

then it is a simple exercise to check that the function

$$v(x,t) = \exp(\mathrm{i}\rho(t)^* x + s(t)) \tag{4.2.7}$$

is the solution of (4.2.1) provided the functions $\rho(t)$, $s(t)$ are chosen as follows:

$$\dot{\rho} + F^*\rho + H^* R^{-1}\beta = 0, \quad \rho(T) = \lambda$$
$$\dot{s} + \mathrm{i}\rho^* f + \mathrm{i}\beta^* R^{-1}h - \frac{1}{2}\rho^* \sigma Q \sigma^* \rho = 0 \tag{4.2.8}$$
$$s(T) = 0.$$

Note that ρ is real, s complex (real if $f, h = 0$). Let us give now some assumptions to guarantee the existence and uniqueness of a pair v_1, v_2 which is a sufficiently smooth solution of (4.2.2).

We can prove the following.

Proposition 4.2.1 *Assume that*

$$\text{a.e. } t, \ a_{ij,\lambda}, \ a_{ij,\lambda\mu}; \ g_{i,\lambda}, \ g_{i,\lambda\mu}; \ h_{i,\lambda}, \ h_{ij,\lambda\mu} \tag{4.2.9}$$

are bounded†, by a constant C_0, $\forall x$. Then there exists a unique solution of (4.2.2) v_1, v_2 such that $v_1, v_2; v_{1,\lambda}, v_{2,\lambda}; v_{1,\lambda\mu}; v_{2,\lambda\mu}$ are bounded by a constant C_1, $\forall x, t$.

Proof

We set

$$f(x,t) = h^*(x,t)R^{-1}(t)\beta(t).$$

From the assumption (4.2.9), we have f_λ, $f_{\lambda\mu}$ bounded. Note that f is not itself bounded, nor is f_t. Let us derive an a priori estimate on v_1, v_2 and their derivatives $v_{1,t}, v_{2,t}; v_{1,\lambda}, v_{2,\lambda}; v_{1,\lambda\mu}; v_{2,\lambda\mu}$.

Set to begin with

$$z = \frac{1}{2}(v_1^2 + v_2^2).$$

After some easy computation we can evaluate

$$-\frac{\partial z}{\partial t} + A(t)z = -a_{\lambda\mu}(v_{1,\lambda}v_{1,\mu} + v_{2,\lambda}v_{2,\mu}) \tag{4.2.10}$$
$$\leq 0.$$

Therefore from the maximum principle we deduce the estimate‡

$$\|v_1\|^2 + \|v_2\|^2 \leq \|\phi_1\|^2 + \|\phi_2\|^2. \tag{4.2.11}$$

Define next

$$z = \sum_\lambda \frac{1}{2}(v_{1,\lambda}^2 + v_{2,\lambda}^2)$$

using the same notation for the intermediary function z (to save notation). Similarly we deduce (using the summation convention)

$$\begin{aligned}
-\frac{\partial z}{\partial t} + Az = {}& -a_{\lambda\mu}(v_{1,\lambda\nu}v_{1,\mu\nu} + v_{2,\lambda\nu}v_{2,\mu\nu}) \\
& + a_{\lambda\mu,\nu}v_{1,\lambda\mu}v_{1,\nu} + a_{\lambda\mu,\nu}v_{2,\lambda\mu}v_{2,\nu} \\
& + g_{\lambda,\mu}v_{1,\lambda}v_{1,\mu} + g_{\lambda,\mu}v_{2,\lambda}v_{2,\mu} \\
& - v_{1,\lambda}f_\lambda v_2 + v_{2,\lambda}f_\lambda v_1.
\end{aligned} \tag{4.2.12}$$

We make use of the estimation (due to Oleinik and Radkevic, see Stroock and Varadhan 1979)

$$\sum_\lambda |a_{\lambda\mu,\nu}v_{1,\lambda\mu}|^2 \leq C a_{\lambda\mu}v_{1,\lambda\nu}v_{1,\mu\nu} \tag{4.2.13}$$

where C depends only on the norm $\|a_{\lambda\mu}\|_{W^{2,\infty}}$. It follows that

$$-\frac{\partial z}{\partial t} + Az \leq C_0 z + C_1$$

---

† We denote $a_{ij,\lambda} = \partial a_{ij}/\partial x_\lambda$. Note that a_{ij} is not necessarily elliptic.

‡ Where $\|v\| = \sup_{x,t}|v(x,t)|$.

where C_0, C_1 depend only on the constants arising in (4.2.9).

From the maximum principle it follows that

$$\sup(v_{1,\lambda}^2 + v_{2,\lambda}^2) \le e^{C_0 T}\left[\sup(\phi_{1\lambda}^2 + \phi_{2,\lambda}^2) + 2\frac{C_1}{C_0}\right].$$

We do a similar calculation for

$$z = \sum_{\lambda\mu}\frac{1}{2}(v_{1,\lambda\mu}^2 + v_{2,\lambda\mu}^2).$$

It follows that

$$
\begin{aligned}
-\frac{\partial z}{\partial t} + A(t)z = &- a_{\lambda\mu}(v_{1,ij\lambda}v_{1,ij\mu} + v_{2,ij\lambda}v_{2,ij\mu})\\
&+ a_{\lambda\mu,ij}(v_{1,\lambda\mu}v_{1,ij} + v_{2,\lambda\mu}v_{2,ij})\\
&+ 2a_{\lambda\mu,i}(v_{1,\lambda\mu\nu}v_{1,i\nu} + v_{2,\lambda\mu\nu}v_{1,i\nu})\\
&+ g_{i,\lambda\mu}(v_{1,i}v_{1,\lambda\mu} + v_{2,i}v_{2,\lambda\mu})\\
&+ 2g_{i,\lambda}(v_{1,i\mu}v_{1,\lambda\mu} + v_{2,i\mu}v_{2,\lambda\mu})\\
&- 2v_{2,\lambda}f_\mu v_{1,\lambda\mu} - v_2 f_{\lambda\mu}v_{1,\lambda\mu}\\
&+ 2v_{1,\lambda}f_\mu v_{2,\lambda\mu} + v_1 f_{\lambda\mu}v_{2,\lambda\mu}\\
\le &\, C_0 z + C_1
\end{aligned}
\tag{4.2.14}
$$

with constants depending only on the bounds (4.2.9).

We deduce again that

$$\sup(v_{1,\lambda\mu}^2 + v_{2,\lambda\mu}^2) \le e^{C_0 T}\left[\sup(\phi_{1,\lambda\mu}^2 + \phi_{2,\lambda\mu}^2) + 2\frac{C_1}{C}\right]. \tag{4.2.15}$$

The above estimates provide immediately (from equation (4.2.2))

$$|v_{1,t}|, \; |v_{2,t}| \le C(1 + |x|). \tag{4.2.16}$$

Once these estimates are known, it is easy to prove the existence and uniqueness of the solution of (4.2.2), in adequate functional spaces. By linearity and the estimate (4.2.12) the uniqueness is clear. The existence is proved by approximating the coefficients by bounded ones; the situation is then classical. These estimates allow us to pass to the limit. ∎

4.2.2   Uniqueness

We shall define some functional spaces in which we seek the solution of (4.1.50). Let

B = the space of Borel bounded functions on R^n with the norm

$$\|\psi\| = \sup_x |\psi(x)|$$

and let B_1, B_2 be the spaces of Borel functions on R^n such that

$$\frac{\psi}{(1+|x|^2)^{1/2}}, \quad \frac{\psi}{1+|x|^2} \quad \text{belong to } B.$$

The norms on B_1, B_2 are

$$\|\psi\|_1 = \sup_x \frac{|\psi(x)|}{(1+|x|^2)^{1/2}}, \quad \|\psi\|_2 = \sup_x \frac{|\psi(x)|}{1+|x|^2}.$$

We define the weak topology on B by

$$\psi_n \to \psi \text{ in } B \text{ weakly, if } \|\psi_n\| \text{ remains bounded,}$$
$$\psi_n(x) \to \psi(x) \, \forall x. \tag{4.2.17}$$

We denote by B^w the space B equipped with the weak topology. Similarly we shall consider B_1^w, B_2^w.

Next consider the spaces $L^\infty(0,T;B)$, $L^\infty(0,T;B_1)$, $L^\infty(0,T;B_2)$ and define a weak topology on these spaces as follows

$$\psi_n \to \psi \text{ in } L^\infty(0,T;B) \text{ weakly, if}$$
$$\|\psi_n\|_{L^\infty(0,T;B)} \leq C, \quad \text{and} \tag{4.2.18}$$
$$\text{for almost all } t, \quad \psi_n(x,t) \to \psi(x,t) \, \forall x,$$

and similar definitions for the weak topology on $L^\infty(0,T;B_1)$, $L^\infty(0,T;B_2)$. We shall denote by $(L^\infty(0,T;B))^w$, the space $L^\infty(0,T;B)$ equipped with the weak topology.

Consider also the spaces $L^1(\Omega, \mathcal{Z}^t, \tilde{P})$ and $\tilde{L}_Z^1(0,T)$. When equipped with the weak topology $\sigma(L^1(\Omega, \mathcal{Z}^t, \tilde{P}); L^\infty(\Omega, \mathcal{Z}^t, \tilde{P}))$ or $\sigma(\tilde{L}_Z^1(0,T); \tilde{L}_Z^\infty(0,T))$ we shall denote these spaces by $(L^1(\Omega, \mathcal{Z}^t, \tilde{P}))^w$ and $(\tilde{L}_Z^1(0,T))^w$. We search for a solution of (4.1.50) in the space

$$\forall t, p(t) \in \mathcal{L}(B^w; (L^1(\Omega, \mathcal{Z}^t, \tilde{P}))^w)$$
$$\text{and } p(\cdot) \in \mathcal{L}((L^\infty(0,T;B_2))^w; (\tilde{L}_Z^1(0,T))^w). \tag{4.2.19}$$

Naturally the unnormalized conditional probability belongs to (4.2.19), as is easily checked.

Theorem 4.2.1 *We make the assumptions of Theorem 4.1.2, and Proposition 4.2.1. Then there exists a unique solution of (4.1.50), in the functional space (4.2.19).*

Proof

By linearity, (4.1.50) holds with ψ complex valued. We know by Proposition 4.2.1 that, for the solution of (4.2.1),

$$|v(x,t)|, \quad |Dv(x,t)|, \quad |D^2 v(x,t)| \leq C, \quad \forall x,t$$
$$\text{for almost all } t, \quad \left|\frac{\partial v}{\partial t}\right| \leq C(1+|x|) \, \forall x.$$

We can construct a sequence $v^k(x,t)$ which is $C_b^{2,1}$ (complex valued) and satisfies the same estimates uniformly in k, i.e.

$$|v_k(x,t)|, \quad |Dv_k(x,t)|, \quad |D^2 v_k(x,t)| \leq C, \quad \forall x,t$$
$$\text{for almost all } t, \quad \left|\frac{\partial v_k}{\partial t}\right| \leq C(1+|x|) \, \forall x. \tag{4.2.20}$$

and converges to v pointwise as well as its derivatives. Indeed, we can take

$$v_k(x,t) = \frac{k^{n+1}}{\Pi^{(n+1)/2}} \int_{R^{n+1}} \exp\left[-k^2\left(|x-y|^2 + |t-s|^2\right)\right] v(y,s)\,dy ds$$

where $v(y,s)$ is extended by $v(y,s) = v(y,T)$ for $s \geq T$, and $v(y,s) = v(y,0)$ for $s \leq 0$.

It follows among other things that

$$v_k(\cdot,T) \to v(\cdot,T) = \phi \text{ in } B^w$$
$$\frac{\partial v_k}{\partial t} - A(t)v_k + iv_k h^* R^{-1}\beta \to \frac{\partial v}{\partial t} - A(t)v + ivh^* R^{-1}\beta = 0 \qquad (4.2.21)$$
$$\text{in } (L^\infty(0,T;B_2))^w.$$

On the other hand, applying (4.1.50) with $\psi = v$ yields

$$\tilde{E}\theta(T)p(T)(v_k(T)) = \Pi_0(v_k(0))$$
$$+ \tilde{E}\int_0^T \theta(t)p(t)\left(\frac{\partial v_k}{\partial t} - A(t)v_k + iv_k h^* R^{-1}\beta\right)dt. \qquad (4.2.22)$$

By virtue of (4.2.21), and the definition of the functional space (4.2.19), we can pass to the limit in (4.2.22) and thus deduce

$$\tilde{E}\theta(T)p(T)(\phi) = \Pi_0(v(0)). \qquad (4.2.23)$$

A priori in (4.2.23), ϕ is smooth bounded (we take it real). Note that from (4.2.11)

$$|\Pi_0(v(0))| \leq \|\phi\|$$

hence

$$|\tilde{E}\theta(T)p(T)(\phi)| \leq \|\phi\|. \qquad (4.2.24)$$

Therefore if $p_1(t)$, $p_2(t)$ are two solutions, we have

$$\tilde{E}\theta(T)p_1(T)(\phi) = \tilde{E}\theta(T)p_2(T)(\phi)$$

and, from (4.2.24), this equality extends first to ϕ continuous bounded. It also extends to ϕ in B, since if $\phi \in B$ it can be approximated in B^w by a sequence of continuous bounded functions.

Now, from Lemma 4.1.4, we deduce

$$p_1(T)(\phi) = p_2(T)(\phi) \quad \forall \phi \in B.$$

Necessarily $p_1(T) = p_2(T)$, as elements of $L^1(\Omega, \mathscr{Z}^T, \tilde{P})$. Since T can be replaced by any $t \in [0,T]$, this proves uniqueness. ∎

Remark 4.2.1 We shall consider that the unnormalized conditional probability satisfies both equations (4.1.37) and (4.1.50), and belongs to the functional space (4.2.19). It is naturally the unique element of (4.2.19) satisfying both equations, since there is uniqueness already for (4.1.50). We do not try to study the uniqueness of (4.1.37) only, since the stochastic integral on the right hand side of (4.1.37) is not easily manipulable.

4.3   Equation of the conditional probability

We now derive an equation for the conditional probability $\Pi(t)$ itself; see (4.1.9).

4.3.1   Preliminaries

We consider in this paragraph the process $p(t)(1)$. Using the relation (4.1.37) we have

$$p(t)(1) = 1 + \int_0^t p(s)(h^*(s))R_s^{-1}\mathrm{d}z. \tag{4.3.1}$$

Lemma 4.3.1 *The process $p(t)(1)$ is explicitly given by*

$$p(t)(1) = \exp\left[\int_0^t \Pi(s)(h^*(s))R_s^{-1}\mathrm{d}z - \frac{1}{2}\int_0^t \Pi(s)(h^*(s))R_s^{-1}\Pi(s)(h(s))\mathrm{d}s\right]. \tag{4.3.2}$$

Proof

We can write (4.3.1) as follows

$$p(t)(1) = 1 + \int_0^t p(s)(1)\Pi(s)(h^*(s))R_s^{-1}\mathrm{d}z. \tag{4.3.3}$$

This equation has the right hand side of (4.3.2) as a solution. However, since we cannot guarantee that (4.3.3) has a unique solution, it does not immediately follow that (4.3.2) holds. Note also that $p(t)(1) > 0$.

We apply Ito's formula to the process $\log\left[\epsilon + (p(t)(1))^2\right]^{1/2}$ and deduce

$$\mathrm{d}\log\left\{\epsilon + [p(t)(1)]^2\right\}^{1/2} = \frac{[p(t)(1)]^2}{\epsilon + [p(t)(1)]^2}\Pi(t)(h^*(t))R_t^{-1}\mathrm{d}z$$

$$+ \frac{1}{2}\frac{\epsilon - [p(t)(1)]^2}{\left\{\epsilon + [p(t)(1)]^2\right\}^2}[p(t)(1)]^2\Pi(t)(h^*(t))R_t^{-1}\Pi(t)(h(t))\mathrm{d}t. \tag{4.3.4}$$

It is worthwhile to replace $\mathrm{d}z$ by its value given by formula (4.1.7). The stochastic differential becomes

$$\frac{[p(t)(1)]^2}{\epsilon + [p(t)(1)]^2}\Pi(t)(h^*(t))R_t^{-1}\mathrm{d}b.$$

We note that

$$E\int_0^T \Pi(t)(h^*(t))R_t^{-1}\Pi(t)(h(t))\mathrm{d}t \leq CE\int_0^T |h(x_t)|^2\mathrm{d}t$$

$$\leq CE\int_0^T (1 + |x_t|^2)\mathrm{d}t. \tag{4.3.5}$$

Integrating (4.3.4), it is then possible to let ϵ tend to 0 on both sides. The result (4.3.2) obtains. ∎

4.3.2 Kushner–Stratonovitch equation

We can now derive the equation for the conditional probability, called the Kushner–Stratonovitch equation.

Theorem 4.3.1 *We make the assumptions of Theorem 4.1.1. Let*

$$\psi(x,t) \in C_b^{2,1}(R^n \times [0,\infty));$$

then one has the relation

$$
\begin{aligned}
\Pi(t)(\psi(t)) =& \Pi_0(\psi(0)) \\
&+ \int_0^t \Pi(s)\left(\frac{\partial \psi}{\partial s} - A(s)\psi(s)\right) ds + \int_0^t \left[\Pi(s)(h^*(s)\psi(s)) \right.\\
&\left. - \Pi(s)(\psi(s))\Pi(s)(h^*(s))\right] R_s^{-1}[dz - \Pi(s)(h(s))ds] \\
& \text{a.s., } \forall t.
\end{aligned}
\tag{4.3.6}
$$

Proof

From (4.3.2) we deduce

$$\frac{1}{p(t)(1)} = \exp\left[-\int_0^t \Pi(s)(h^*(s))R_s^{-1}dz + \frac{1}{2}\int_0^t \Pi(s)(h^*(s))R_s^{-1}\Pi(s)(h(s))\right] ds$$

which proves that $1/p(t)(1) \in L^\infty(0,T)$ a.s.

Moreover

$$d\frac{1}{p(t)(1)} = \frac{1}{p(t)(1)}\left[-\Pi(t)(h^*(t))R_t^{-1}dz + \Pi(t)(h^*(t))R_t^{-1}\Pi(t)(h(t))dt\right].$$

However, we have

$$\Pi(t)(\psi(t)) = p(t)(\psi(t)) \times \frac{1}{p(t)(1)}$$

therefore we can use Ito's calculus to conclude that

$$
\begin{aligned}
d\Pi(t)(\psi(t)) =& \Pi(t)\left(\frac{\partial \psi}{\partial t} - A(t)\psi(t)\right) dt + \Pi(t)(h^*(t)\psi(s))R_t^{-1}dz \\
&+ \Pi(t)(\psi(t))\left[-\Pi(t)(h^*(t))R_t^{-1}dz + \Pi(t)(h^*(t))R_t^{-1}\Pi(t)(h(t))dt\right] \\
&- \Pi(t)(h^*(t)\psi(t))R_t^{-1}\Pi(t)(h(t))dt
\end{aligned}
$$

and the desired result (4.3.6) obtains. ∎

The uniqueness of the solution of the Kushner equation is a consequence of the uniqueness of the solution of the Zakai equation. In fact the key point is that the integrated forms of the Zakai and Kushner equations will be identical, except for the role of $\tilde{P}$ in (4.1.50) replaced by P.

We begin with the following result, interesting in itself.

Lemma 4.3.2 *The process ρ_t defined in (4.1.11) and its inverse η_t (see (4.1.19)) verify*

$$E[\rho_t|\mathcal{Z}^t]\tilde{E}[\eta_t|\mathcal{Z}^t] = 1 \quad \text{a.s.} \tag{4.3.7}$$

Proof

Let ξ_t be any $\mathcal{Z}^t$ measurable random variable, which is bounded a.s., then

$$E\{E[\rho_t|\mathcal{Z}^t]\xi_t\} = E\rho_t\xi_t.$$

Hence

$$\tilde{E}\left\{E[\rho_t|\mathcal{Z}^t]\xi_t\eta_t\right\} = \tilde{E}\rho_t\xi_t\eta_t$$
$$= \tilde{E}\xi_t.$$

Since $E[\rho_t|\mathcal{Z}^t]\eta_t$ belongs to $L^1(\Omega, \mathcal{A}, \tilde{P})$, we can take its conditional expectation with respect to $\mathcal{Z}^t$, which is $E[\rho_t|\mathcal{Z}^t]\tilde{E}[\eta_t|\mathcal{Z}^t]$. But from the preceding equality it is equal to 1, a.s., hence the desired result. ∎

We can now derive the integrated form of Kushner's equation, namely

Theorem 4.3.2 *We make the assumptions of Theorem 4.1.2. Let $\theta(t)$ be the process defined in (4.1.28), then one has*

$$E\theta(t)\Pi(t)(\psi(t)) = \Pi_0(\psi(0)) + E\int_0^t \theta(s)\Pi(s)\left(\frac{\partial\psi}{\partial s} - A(s)\psi(s)\right.$$
$$\left. + i\psi(s)h^*(s)R_s^{-1}\beta(s)\right)\mathrm{d}s. \tag{4.3.8}$$

Proof

We have

$$\tilde{E}\theta(t)p(t)(\psi(t)) = E\theta(t)\rho(t)p(t)(\psi(t))$$
$$= E\left\{\theta(t)p(t)(\psi(t))E[\rho(t)|\mathcal{Z}^t]\right\}$$

and from Lemma 4.3.2

$$E\{\theta(t)p(t)(\psi(t))E[\rho(t)|\mathcal{Z}^t]\} = E\left\{\theta(t)p(t)(\psi(t))\frac{1}{\tilde{E}[\eta_t|\mathcal{Z}^t]}\right\}$$
$$= E\left[\theta(t)\frac{p(t)(\psi(t))}{p(t)(1)}\right]$$
$$= E\left[\theta(t)\Pi(t)(\psi(t))\right].$$

A similar argument holds for the right hand side of (4.3.9), hence (4.3.8) follows immediately from (4.1.50). ∎

From the method of uniqueness of the Zakai equation, we get immediately a uniqueness theorem for the integrated form of the Kushner equation.

We define the functional space

$$\forall t, \ \Pi(t) \in \mathcal{L}(B^w; (L^1(\Omega, \mathcal{Z}^t, P))^w) \tag{4.3.9}$$

and

$$\Pi(\cdot) \in \mathcal{L}((L^\infty(0, T; B_2))^w; (L_Z^1(0, T))^w).$$

We can state

Theorem 4.3.3 *We make the assumptions of Theorem 4.1.2 and Proposition 4.2.1. Then there exists a unique solution of (4.3.8), in the functional space (4.3.9).*

Proof

Identical to that of Theorem 4.2.1, thanks to Corollary 4.1.1. ∎

Remark 4.3.1 Kurtz and Ocone (1988) have derived uniqueness theorems for conditional distributions in non linear filtering, in a general framework. Applied to diffusions, their results bear some similarities to ours, especially with respect to the use of duality.

Our approach uses PDE techniques. It does not extend to the case of coefficients g, σ which are random, except to the situation of Section 4.5. In the general random case, a recent paper of Chaleyat-Maurel, Michel and Pardoux (1989) overcomes this restriction, using techniques of pseudo differentials. However they need to assume bounded coefficients. ∎

4.3.3 The innovation approach

We shall develop in this section some interesting ideas related to the concept of innovation. In fact it will be shown that this leads to a direct approach to the Kushner–Stratonovitch equation (4.3.6). The concept of innovation was introduced by Kailath (1968, 1970) first in the context of linear filtering theory. The approach developed here is mainly that of Fujisaki, Kallianpur and Kunita (1972). We follow the presentation of Liptser and Shiryaev (1977).

4.3.3.1 Definition of the innovation

The innovation is the process

$$\nu_t = z_t - \int_0^t \Pi(s)(h(s)) \mathrm{d}s. \tag{4.3.10}$$

The first important result is the following.

Theorem 4.3.4 *We make the assumptions of Theorem 4.1.1. Then the innovation process is a P, $\mathcal{Z}^t$ Wiener process, with covariance $R(\cdot)$.*

Proof

From the definition (4.3.10) ν_t is clearly adapted to $\mathcal{Z}^t$. Moreover one has

$$\mathrm{d}\nu_t = \mathrm{d}b_t + (h(x_t) + \Pi_t(h))\mathrm{d}t \tag{4.3.11}$$

where $\Pi_t(h) = \Pi(t)(h(t)) = E\left[h(x_t, t)|\mathcal{Z}^t\right]$.

Let $\phi \in L^2(0, T; R^m)$ (non random) and set

$$\zeta_t = \exp\left(\mathrm{i} \int_0^t \phi_s^* \mathrm{d}\nu_s\right).$$

By Ito's formula, one has

$$\begin{aligned}
d\zeta_t &= \zeta_t(\mathrm{i}\phi_t^* \mathrm{d}\nu_t - \phi_t^* R_t \phi_t \mathrm{d}t) \\
&= \zeta_t \left\{\mathrm{i}\phi_t^* \mathrm{d}b_t + [\mathrm{i}\phi_t^*(h(x_t) - \Pi_t(h)) - \phi_t^* R_t \phi_t]\mathrm{d}t\right\}.
\end{aligned}$$

It follows that

$$E\left[\zeta_t | \mathcal{Z}^s\right] = \zeta_s - \frac{1}{2}\int_s^t E\left[\zeta_\lambda | \mathcal{Z}^s\right]\phi_\lambda^* R_\lambda \phi_\lambda d\lambda \qquad (4.3.12)$$

since

$$\int_s^t \phi_\lambda^* E\left[\zeta_\lambda(h(x_\lambda) - \Pi_\lambda(h)) | \mathcal{Z}^s\right] d\lambda = 0.$$

But (4.3.12) implies

$$E[\zeta_t | \mathcal{Z}^s] = \zeta_s \exp\left\{-\frac{1}{2}\int_s^t \phi_\lambda^* R_\lambda \phi_\lambda d\lambda\right\}$$

or

$$E\left[\exp\left(i\int_s^t \phi_\lambda^* d\nu_\lambda\right) \Big| \mathcal{Z}^s\right] = \exp\left(-\frac{1}{2}\int_s^t \phi_\lambda^* R_\lambda \phi_\lambda d\lambda\right)$$

and this proves the desired result.   ∎

Remark 4.3.2 We can write (4.3.6) as

$$\begin{aligned}\Pi(t)(\psi(t)) =&\, \Pi_0(\psi(0)) + \int_0^t \Pi(s)\left(\frac{\partial \psi}{\partial s} - A(s)\psi(s)\right) ds \\ &+ \int_0^t \left[\Pi(s)(h^*(s)\psi(s)) - \Pi(s)(\psi(s))\Pi(s)(h^*(s))\right] R_s^{-1} d\nu_s.\end{aligned} \qquad (4.3.13)$$

Unlike the case of the Zakai equation, the stochastic integral in (4.3.13) is easily manipulable, since the integrand is in $L_Z^2(0,T)$ (and not only in $L_Z^1(0,T)$). In particular noting that $\theta(t)$ admits the differential

$$d\theta(t) = \theta(t)i\beta_t^* R_t^{-1}[\Pi(t)(h(t))dt + d\nu_t] \qquad (4.3.14)$$

it is easy to derive from (4.3.13) and (4.3.14) the integrated form (4.3.8).   ∎

4.3.3.2   A representation formula

Note that if $N^t = \sigma(\nu_s, s \leq t)$ then $N^t \subset \mathcal{Z}^t$. These σ-algebras do not coincide in general but, as we have noticed, the observation has the Ito differential

$$dz = \Pi_t(h)dt + d\nu \qquad (4.3.15)$$

where $\Pi_t(h)$ is adapted to $\mathcal{Z}^t$ and

$$E|\Pi_t(h)|^2 < C, \quad \forall t \in [0,T].$$

The theory of Kunita and Watanabe (1967) asserts that any square integrable P, N^t martingale μ_t has a representation of the form (up to a modification)

$$\mu_t = \mu_0 + \int_0^t \lambda_s^* R_s^{-1} d\nu_s, \quad E\int_0^T |\lambda s_t|^2 dt < \infty. \qquad (4.3.16)$$

But because of the special form (4.3.15), this important result *extends to square integrable P, $\mathcal{Z}^t$ martingales*. Thus any square integrable P, $\mathcal{Z}^t$ martingale will have a representation of the form (4.3.16) and λ_t is adapted to $\mathcal{Z}^t$. For the proof we refer the reader to Liptser and Shiryaev (1977, p. 193).

4.3.3.3   A new proof of Theorem 4.3.1

We are now equipped to derive a new proof of Theorem 4.3.1. Let us define

$$M_t = \Pi(t)(\psi(t)) - \int_0^t \Pi(s)\left(\frac{\partial \psi}{\partial s} - A(s)\psi(s)\right) ds$$

then it is an L^2 process adapted to $\mathcal{Z}^t$. Let us prove that M_t is a P, $\mathcal{Z}^t$ martingale. But

$$E\left[M_t - M_s | \mathcal{Z}^s\right] = E\left[\psi(x_t, t) - \psi(x_s, s)\right.$$
$$\left. - \int_s^t \left(\frac{\partial \psi}{\partial \lambda} - A(\lambda)\psi(\lambda)\right)(x(\lambda), \lambda)d\lambda \middle| \mathcal{Z}^s\right]$$
$$= E\left[\int_s^t D\psi(x_\lambda, \lambda)\sigma(x_\lambda, \lambda)dw_\lambda \middle| \mathcal{Z}^s\right] = 0.$$

But then by the result mentioned in the preceding paragraph, we can state that

$$\Pi(t)(\psi(t)) = \Pi_0(\psi(0)) + \int_0^t \Pi(s)\left(\frac{\partial \psi}{\partial s} - A(s)\psi(s)\right) ds$$
$$+ \int_0^t \lambda_s^* R_s^{-1}[dz_s - \Pi(s)(h(s))ds] \tag{4.3.17}$$

where λ_t is adapted to $\mathcal{Z}^t$ and $E \int_0^T |\lambda_t|^2 dt < \infty$.

It thus remains to identify the process λ_t.

Let us consider a process of the form

$$\xi_t = \int_0^t \nu_s^* R_s^{-1} d\nu_s$$

where ν is $\mathcal{Z}^t$ adapted and bounded.

Clearly

$$EM_t\xi_t = E\int_0^t \nu_s^* R_s^{-1}\lambda_s ds. \tag{4.3.18}$$

We are going to compute $EM_t\xi_t$ differently using the definition of M_t; hence, as is easily seen,

$$EM_t\xi_t = E\psi(x_t, t)\xi_t - E\int_0^t \xi_s\left(\frac{\partial \psi}{\partial s} - A(s)\psi\right) ds.$$

But

$$d\xi_t = \nu_t^* R_t^{-1}[db + (h(x_t) - \Pi_t(h))dt]$$

and from Ito's formula

$$d\psi(x_t, t)\xi_t = \left[\left(\frac{\partial \psi}{\partial t} - A(t)\psi\right)\xi_t + \nu_t^* R_t^{-1}(h(x_t) - \Pi_t(h))\psi(x_t)\right] dt + \xi_t D\psi\sigma dw$$

hence

$$EM_t\xi_t = E\int_0^t \nu_s^* R_s^{-1}\left[\Pi(s)(h(s)\psi(s)) - \Pi(s)(h(s))\Pi(s)(\psi(s))\right]ds.$$

Comparing with (4.3.18), we easily deduce since ν_t is arbitrary that

$$\lambda_t = \Pi(t)(h(t)\psi(t)) - \Pi(t)(h(t))\Pi(t)(\psi(t))$$

which is the desired result.   ∎

Remark 4.3.3 Since we have proven the result of Theorem 4.3.1 directly, in other words without using the Zakai equation first, it is reasonable to wonder whether one can derive the Zakai equation from the Kushner equation. The integrated form (4.1.50) follows easily from (4.3.8).

As far as the full equation (4.1.37) is concerned, we proceed as follows.

Let (cf. (4.3.3)) ζ_t be defined by

$$\zeta_t = \exp\left[\int_0^t \Pi(s)(h^*(s))R_s^{-1}dz - \frac{1}{2}\int_0^t \Pi(s)(h^*(s))R_s^{-1}\Pi(s)(h(s))ds\right]. \quad (4.3.19)$$

Let us check that

$$\zeta_t = \tilde{E}\left[\eta_t|\mathcal{Z}^t\right] = p(t)(1). \quad (4.3.20)$$

For that purpose consider

$$\mu_t = \frac{1}{\zeta_t} = \exp\left[-\int_0^t \Pi(s)(h^*(s))R_s^{-1}dz + \frac{1}{2}\int_0^t \Pi(s)(h^*(s))R_s^{-1}\Pi(s)(h(s))ds\right]$$

$$= \exp\left[-\int_0^t \Pi(s)(h^*(s))R_s^{-1}d\nu_s - \frac{1}{2}\int_0^t \Pi(s)(h^*(s))R_s^{-1}\Pi(s)(h(s))ds\right]$$

hence

$$d\mu_t = -\mu_t\Pi(t)(h^*(t))R_t^{-1}d\nu_t.$$

Recall that

$$d\rho_t = -\rho_t h^*(x_t)R_t^{-1}db_t;$$

one easily checks that

$$d\theta_t\rho_t = \theta_t\rho_t(-h^*(x_t) + i\beta^*)R_t^{-1}db_t$$

$$d\theta_t\mu_t = \theta_t\mu_t(-\Pi(t)(h^*(t)) + i\beta^*)R_t^{-1}d\nu_t$$

and thus (reasoning as for Lemma 4.1.1)

$$E\theta_t\rho_t = E\theta_t\mu_t = 1.$$

Therefore we have proven that

$$E\theta_t\{\mu_t - E\left[\rho_t|\mathcal{Z}^t\right]\} = 0$$

which implies, thanks to Corollary 4.1.1,

$$\mu_t = E\left[\rho_t|\mathcal{Z}^t\right]$$

and, from Lemma 4.3.2, we deduce (4.3.20). But then from (4.3.19) it follows that ζ_t has the Ito differential

$$d\zeta_t = \zeta_t\Pi(t)(h^*(t))R_t^{-1}[d\nu_t + \Pi(t)(h(t))dt]. \quad (4.3.21)$$

Since $p(t)(\psi) = \Pi(t)(\psi)\zeta_t$, we can use stochastic calculus between (4.3.13) and (4.3.21) to derive the Ito differential of $p(t)(\psi)$, and we recover the Zakai equation.   ∎

4.4 An explicit solution

4.4.1 Notation

We consider the linear case

$$
\begin{aligned}
g(x,t) &= F(t)x + f(t) \\
\sigma(x,t) &= I \\
h(x,t) &= H(t)x + h(t).
\end{aligned}
\tag{4.4.1}
$$

We know from Sections 4.2 and 4.3 that the solutions of the Zakai and Kushner equations (4.1.37) and (4.3.6) exist and are unique in the functional spaces (4.2.19) and (4.3.9). We also assume that the initial data Π_0 is given by

$$
\Pi_0(\psi) = \int \psi(x_0 + P_0^{1/2}\xi)\frac{\exp\left(-\frac{1}{2}|\xi|^2\right)}{(2n)^{n/2}}\,\mathrm{d}\xi
\tag{4.4.2}
$$

where x_0, P_0 are given, P_0 symmetric non negative; hence Π_0 is a gaussian with mean x_0 and covariance matrix P_0.

4.4.2 Solution of the Zakai equation

We shall prove

Theorem 4.4.1 *We make the assumptions (4.4.1), (4.4.2). Then the solution of the Zakai equation (4.1.37) (unique in the sense of Theorem 4.2.1) is given by*

$$
p(t)(\psi) = \left[\int \psi(\hat{x}_t + P_t^{1/2}\xi)\frac{\exp(-\frac{1}{2}|\xi|^2)}{(2n)^{n/2}}\,\mathrm{d}\xi\right] s_t
\tag{4.4.3}
$$

where $P(t)$ is the solution of the Riccati equation

$$
\begin{aligned}
\dot{P} + PH^*R^{-1}HP - Q - FP - PF^* &= 0 \\
P(0) &= P_0
\end{aligned}
\tag{4.4.4}
$$

and $\hat{x}_t$ is the Kalman filter solution of

$$
\begin{aligned}
\mathrm{d}\hat{x}_t &= (F\hat{x}_t + f)\mathrm{d}t + PH^*R^{-1}[\mathrm{d}z - (H\hat{x} + h)\mathrm{d}t] \\
\hat{x}(0) &= x_0.
\end{aligned}
\tag{4.4.5}
$$

The process s_t is given by

$$
s_t = \exp\left[\int_0^t (\hat{x}^*H^* + h^*)R^{-1}\mathrm{d}z - \frac{1}{2}\int_0^t (\hat{x}^*H^* + h^*)R^{-1}(H\hat{x} + h)\mathrm{d}s\right].
\tag{4.4.6}
$$

Proof

By a proof similar to that of Lemma 4.1.1, one checks that

$$\tilde{E}s_t = 1. \tag{4.4.7}$$

From this property it is easy to check that $p(\cdot)$ belongs to the functional space (4.2.19).

Let us check (4.1.37). Let us set

$$\Gamma_t = P_t^{1/2}$$

and assume for a while that P_t is invertible the moment and that Γ_t is differentiable and $\dot{\Gamma}_t$ is the solution of the Liapunov equation

$$\dot{P}_t = \Gamma_t\dot{\Gamma}_t + \dot{\Gamma}_t\Gamma_t.$$

An easy calculation shows that

$$d[\psi(\hat{x}_t + P_t^{1/2}\xi, t)s_t] = s_t\left[\frac{\partial\psi}{\partial t} + D\psi^*(F\hat{x} + f) + D\psi^*\dot{\Gamma}_t\xi\right.$$
$$\left. + \frac{1}{2}\text{tr } D^2\psi PH^*R^{-1}HP\right]dt$$
$$+ s_t\left[D\psi^*PH^* + \psi(\hat{x}^*H^* + h^*)\right]R^{-1}dz$$

where the space argument of $\partial\psi/\partial t$, $D\psi$, $D^2\psi$ on the right hand side is evaluated at $\hat{x}_t + P_t^{1/2}\xi$. One then notices that

$$\int D\psi^*(\hat{x}_t + P_t^{1/2}\xi, t)\dot{\Gamma}_t\xi \exp\left(-\frac{1}{2}|\xi|^2\right) d\xi$$
$$= -\int D\psi^*(\hat{x}_t + P_t^{1/2}\xi, t)\dot{\Gamma}_t \text{ grad}\left[\exp\left(-\frac{1}{2}|\xi|^2\right)\right] d\xi \tag{4.4.8}$$
$$= \frac{1}{2}\int \text{tr } D^2\psi(\hat{x}_t + P_t^{1/2}\xi, t)\dot{P}_t \exp\left(-\frac{1}{2}|\xi|^2\right) d\xi$$

and thus it follows that

$$dp(t)(\psi(t)) = s_t\int\left[\frac{\partial\psi}{\partial t} + D\psi^*(F\hat{x} + f)\right.$$
$$\left. + \frac{1}{2}\text{tr } D^2\psi(\dot{P}_t + PH^*R^{-1}HP)\right]\frac{\exp(-\frac{1}{2}|\xi|^2)}{(2n)^{n/2}}d\xi$$
$$+ s_t\left\{\int\left[D\psi^*PH^* + \psi(\hat{x}^*H^* + h^*)\right]\right. \tag{4.4.9}$$
$$\left.\cdot\frac{\exp(-\frac{1}{2}|\xi|^2)}{(2n)^{n/2}}d\xi\right\}R^{-1}dz.$$

Obviously this relation extends even when P_t is not invertible. Notice then from (4.4.4) that

$$\frac{1}{2}\int \text{tr } D^2\psi(\dot{P}_t + PH^*R^{-1}HP)\exp\left(-\frac{1}{2}|\xi|^2\right) d\xi$$
$$= \frac{1}{2}\int \text{tr } D^2\psi(Q + FP + PF^*)\exp\left(-\frac{1}{2}|\xi|^2\right) d\xi \tag{4.4.10}$$
$$= \int\left(\frac{1}{2}\text{tr } D^2\psi Q + D\psi^*FP^{1/2}\xi\right)\exp\left(-\frac{1}{2}|\xi|^2\right) d\xi.$$

Since also

$$\int D\psi^* PH^* \exp\left(-\frac{1}{2}|\xi|^2\right) d\xi = \int \psi\xi^* P^{1/2} H^* \exp\left(-\frac{1}{2}|\xi|^2\right) d\xi \qquad (4.4.11)$$

we can use (4.4.10) and (4.4.11) to evaluate the right hand side of (4.4.9) and obtain

$$dp(t)(\psi(t)) = p(t)\left[\frac{\partial\psi}{\partial t} + D\psi^*(Fx+f) + \frac{1}{2}\mathrm{tr}\,D^2\psi Q\right]dt$$
$$+ p(t)\left[\psi(x^* H^* + h^*)\right]R^{-1}dz$$

which is exactly the desired result (4.1.37). ∎

Corollary 4.4.1 *Under the assumptions (4.4.1), (4.4.2), the unique solution of Kushner's equation (4.3.6), satisfying (4.3.9), is given by*

$$\Pi(t)(\psi) = \int \psi(\hat{x}_t + P_t^{1/2}\xi)\frac{\exp(-\frac{1}{2}|\xi|^2)}{(2n)^{n/2}}d\xi. \qquad (4.4.12)$$

Proof

From (4.4.3)

$$p(t)(1) = s_t$$

hence

$$\Pi(t)(\psi) = \frac{p(t)\psi)}{p(t)(1)}$$

which with the uniqueness result of Theorem 4.3.3 completes the proof. ∎

4.4.3 Some remarks on the moments

The result of Corollary 4.4.1 implies that the process x_t has a conditional probability law (given $\mathcal{Z}^t$) which is gaussian with mean $\hat{x}_t$ and variance P_t. This implies at once

$$\hat{x}_t = E\left[x_t|\mathcal{Z}^t\right] \qquad (4.4.13)$$

$$P_t = E\left[(x_t - \hat{x}_t)(x_t - \hat{x}_t)^*|\mathcal{Z}^t\right] \qquad (4.4.14)$$

which is what the Kalman filter theory gives (see Chapter 2). The non linear filtering theory applied to the linear case thus gives a stronger result than the Kalman filter theory.

Let us show that we can derive the equations for $\hat{x}_t$ and P_t directly from the Kushner equation (4.3.6). We do it formally to avoid lengthy details. Usually (4.3.6) with $\psi(x) = x$ (a vector) yields

$$d\hat{x}_t = (F\hat{x}_t + f)dt + PH^*R^{-1}[dz - (H\hat{x} + h)dt] \qquad (4.4.15)$$

where P_t is defined by (4.4.14) (*a priori* not deterministic).

To derive an equation for P_t we use (4.3.6) with $\psi(x) = xx^*$ (in fact we do it component by component) and notice that

$$P_t = E[x_t x_t^* | \mathcal{Z}^t] - \hat{x}_t \hat{x}_t^*.$$

After some calculations, one can check that the following relation holds

$$
\begin{aligned}
dP_t =& (FP + PF^* + Q - PH^*R^{-1}H)dt \\
& + E^{\mathcal{Z}^t}\left[(x_t - \hat{x}_t)(x_t - \hat{x}_t)^*(x_t - \hat{x}_t)^*\right]H^*R^{-1}[dz - (H\hat{x}_t + h)dt]
\end{aligned}
\tag{4.4.16}
$$

and everything amounts to showing that

$$E^{\mathcal{Z}^t}\left[(x_t - \hat{x}_t)_i(x_t - \hat{x}_t)_j(x_t - \hat{x}_t)_k\right] = 0 \quad \forall i, j, k = 1, \dots, n \tag{4.4.17}$$

and this is a consequence of the property

$$\int \xi_i \xi_j \xi_k \left(\exp - \frac{1}{2}|\xi|^2\right) d\xi = 0 \quad \forall i, j, k$$

hence P_t is deterministic and is the solution of the Riccati equation (4.4.4).

4.5 Correlation between the signal noise and the observation noise

4.5.1 Setting of the problem

In this section we extend the problem considered in Section 4.1 as follows. The model is now given by

$$
\begin{aligned}
\mathrm{d}x &= g(x_t, t)\mathrm{d}t + \sigma(x_t)\mathrm{d}w_t \\
x(0) &= \xi \\
\mathrm{d}z &= h(x_t, t)\mathrm{d}t + \alpha(t)\mathrm{d}w_t + \mathrm{d}b_t \\
z(0) &= 0.
\end{aligned}
$$

(4.5.1)

(4.5.2)

The notation and assumptions are identical to those in § 4.1.1; the only new term is the deterministic matrix $\alpha(t)$ in the observation process, creating a correlation between the signal noise and the observation noise.

The problem is the same, namely to compute the process

$$
\Pi(t)(\psi) = E\left[\psi(x_t)|\mathcal{Z}^t\right].
$$

(4.5.3)

We shall describe the modifications which arise in the non linear filtering equations, skipping some details.

Define the process ρ_t now by

$$
\begin{aligned}
\mathrm{d}\rho &= -\rho h^*(x_t)(\alpha Q\alpha^* + R)^{-1}(\alpha \mathrm{d}w + \mathrm{d}b) \\
\rho(0) &= 1
\end{aligned}
$$

(4.5.4)

explicitly

$$
\rho_t = \exp\left[\int_0^t -h^*(x_s)(\alpha Q\alpha^* + R)^{-1}(\alpha \mathrm{d}w + \mathrm{d}b) \right.
$$
$$
\left. -\frac{1}{2}\int_0^t h^*(x_s)(\alpha Q\alpha^* + R)^{-1}h(x_s)\mathrm{d}s\right]
$$

(4.5.5)

and the change of probability

$$
\left.\frac{\mathrm{d}\tilde{P}}{\mathrm{d}P}\right|_{\mathcal{F}^t} = \rho_t
$$

(4.5.6)

$$
\left.\frac{\mathrm{d}P}{\mathrm{d}\tilde{P}}\right|_{\mathcal{F}^t} = \eta_t = \frac{1}{\rho_t}.
$$

(4.5.7)

Consider next the process

$$\tilde{w}(t) = w(t) + \int_0^t Q\alpha^*(\alpha Q\alpha^* + R)^{-1}h(x_s)ds \qquad (4.5.8)$$

then we can state

Lemma 4.5.1 *For the system* $(\Omega, \mathcal{A}, \tilde{P}, \mathcal{F}^t)$ *the processes* $\tilde{w}(\cdot)$ *and* $z(\cdot)$ *are* $\mathcal{F}^t$ *Wiener processes with covariance matrices* $Q(\cdot)$ *and* $\alpha Q\alpha^* + R$ *respectively. The correlation is given by*

$$\tilde{E}\left[(z(t) - z(s))(\tilde{w}(t) - \tilde{w}(s))^*\right] = \int_s^t \alpha(\tau)Q(\tau)d\tau. \qquad (4.5.9)$$

Proof

Consider the stochastic integral

$$I_t(\phi, \psi) = \int_t^T \phi^* dz + \int_t^T \psi^* d\tilde{w}$$

where ϕ, ψ are L^2 deterministic functions. The result is an immediate consequence of the formula

$$\tilde{E}\left[\exp iI_t(\phi, \psi)|\mathcal{F}^t\right] = \exp\left[-\frac{1}{2}\int_t^T \phi^*(\alpha Q\alpha^* + R)\phi ds\right.$$
$$\left. -\frac{1}{2}\int_t^T \psi^* Q\psi ds - \int_t^T \phi^*\alpha Q\psi ds\right]. \qquad (4.5.10)$$

Details are left for the reader (see also Bensoussan 1982). ∎

For the system $(\Omega, \mathcal{A}, \tilde{P}, \mathcal{F}^t)$ the process $x(t)$ appears as the solution of

$$dx = \left[g(x_t, t) - \sigma(x_t)Q_t\alpha_t^*(\alpha Q\alpha^* + R)^{-1}h(x_t)\right]dt + \sigma(x_t)d\tilde{w}$$
$$x(0) = \xi \qquad (4.5.11)$$

where $\tilde{w}$ is a $\mathcal{F}^t$ Wiener process, the observation z being also a $\mathcal{F}^t$ Wiener process, with covariance and correlation as described in Lemma 4.5.1.

It will be convenient in what follows to use the notation

$$\tilde{g}(x, t) = g(x, t) - \sigma(x, t)Q_t\alpha_t^*(\alpha_t Q_t\alpha_t^* + R_t)^{-1}h(x, t). \qquad (4.5.12)$$

4.5.2   Equations for the unnormalized conditional probability

We consider as in Section 4.1

$$p(t)(\psi) = \tilde{E}[\psi(x_t)\eta_t|\mathcal{Z}^t] \qquad (4.5.13)$$

and of course

$$\Pi(t)(\psi) = \frac{p(t)(\psi)}{p(t)(1)}. \qquad (4.5.14)$$

The generalization of Theorem 4.1.1 is expressed as follows.

Theorem 4.5.1 *We make the assumptions of Theorem 4.1.1 and $\alpha(t)$ belongs to $L^\infty(0,T;\mathcal{L}(R^n;R^m))$. Let $\psi(x,t) \in C_b^{2,1}(R^n \times [0,\infty))$, then one has the relation*

$$
p(t)(\psi(t)) = \Pi_0(\psi(0)) + \int_0^t p(s)\left(\frac{\partial\psi}{\partial s} - A(s)\psi(s)\right)\mathrm{d}s
$$

$$
+ \int_0^t p(s)\left[h^*(s)\psi(s) + \mathrm{D}\psi^*(s)\sigma(s)Q(s)\alpha^*(s)\right](\alpha Q\alpha^* + R)_s^{-1}\mathrm{d}z
$$

a.s., $\forall t$.

$$(4.5.15)$$

Proof

We do not repeat all the details of the proof of Theorem 4.1.1. We only mention the formal calculation leading to (4.5.15). As we have seen in Theorem 4.1.1, the technicalities arise in justifying the treatment of the stochastic integral (for instance justifying that one can take the mathematical expectation and that it is equal to 0).

Note that η_t defined in (4.5.7) satisfies

$$
\mathrm{d}\eta_t = \eta_t h^*(x_t)(\alpha Q\alpha^* + R)^{-1}\mathrm{d}z_t \tag{4.5.16}
$$

and

$$
\mathrm{d}\eta_t\psi(x_t) = \eta\left(\frac{\partial\psi}{\partial t} - A(t)\psi\right) + \eta_t\mathrm{D}\psi^*(x_t)\sigma(x_t)\mathrm{d}\tilde{w}
$$

$$
+ \eta_t\psi(x_t)h^*(x_t)(\alpha Q\alpha^* + R)^{-1}\mathrm{d}z. \tag{4.5.17}
$$

Consider now the process θ_t defined as in (4.1.28) by

$$
\mathrm{d}\theta = \theta i\beta^*(\alpha Q\alpha^* + R)^{-1}\mathrm{d}z, \quad \theta(0) = 1 \tag{4.5.18}
$$

where $\beta \in L^\infty(0,T;R^m)$. We deduce from (4.5.17), (4.5.18)

$$
\mathrm{d}\theta_t\eta_t\psi(x_t)
$$

$$
= \theta\eta\left[\left(\frac{\partial\psi}{\partial t} - A(t)\psi\right) + i\beta^*(\alpha Q\alpha^* + R)^{-1}(h(x_t)\psi(x_t)) + \alpha_t Q_t\sigma^*(x_t)\mathrm{D}\psi(x_t)\right]\mathrm{d}t
$$

$$
+ \theta_t\eta_t\psi(x_t)(h^*(x_t) + i\beta_t^*)(\alpha Q\alpha^* + R)^{-1}\mathrm{d}z + \theta_t\eta_t\mathrm{D}\psi^*(x_t)\sigma(x_t)\mathrm{d}\tilde{w}.
$$

$$(4.5.19)$$

Integrating between 0 and t and taking the expectation with respect to $\tilde{P}$ yields

$$
\tilde{E}\theta_t\eta_t\psi(x_t) = \Pi_0(\psi(0)) + \tilde{E}\int_0^t \theta_s\eta_s\left\{\frac{\partial\psi}{\partial s} - A(s)\psi\right.
$$

$$
\left. + i\beta_s^*(\alpha Q\alpha^* + R)^{-1}[h(x_s)\psi(x_s) + \alpha_s Q_s\sigma^*(x_s)\mathrm{D}\psi(x_s)]\right\}\mathrm{d}s
$$

$$(4.5.20)$$

hence

$$
\tilde{E}\theta_t p(t)(\psi(t)) = \Pi_0(\psi(0)) + \tilde{E}\int_0^t \theta_s\left\{p(s)\left(\frac{\partial\psi}{\partial s} - A(s)\psi\right)\right.
$$

$$
\left. + i\beta_s^*(\alpha Q\alpha^* + R)^{-1}p(s)\left[h(s)\psi(s) + \alpha(s)Q(s)\sigma^*(s)\mathrm{D}\psi(s)\right]\right\}\mathrm{d}s
$$

$$
= \tilde{E}\theta_t\int_0^t p(s)\left(\frac{\partial\psi}{\partial s} - A(s)\psi\right)\mathrm{d}s
$$

$$
+ \tilde{E}\theta_t\int_0^t p(s)\left[h^*(s)\psi(s)\mathrm{D}\psi^*(s)\sigma(s)Q(s)\alpha^*(s)\right](\alpha Q\alpha^* + R)^{-1}\mathrm{d}z
$$

and the result (4.5.15) follows easily.   ∎

The result of Theorem 4.1.2 has been included in the proof of Theorem 4.5.1.

As far as uniqueness is concerned, one can follow the method developed in Section 4.2. This approach simplifies the studying of the PDE (whose unknown is a complex valued function)

$$\frac{\partial v}{\partial t} - A(t)v + i\beta^*(\alpha Q\alpha^* + R)^{-1}(vh + \alpha Q\sigma^* Dv) = 0$$
$$v(x,t) = \phi(x) \tag{4.5.21}$$

which decomposes into the system

$$\frac{\partial v_1}{\partial t} - A(t)v_1 - \beta^*(\alpha Q\alpha^* + R)^{-1}(hv_2 + \alpha Q\sigma^* Dv_2) = 0$$
$$\frac{\partial v_2}{\partial t} - A(t)v_2 + \beta^*(\alpha Q\alpha^* + R)^{-1}(hv_1 + \alpha Q\sigma^* Dv_1) = 0 \tag{4.5.22}$$
$$v_1(x,T) = \phi_1(x), \quad v_2(x,T) = \phi_2(x).$$

We need to obtain the same results as in Proposition 4.2.1.

Proposition 4.5.1 *We make the assumptions of Proposition 4.2.1, then the same result holds as in Proposition 4.2.1.*

Proof

We set

$$f(x,t) = \beta^*(t)[\alpha(t)Q(t)\alpha^*(t) + R(t)]^{-1}h(x,t)$$
$$c(x,t) = \sigma(x,t)Q(t)\alpha^*(t)[\alpha(t)Q(t)\alpha^*(t) + R(t)]^{-1}\beta(t)$$

and (4.5.22) reads

$$-\frac{\partial v_1}{\partial t} + Av_1 + fv_2 + c^* Dv_2 = 0$$
$$-\frac{\partial v_2}{\partial t} + Av_2 - fv_1 - c^* Dv_1 = 0. \tag{4.5.24}$$

By virtue of (4.5.23) and the general assumptions we have†

$$c_\lambda, \ c_{\lambda,\mu}, \ c_{\lambda,\mu\nu} \text{ are bounded.} \tag{4.5.25}$$

We derive the same estimates as in Proposition 4.2.1.

We begin with

$$z = \frac{1}{2}(v_1^2 + v_2^2)$$

hence

$$-\frac{\partial z}{\partial t} + Az = -a_{\lambda\mu}(v_{1,\lambda}v_{1,\mu} + v_{2,\lambda}v_{2,\mu})$$
$$+ c_\lambda(-v_1 v_{2,\lambda} + v_2 v_{1,\lambda}) \tag{4.5.26}$$
$$\leq c_0 z$$

and the maximum principle yields

$$v_1, \ v_2 \text{ are bounded.} \tag{4.5.27}$$

―――――――――――――
† With the notation of Proposition 4.2.1 for the derivations.

Define next

$$z = \frac{1}{2} \sum_\lambda (v_{1,\lambda}^2 + v_{2,\lambda}^2)$$

we deduce

$$
\begin{aligned}
-\frac{\partial z}{\partial t} + Az = &- a_{\lambda\mu}(v_{1,\lambda\nu} v_{1,\mu\nu} + v_{2,\lambda\nu} v_{2,\mu\nu}) \\
&+ a_{\lambda\mu,\nu} v_{1,\lambda\mu} v_{1,\nu} + a_{\lambda\mu,\nu} v_{2,\lambda\mu} v_{2,\nu} + g_{\lambda,\mu} v_{1,\lambda} v_{1,\mu} \\
&+ g_{\lambda,\mu} v_{2,\lambda} v_{2,\mu} - v_{1,\lambda} f_\lambda v_2 + v_{2,\lambda} f_\lambda v_1 \\
&- c_{\lambda,\mu}(v_{2,\lambda} v_{1,\mu} - v_{1,\lambda\mu} v_{2,\mu}) - c_\lambda(v_{2,\lambda\mu} v_{1,\mu} - v_{1,\lambda} v_{2,\mu}) \\
\leq &\, c_0 z + c_1
\end{aligned}
\tag{4.5.28}
$$

hence again

$$v_{1,\lambda}, \; v_{2,\lambda} \text{ are bounded.} \tag{4.5.29}$$

A similar calculation is done to estimate the quantity

$$z = \frac{1}{2} \sum_{\lambda\mu}(v_{1,\lambda\mu}^2 + v_{2,\lambda\mu}^2)$$

(see (4.2.14)); we leave it to the reader to make the details explicit.

The remainder of the proof is the same as in Proposition 4.2.1. ∎

Remark 4.5.1 In the linear case

$$
\begin{aligned}
g(x,t) &= F(t)x + f(t) \\
\sigma(x,t) &= \sigma(t) \\
h(x,t) &= H(t)x + h(t)
\end{aligned}
$$

and initial data

$$\phi(x) = \exp(i\lambda^* x),$$

the solution of (4.5.21) is still given by

$$v(x,t) = \exp(i\rho(t)^* x + s(t)) \tag{4.5.30}$$

with $\rho(t)$, $s(t)$ solutions of

$$
\begin{aligned}
&\dot{\rho} + F^*\rho + H^*(\alpha Q\alpha^* + R)^{-1}\beta = 0, \quad \rho(T) = \lambda \\
&\dot{s} + i\rho^* f + i\beta^*(\alpha Q\alpha^* + R)^{-1}h - \frac{1}{2}\rho^*\sigma Q\sigma^*\rho - \beta^*(\alpha Q\alpha^* + R)^{-1}\alpha Q\sigma^*\rho = 0 \\
&s(T) = 0.
\end{aligned}
\tag{4.5.31}
$$

∎

The uniqueness result for the solution of the Zakai equation (4.5.15) can then be stated in the same way as in § 4.2.2; see Theorem 4.2.1.

4.5.3 Equation for the normalized conditional probability

We can state

Theorem 4.5.2 *We make the assumptions of Theorem 4.5.1. Let*

$$\psi(x,t) \in C_b^{2,1}(R^n \times [0,\infty)),$$

then one has the relation

$$
\begin{aligned}
\Pi(t)(\psi(t)) ={}& \Pi_0(\psi(0)) + \int_0^t \Pi(s)\left(\frac{\partial\psi}{\partial s} - A(s)\psi(s)\right) ds \\
& + \int_0^t [\Pi(s)(h^*(s)\psi(s) + D\psi^*(s)\sigma(s)Q(s)\alpha^*(s)) \\
& - \Pi(s)(\psi(s))\Pi(s)(h^*(s))] (\alpha Q\alpha^* + R)^{-1}(dz - \Pi(s)(h(s))ds),
\end{aligned}
$$ (4.5.32)

a.s., $\forall t$.

Proof

Note that

$$dp(t)(1) = p(t)(1)\Pi(t)(h^*(t))(\alpha Q\alpha^* + R)^{-1}dz$$

hence

$$
\begin{aligned}
d\frac{1}{p(t)(1)} = \frac{1}{p(t)(1)} \Big[& -\Pi(t)(h^*(t))(\alpha Q\alpha^* + R)^{-1}dz \\
& + \Pi(t)(h^*(t))(\alpha Q\alpha^* + R)^{-1}\Pi(t)(h(t))dt \Big]
\end{aligned}
$$

and the result (4.5.32) follows easily. ∎

The uniqueness is studied as in Theorem 4.3.3.

4.5.4 Explicit solution in the linear case

We consider the situation (4.4.1), (4.4.2) and look for the solution of the Zakai equation. We can check by calculations similar to those of Theorem 4.4.1 that

$$p(t)(\psi) = \left[\int \psi(\hat{x}_t + P_t^{1/2}\xi)\frac{\exp(-\frac{1}{2}|\xi|^2)}{(2n)^{n/2}}d\xi\right] s_t$$ (4.5.33)

where P is the solution of the Riccati equation.

$$\dot{P} + (PH^* + Q\alpha^*)(\alpha Q\alpha^* + R)^{-1}(HP + \alpha Q) - Q - FP - PF^* = 0$$

$$P(0) = P_0$$ (4.5.34)

and the Kalman filter $\hat{x}_t$ is given by the expression

$$d\hat{x}_t = (F\hat{x}_t + f)dt + (PH^* + Q\alpha^*)(\alpha Q\alpha^* + R)^{-1}[dz - (H\hat{x} + h)dt]$$

$$\hat{x}(0) = x_0.$$ (4.5.35)

The process s_t is given by the expression

$$
\begin{aligned}
s_t ={}& \exp\Big[\int_0^t (\hat{x}^*H^* + h^*)(\alpha Q\alpha^* + R)^{-1}dz \\
& -\frac{1}{2}\int_0^t (\hat{x}^*H^* + h^*)(\alpha Q\alpha^* + R)^{-1}(H\hat{x} + h)ds\Big].
\end{aligned}
$$ (4.5.36)

4.5.5 Further correlation

We extend the model (4.5.1), (4.5.2) as follows. We proceed formally to avoid lengthy statements and proofs.

We thus consider

$$\mathrm{d}x = g(x_t, z_t)\mathrm{d}t + \sigma(x_t, z_t)\mathrm{d}w$$
$$x(0) = \xi \tag{4.5.37}$$

$$\mathrm{d}z = h(x_t, z_t)\mathrm{d}t + \alpha(z_t)\mathrm{d}w + \mathrm{d}b$$
$$z(0) = 0. \tag{4.5.38}$$

The Wiener processes w, b are the same as in (4.5.1). We have omitted the dependence of g, σ, h, α on t; it is implicit. We see that g, σ, h depend on the signal as well as on the observation. However α depends on the observation only. The case when α depends also on x is open, at least with the present techniques (see Szpirglas and Mazzioto 1978; Kunita 1982; Pardoux 1982). To simplify the notation we set

$$D(z) = R + \alpha(z)Q\alpha^*(z). \tag{4.5.39}$$

Define the process ρ_t by

$$\mathrm{d}\rho = -\rho h^*(x_t, z_t)D(z_t)^{-1}(\alpha\mathrm{d}w + \mathrm{d}b), \quad \rho(0) = 1. \tag{4.5.40}$$

Next define $\tilde{z}_t$ by

$$\mathrm{d}\tilde{z}_t = D(z_t)^{-1/2}\mathrm{d}z_t, \quad \tilde{z}(0) = 0$$
$$= D(z_t)^{-1/2}[h(x_t, z_t)\mathrm{d}t + \alpha(z_t)\mathrm{d}w + \mathrm{d}b]. \tag{4.5.41}$$

Let also

$$B(z) = Q - Q\alpha^*(z)D(z)^{-1}\alpha(z)Q \tag{4.5.42}$$

which is positive definite, as is easily seen.† We also define the process $\tilde{w}_t$ by

$$d\tilde{w}_t = B(z_t)^{-1/2}\left[\mathrm{d}w - Q\alpha^*(z_t)D(z_t)^{-1}\mathrm{d}z_t\right.$$
$$\left. + Q\alpha^*(z_t)D(z_t)^{-1}h(x_t, z_t)\mathrm{d}t\right] \tag{4.5.43}$$
$$\tilde{w}(0) = 0.$$

Let $\tilde{P}$ be defined as in (4.5.6) by

$$\left.\frac{\mathrm{d}\tilde{P}}{\mathrm{d}P}\right|_{\mathcal{F}^t} = \rho_t \tag{4.5.44}$$

We can state

Lemma 4.5.2 *For the system $(\Omega, \mathcal{A}, \tilde{P}, \mathcal{F}^t)$ the processes $\tilde{w}_t$ and z_t are independent standard Wiener processes.*

Proof

Proceed as in Lemma 4.5.1. ∎

---

† If Q is positive definite, which we assume to avoid technical difficulties.

For the system $(\Omega, \mathcal{A}, \tilde{P}, \mathcal{F}^t)$ the process x_t appears as the solution of

$$
\mathrm{d}x = \left[g(x_t, z_t) - \sigma(x_t, z_t) Q \alpha^*(z_t) D(z_t)^{-1} h(x_t, z_t) \right] \mathrm{d}t
$$
$$
+ \sigma(x_t, z_t) Q_\alpha^*(z_t) D(z_t)^{-1/2} \mathrm{d}\tilde{z}_t + \sigma(x_t, z_t) B(z_t)^{1/2} \mathrm{d}\tilde{w} \qquad (4.5.45)
$$
$$
x(0) = \xi
$$

and $\tilde{z}_t$, $\tilde{w}$ are independent Wiener processes. Since $D(z_t)$ depends only on z and is invertible, we can assert that

$$
\mathcal{Z}^t = \tilde{\mathcal{Z}}^t = \sigma(\tilde{z}_s, \ s \le t) \qquad (4.5.46)
$$

provided of course some smoothness on $D(z)$ is available which permits us to solve the stochastic differential equation

$$
\mathrm{d}z = D(z_t)^{1/2} \mathrm{d}\tilde{z}_t, \quad z(0) = 0
$$

in which $\tilde{z}_t$ is viewed as a Wiener process (for $\tilde{P}$).

Introduce $\eta_t = 1/\rho_t$ so that

$$
\mathrm{d}\eta_t = \eta_t h^*(x_t, z_t) D^{-1/2}(z_t) \mathrm{d}\tilde{z}, \quad \eta(0) = 1. \qquad (4.5.47)
$$

We are going to derive the equations for the unnormalized and for the normalized conditional probabilities. We proceed as in the proof of Theorem 4.5.1, and get the equation

$$
p(t)(\psi(t)) = \Pi_0(\psi(0)) + \int_0^t p(s) \left(\frac{\partial \psi}{\partial s} - A(s)\psi(s) \right) \mathrm{d}s
$$
$$
+ \int_0^t \left[p(s)(\mathrm{D}\psi^*(s)\sigma(s)) Q_s \alpha^*(z_s) + p(s)(\psi(s)h^*(s)) \right] D(z_s)^{-1} \mathrm{d}z_s. \qquad (4.5.48)
$$

Note that in this equation σ and h are functions of x and z. Therefore more explicitly

$$
p(s)(\psi(s)h(s)) = \tilde{E}[\psi(x_s)h(x_s, z_s)\eta_s | \mathcal{Z}^s].
$$

In fact (4.5.48) will not be sufficient to characterize in a unique way the process $p(t)$. That is why one considers a test function $\psi(x, z, t)$ depending on x and z, which is $C_b^{2,2,1}(R^n \times R^m \times [0, \infty))$.

By arguments similar to those needed to derive (4.5.48) one obtains the relation generalizing (4.5.48)

$$
p(t)(\psi(t)) = \Pi_0(\psi(\cdot, 0, 0))
$$
$$
+ \int_0^t p(s) \left(\frac{\partial \psi}{\partial s} - A\psi + h^* \mathrm{D}_z \psi + \mathrm{tr} \ \sigma Q \alpha^* \mathrm{D}_{zx}^2 \psi + \frac{1}{2} \mathrm{tr} \ D(z) \mathrm{D}_z^2 \psi \right) \mathrm{d}s
$$
$$
+ \int_0^t \left[p(s)(\mathrm{D}_x \psi^*(s)\sigma(s)) Q_s \alpha^*(z_s) \right.
$$
$$
\left. + p(s)(\psi(s)h^*(s)) + p(s)(\mathrm{D}_z \psi^*(s)) D(z_s) \right] D^{-1}(z_s) \mathrm{d}z_s. \qquad (4.5.49)
$$

We may also write an integrated form of (4.5.49). For uniqueness, one needs to study the complex valued PDE

$$
\begin{aligned}
&\frac{\partial v}{\partial t} - A(t)v + h^* \mathrm{D}_z v + \frac{1}{2} \mathrm{tr}\, D(z) \mathrm{D}_z^2 v + \mathrm{tr}\, \sigma Q \alpha^* \mathrm{D}_{zx}^2 v \\
&+ \mathrm{i} \beta^* D^{-1}(hv + D\mathrm{D}_z v + \alpha Q \sigma^* \mathrm{D}_x v) = 0 \\
&v(x, z, T) = \phi(x).
\end{aligned}
\tag{4.5.50}
$$

Clearly, when the coefficients do not depend on z, (4.5.50) has for its solution a function which does not depend on z as well. One easily extends the results of Proposition 4.5.1 to (4.5.50). Details are left to the reader.

4.6   Some representation formulas for the conditional probability

4.6.1   Setting of the problem

In presenting the ideas of this section, we shall not try to make the most general statements possible, and this allows convenient technical assumptions, which will be discussed later on. We thus consider the model

$$dx = g(x_t, t)dt + \sigma(x_t, t)dw$$
$$x(0) = \xi \tag{4.6.1}$$

$$dz = h(x_t, t)dt + db_t$$
$$z(0) = 0 \tag{4.6.2}$$

assuming

$$g, \sigma \text{ are bounded Borel functions} \tag{4.6.3}$$

$$|g(x_1, t) - g(x_2, t)| \le k|x_1 - x_2| \tag{4.6.4}$$

$$\|\sigma(x_1, t) - \sigma(x_2, t)\| \le k|x_1 - x_2| \tag{4.6.5}$$

and

$$h(x, t) \text{ is bounded.} \tag{4.6.6}$$

The probability law of ξ is Π_0; it has a density $p_0(x)$ with respect to Lebesgue measure and $\int |x|^2 p_0(x)dx < \infty$. We also assume that $p_0 \in L^2(R^n)$. $\tag{4.6.7}$

w, b are independent standard Wiener processes in R^n, R^m respectively, with respect to a filtration $\mathcal{F}^t$. $\tag{4.6.8}$

Moreover ξ, w, b are mutually independent.

Therefore all the assumptions of Section 4.1 are satisfied. We have assumed in addition boundedness of the functions, and strong regularity of the observation function h. For simplicity we have taken $Q = R = I$.

Recalling the notation

$$a(x, t) = \frac{1}{2}\sigma(x, t)\sigma^*(x, t) \tag{4.6.9}$$

we also assume the uniform definiteness

$$a(x,t) \geq \delta I, \quad \delta > 0. \tag{4.6.10}$$

Thus the second order differential operator

$$A(t) = -\sum_{i,j} a_{ij} \frac{\partial^2}{\partial x_i \partial x_j} - \sum_i g_i \frac{\partial}{\partial x_i}. \tag{4.6.11}$$
$$= -g^*D - \operatorname{tr} aD^2$$

is strongly elliptic.

The problem we are concerned with is the existence of a density (with respect to Lebesgue measure) for the conditional probability; namely, can we write

$$p(t)(\psi) = \int p(x,t)\psi(x)\mathrm{d}x \tag{4.6.12}$$

where $p(x,t)$ is a stochastic function of x,t.

To guess what the equation could be for which $p(x,t)$ is the solution, we can proceed formally with (4.1.37), writing

$$\mathrm{d} \int p(x,t)\psi(x,t)\mathrm{d}x = \int p(x,t) \left(\frac{\partial\psi}{\partial t} - A(t)\psi \right) \mathrm{d}x$$
$$+ \left[\int p(x,t)\psi(x,t)h^*(x,t)\mathrm{d}x \right] \mathrm{d}z$$

or

$$\int (\mathrm{d}p + A^*p\mathrm{d}t - ph^*\mathrm{d}z)\psi\mathrm{d}x = 0$$

and we are naturally led to the equation to be studied

$$\mathrm{d}p + A^*p\mathrm{d}t = ph^*\mathrm{d}z, \quad p(x,0) = p_0(x) \tag{4.6.13}$$

where A^* is the formal adjoint of A, namely

$$A^* = -\sum_{i,j} \frac{\partial^2}{\partial x_i \partial x_j} (a_{ij} \ \cdot) + \sum_i \frac{\partial}{\partial x_i} (g_i \ \cdot). \tag{4.6.14}$$

We shall in a subsequent section study (4.6.13) rigorously (see Section 4.7). However when h is smooth, it is possible to reduce (4.6.13) to an ordinary PDE with random coefficients. This is called the robust form.

4.6.2 Robust form

We shall assume here that the function is smooth, namely

$$\frac{\partial h}{\partial t}, \quad Dh, \quad D^2h \quad \text{are bounded.} \tag{4.6.15}$$

We make the transformation

$$q(x,t) = p(x,t) \exp[-h^*(x,t)z(t)] \tag{4.6.16}$$

in equation (4.6.13). This change of unknown function (known as robust form) was first introduced by Doss (1977) and Sussmann (1978), in order to reduce Ito equations to ordinary differential equations. It has been used extensively in the context of non linear filtering (see for example Liptser and Shiryaev 1977; Clark 1978; Davis 1981) and permits us to reduce (4.6.13) to an ordinary PDE with random coefficients (i.e. depending on the path $z(\cdot)$). Indeed write

$$\chi_t = \exp[-h^*(x,t)z(t)]$$

then

$$d\chi_t = \chi\left[-h^*dz + \left(-h_t^*z + \frac{1}{2}|h|^2\right)dt\right]$$

and thus

$$dq = \chi dp + pd\chi - p\chi|h|^2 dt$$

and we see that the term in dz cancels out. Hence

$$\frac{\partial q}{\partial t} = -\chi A^* p + q\left(-\frac{\partial h^*}{\partial t}z - \frac{1}{2}|h|^2\right).$$

To simplify the notation we use the scalar product hz instead of h^*z (omitting the dot),

$$a_{ij,k} = \frac{\partial a_{ij}}{\partial x_k}, \quad [\mathrm{D}(hz)]_i = \frac{\partial hz}{\partial x_i}, \quad \text{etc.}$$

We also define

$$f_i = g_i - a_{ij,j} \tag{4.6.17}$$

then we have

$$\chi A^* p = A^* q - 2a\mathrm{D}q\mathrm{D}(hz) + q\left[A(hz) + 2f\mathrm{D}(hz) - a\mathrm{D}(hz)\mathrm{D}(hz)\right]$$

therefore q is the solution of the PDE

$$\frac{\partial q}{\partial t} + A^* q - 2a\mathrm{D}q\mathrm{D}(hz) + q\left[\frac{\partial h}{\partial t}z + A(hz)\right.$$

$$\left. +2f\mathrm{D}(hz) - a\mathrm{D}(hz)\mathrm{D}(hz) + \frac{1}{2}|h|^2\right] = 0.$$

Define the quantity†

$$e_t(h,z) = \frac{\partial h}{\partial t}z_t + A(t)(h_t z_t) + ef_t\mathrm{D}(h_t z_t)$$

$$- a(t)\mathrm{D}(h_t z_t)\mathrm{D}(h_t z_t) + \frac{1}{2}|h_t|^2 \tag{4.6.18}$$

and thus the PDE can be written as follows

$$\frac{\partial q}{\partial t} + A^* q - 2a\mathrm{D}q\mathrm{D}(hz) + qe_t(h,z) = 0 \tag{4.6.19}$$

$$q(x,0) = p_0(x).$$

---

† $h_t \equiv h(x,t)$, $f_t = f(x,t)$, etc. Note that when h is not bounded, the situation becomes much more complicated.

For any sample, $z(t)$ is a continuous function, and thus, by the assumptions made on h, $e_t(h, z)$ is bounded. Therefore (4.6.19) appears for any sample as an ordinary parabolic PDE. Since $p_0 \in L^2(R^n)$, it has, by standard theory, one and only one solution in the space

$$q \in L^2(0, T; H^1(R^n)), \quad \frac{\partial q}{\partial t} \in L^2(0, T; H^{-1}(R^n)). \qquad (4.6.20)$$

Remark 4.6.1 The above approach leads to a solution defined pathwise. We do not obtain information concerning expected values. This requires a direct approach to (4.6.18). ∎

4.6.3 A first representation formula for the conditional probability

Our objective is to give an interesting representation formula for $p(x, T)$, in the spirit of the probabilistic interpretation of the solution of parabolic equations.

Let us introduce some notation. If $\phi(x, t)$ is any function, we shall write

$$\bar{\phi}(x, t) = \phi(x, T - t);$$

similarly we write $\bar{z}_t = z(T - t)$, $\bar{A}(t) = A(T - t)$, and define

$$\begin{aligned}
\bar{e}_t(\bar{h}, \bar{z}) = &-\frac{\partial \bar{h}}{\partial t} \bar{z}_t + \bar{A}(t)(\bar{h}_t \bar{z}_t) + 2\bar{f}_t \mathrm{D}(\bar{h}_t \bar{z}_t) \\
&- \bar{a}(t)\mathrm{D}(\bar{h}_t \bar{z}_t)\mathrm{D}(\bar{h}_t \bar{z}_t) + \frac{1}{2}|\bar{h}_t|^2.
\end{aligned} \qquad (4.6.21)$$

Clearly from (4.6.21) the function $\bar{q}(x, t)$ is the solution of

$$\begin{aligned}
&-\frac{\partial \bar{q}}{\partial t} + \bar{A}^* \bar{q} - 2\bar{a}\mathrm{D}\bar{q}\mathrm{D}(\bar{h}\bar{z}) + \bar{q}\bar{e}_t(\bar{h}, \bar{z}) = 0 \\
&\bar{q}(x, T) = p_0(x).
\end{aligned} \qquad (4.6.22)$$

Introduce next

$$\tilde{q}(x, t) = \bar{q}(x, t) \exp[\bar{h}(x, t) z_T] \qquad (4.6.23)$$

and the process

$$\tilde{z}_t = z_t - \bar{z}_t \qquad (4.6.24)$$

(note that $\tilde{z}_T = z_t$, $\tilde{z}_0 = 0$).

We shall justify the introduction of $\tilde{q}$ presently. Note for the moment that

$$\tilde{q}(x, 0) = p(x, T) \qquad (4.6.25)$$

which is the quantity we are interested in.

We leave it to the reader to check that $\tilde{q}(x, t)$ is the solution of

$$\begin{aligned}
&-\frac{\partial \tilde{q}}{\partial t} + \bar{A}\tilde{q} + [2\bar{a}D(\bar{h}\tilde{z}) + 2\bar{f}]\mathrm{D}\tilde{q} + \tilde{q}\left[\mathrm{div}\,\bar{f} + \tilde{e}_t(\bar{h}, \bar{z})\right] = 0 \\
&\tilde{q}(x, T) = p_0(x) \exp[\bar{h}(x, T)\tilde{z}_T]
\end{aligned} \qquad (4.6.26)$$

where we have set

$$\tilde{e}_t(\bar{h}, \tilde{z}) = \frac{\partial \bar{h}}{\partial t} \tilde{z}_t - \bar{A}(t)(\bar{h}_t \tilde{z}_t) - 2\bar{f}_t \mathrm{D}(\bar{h}_t \tilde{z}_t)$$

$$- \bar{a}(t)\mathrm{D}(\bar{h}_t \tilde{z})\mathrm{D}(\bar{h}_t \tilde{z}) + \frac{1}{2}|\bar{h}_t|^2. \tag{4.6.27}$$

It is interesting and important to notice that the coefficients of the equation (4.6.27) depend only on $\tilde{z}$. Moreover $\tilde{z}_t$ is a standard Wiener process adapted to the filtration

$$\tilde{\mathcal{Z}}^t = \sigma(\tilde{z}_s, s \leq t).$$

Since (4.6.26) is a backward parabolic PDE, one can write a representation formula. Consider another probability space $(\Omega_0, \mathcal{A}_0, P_0)$, equipped with a filtration Γ^t on which is defined a standard Γ^t Wiener process $w_{0,t}$, with values in R^n. We take as the probabilistic set up the product space

$$(\Omega \times \Omega_0, \ \mathcal{A} \times \mathcal{A}_0, \ \tilde{P} \otimes P_0) = Q$$

equipped with the filtration $\tilde{\mathcal{Z}}^t \otimes \Gamma^t$. Then the pair $\tilde{z}_t$, $w_{0,t}$ is a pair of independent standard Wiener processes with values in R^m and R^n respectively.

Consider next the stochastic differential equation

$$dX = (\bar{g} - 2\bar{f})(X)dt + \bar{\sigma}(X)dw_0$$

$$X(0) = x \tag{4.6.28}$$

where x is given, and the process

$$d\zeta_t = \zeta_t \left\{ [-\mathrm{div}\, \bar{f}(X_t) - \tilde{e}_t(\bar{h}, \tilde{z})(X_t)]dt \right.$$

$$\left. -\mathrm{D}(\bar{h}\tilde{z})\bar{\sigma}(X_t)dw_0 \right\} \tag{4.6.29}$$

$$\zeta(0) = 1$$

where $\tilde{z}$ is considered as a parameter, and the randomness is due only to the process w_0 (and thus X).

Considering the process $\tilde{q}(X_t, t)\zeta_t$ we can immediately assert that

$$d\tilde{q}(X_t, t)\zeta_t = \zeta_t \left[\mathrm{D}\tilde{q} - \tilde{q}\mathrm{D}(\bar{h}\tilde{z}) \right] \bar{\sigma}dw_0$$

hence

$$E_0 \tilde{q}(X_T, T)\zeta_T = \tilde{q}(x, 0)$$

i.e.

$$\tilde{q}(x, 0) = E_0 \left\{ p_0(X_T)\zeta_T \exp[\bar{h}(X_T, T)\tilde{z}_T] \right\}.$$

Note that from (4.6.28) we deduce

$$\zeta_T = \exp \left\{ \int_0^T \left[-\mathrm{div}\, \bar{f}(X_t) - \tilde{e}_t(\bar{h}, \tilde{z})(X_t) - \bar{a}\mathrm{D}(\bar{h}\tilde{z})\mathrm{D}(\bar{h}\tilde{z}) \right] dt \right.$$

$$\left. - \int_0^T \mathrm{D}(\bar{h}\tilde{z})\bar{\sigma}(X_t)dw_0 \right\}$$

$$= \exp \left\{ \int_0^T \left[-\mathrm{div}\, \bar{f}(X_t) - \frac{\partial \bar{h}}{\partial t}(X_t)\tilde{z}_t + \bar{A}(t)(\bar{h}_t \tilde{z}_t) \right. \right.$$

$$\left. \left. +2\bar{f}_t \mathrm{D}(\bar{h}_t \tilde{z}_t) - \frac{1}{2}|\bar{h}_t|^2 \right] dt - \int_0^T \mathrm{D}(\bar{h}\tilde{z})\bar{\sigma}(X_t)dw_0 \right\}$$

hence

$$
\begin{aligned}
\tilde{q}(x,0) = E_0 &\left[p_0(X_T) \exp\left(-\int_0^T \operatorname{div} \bar{f}(X_t)\mathrm{d}t \right) \right. \\
&\times \exp\left\{ \bar{h}(X_T,T)\tilde{z}_T + \int_0^T \left[-\frac{\partial \bar{h}}{\partial t}\tilde{z}_t + \bar{A}(t)(\bar{h}_t\tilde{z}_t) \right.\right. \\
&\left.\left.\left. +2\bar{f}_t \mathrm{D}(\bar{h}_t\tilde{z}_t) - \frac{1}{2}|\bar{h}_t|^2 \right]\mathrm{d}t - \int_0^T \mathrm{D}(\bar{h}\tilde{z})\bar{\sigma}\mathrm{d}w_0 \right\} \right]
\end{aligned}
\tag{4.6.30}
$$

and because of the independence of X_t and $\tilde{z}_t$

$$
\begin{aligned}
\tilde{q}(x,0) = E^Q &\left[p_0(X_T) \exp\left(-\int_0^T \operatorname{div}\bar{f}(X_t)\mathrm{d}t \right) \right. \\
&\times \exp\left\{ \bar{h}(X_T,T)\tilde{z}_T + \int_0^T \left[-\frac{\partial\bar{h}}{\partial t}\tilde{z}_t + \bar{A}(t)(\bar{h}_t\tilde{z}_t) \right.\right. \\
&\left.\left.\left. +2\bar{f}_t\mathrm{D}(\bar{h}_t\tilde{z}_t) - \frac{1}{2}|\bar{h}_t|^2 \right]\mathrm{d}t - \int_0^T \mathrm{D}(\bar{h}\tilde{z})\bar{\sigma}\mathrm{d}w_0 \right\} \middle| \tilde{\mathcal{Z}}^T \right].
\end{aligned}
$$

But we may compute the Ito differential of $\bar{h}(X_t,t)\tilde{z}_t$ where the stochastic process involved is the pair X_t, $\tilde{z}_t$. We have

$$
\begin{aligned}
\mathrm{d}\bar{h}(X_t,t)\tilde{z}_t = &\left(\frac{\partial\bar{h}}{\partial t}\tilde{z}_t - \bar{A}(t)(\bar{h}_t\tilde{z}_t) - 2\bar{f}_t\mathrm{D}(\bar{h}_t\tilde{z}_t) \right)\mathrm{d}t \\
&+ \mathrm{D}(\bar{h}\tilde{z})\bar{\sigma}\mathrm{d}w_0 + \bar{h}(X_t,t)\mathrm{d}\tilde{z}
\end{aligned}
$$

and thus collecting results we finally obtain the formula

$$
\begin{aligned}
p(x,T) = E^Q &\left[p_0(X_T) \exp\left(-\int_0^T \operatorname{div}\bar{f}(X_t)\mathrm{d}t \right) \right. \\
&\times \exp\left\{ \int_0^T \bar{h}(X_t)\mathrm{d}\tilde{z} - \frac{1}{2}\int_0^T |\bar{h}(X_t,t)|^2\mathrm{d}t \right\} \middle| \tilde{\mathcal{Z}}^T \right].
\end{aligned}
\tag{4.6.31}
$$

We thus have proved

Theorem 4.6.1 *We make assumptions (4.6.3) to (4.6.8) and (4.6.10), (4.6.15). Then $p(x,T)$ is given explicitly by the formula (4.6.31).*

Remark 4.6.2 If $h = 0$, then

$$
p(x,t) = \text{ probability density of the process } x_t
$$

where x_t was defined in (4.6.1).

One recovers the standard probabilistic interpretation of the Fokker–Planck equation (written backwards after a change of time variable). ∎

Remark 4.6.3 Formula (4.6.31) involves only h but not its derivatives. Therefore it can be extended to cases when the regularity assumptions (4.6.15) on h do not hold, and also to cases when h is unbounded. ∎

It is possible and worth while to relate the process X_t introduced in (4.6.28) to the original process x_t, and thus to keep the same probabilistic set up as initially. This follows from the theory of time reversal of diffusions (see Haussmann and Pardoux 1988). We shall just check the main points of this theory, referring the reader to the article of Haussmann and Pardoux for all justifications.

Consider the process $\bar{x}_t = x(T - t)$, then one has

Lemma 4.6.1 *The process $\bar{x}_t$ is Markov.*

Proof

This is a general property of Markov processes. In the case of diffusions like x_t it can be readily seen by the following argument. We want to show that for $\phi \in C_0^\infty(R^n)$

$$E\left[\phi(\bar{x}_t)|\bar{x}_s\right] = E\left[\phi(\bar{x}_t)|\sigma(\bar{x}_r, r \leq s)\right] \quad \forall s \leq t.$$

But this amounts to showing

$$E\left[\phi(x_{T-t})|x_{T-s}\right] = E\left[\phi(x_{T-t})|\sigma(x_{T-s}, \ r \leq s)\right]$$

which is a consequence of

$$E\left[\phi(x_{T-r})|x_{T-s}\right] = E\left[\phi(x_{T-t}|x_{T-s}, \sigma(w_\lambda - w_{T-s}, T - s \leq \lambda \leq T)\right] \qquad (4.6.32)$$

since, for $r \leq s$, x_{T-r} is measurable with respect to the σ-algebra generated by x_{T-r} and $\sigma(w_\lambda - w_{T-s}, T - s \leq \lambda \leq T - t)$. Now the fact that (4.6.32) holds is merely a consequence of the independence of $w_\lambda - w_{T-s}$ for $\lambda \geq T - s$, and x_{T-t}. ∎

Let us denote by $m(x, t)$ the probability density of the process x_t. It is the solution of the Fokker–Planck equation

$$\frac{\partial m}{\partial t} + A^* m = 0, \quad m(x, 0) = p_0(x). \qquad (4.6.33)$$

Consider the second order differential operator

$$A_0 = -\text{tr } \bar{a}D^2 + (-\bar{g} + 2\bar{f} - 2\bar{a}\text{D} \log \bar{m})\text{D} \qquad (4.6.34)$$

and the filtration $\bar{\mathcal{X}}^t = \sigma(\bar{x}_s, \ s \leq t)$, then we can assert

Lemma 4.6.2 *For any $\phi \in C_0^\infty(R^n)$, the process $\phi(\bar{x}_t) + \int_0^t A_0\phi(\bar{x}_s)ds$ is a P, $\bar{\mathcal{X}}^t$ martingale.*

Proof

One has to prove that

$$E\left[\phi(\bar{x}_t) + \int_s^t A_0\phi(\bar{x}_\lambda)d\lambda \,\middle|\, \bar{\mathcal{X}}^s\right] = \phi(\bar{x}_s) \quad \forall s \leq t.$$

By the Markov property of $\bar{x}_t$, this is equivalent to

$$E\left[\phi(\bar{x}_t) + \int_s^t A_0\phi(\bar{x}_\lambda)d\lambda \,\middle|\, \bar{x}_s\right] = \phi(\bar{x}_s). \qquad (4.6.35)$$

If ψ is in $C_0^\infty(R^n)$, the property (4.6.35) is equivalent to

$$E\left\{\left[\phi(\bar{x}_t) + \int_s^t A_0\psi(\bar{x}_\lambda)d\lambda\right]\phi(\bar{x}_s)\right\} = E\phi(\bar{x}_s)\psi(\bar{x}_s).$$

After a reversal of time, the problem reduces to proving that

$$E\left\{\left[\phi(x_s) + \int_s^t \bar{A}_0(\mu)\phi(x_\mu)d\mu\right]\psi(x_t)\right\} = E\phi(x_t)\psi(x_t) \tag{4.6.36}$$

where of course

$$\bar{A}_0 = -\text{tr } aD^2 + (f - aD\log m)D. \tag{4.6.37}$$

Now consider the function $v(x,s)$ which is the solution of

$$-\frac{\partial v}{\partial s} + Av = 0, \quad v(x,t) = \psi(x), \quad s \leq t. \tag{4.6.38}$$

Note that

$$E\phi(x_s)\psi(x_t) = E\phi(x_s)v(x_s,s)$$
$$= \int \phi(x)v(x,s)m(x,s)dx$$

and thus (4.6.36) is equivalent to the analytic relation

$$\int \phi(x)v(x,s)m(x,s)dx + \int_s^t \left\{\left[-\text{tr } aD^2\phi + (-g + 2f - 2aD\log m)D\phi\right]\right.$$
$$\left.\times (x,\mu)v(x,\mu)m(x,\mu)\right\}dxd\mu \tag{4.6.39}$$
$$= \int \phi(x)\psi(x)m(x,t)dx$$

which can be checked easily by combining (4.6.33), (4.6.38) and performing some integration by parts. ∎

From the theory of Stroock and Varadhan (1979), it follows that there exists a $\bar{\mathcal{X}}^t$ standard Wiener process with values in R^n, denoted $\hat{w}$ such that

$$d\bar{x} = \left[\bar{g} - 2\bar{f} + 2\bar{a}D\log\bar{m}\right](\bar{x}_t,t)dt + \bar{\sigma}(\bar{x}_t,t)d\hat{w}$$
$$\bar{x}(0) = x(T) \tag{4.6.40}$$

random variable, $\bar{\mathcal{X}}(0)$ measurable with probability density $m(x,T)$.

So far the probability on $(\Omega, \mathcal{A})$ is the initial one P. But of course everything remains unchanged if one considers $\tilde{P}$ instead of P. Indeed (cf. (4.1.18)), the use of $\tilde{P}$ does not affect the process x_t at all. It affects only the process z_t which becomes a Wiener process.

Comparing (4.6.28) and (4.6.40) we see that there is a difference in the drift and in the initial condition.

To take care of the drift, one can perform a new Girsanov transformation, as follows. Firstly we consider on Ω the σ-algebra $\bar{\mathcal{X}}^T \otimes \tilde{\mathcal{Z}}^T \subset \mathcal{F}^T$, and define

$$\frac{d\tilde{P}^T}{d\tilde{P}} = \exp\left[\int_0^T \left(-D\log\bar{m}\,\bar{\sigma}d\hat{w} - \bar{a}D\log\bar{m}\cdot D\log\bar{m}\right)dt\right]. \tag{4.6.41}$$

Note that the function $\bar{m}$ satisfies the backward equation

$$-\frac{\partial \bar{m}}{\partial t} - \mathrm{tr}\, \mathrm{D}\bar{m}\, \bar{a} - (\bar{g} - 2\bar{f})\mathrm{D}\bar{m} + \bar{m}\, \mathrm{div}\, \bar{f} = 0$$
$$\bar{m}(x, T) = p_0(x)$$

and applying Ito's calculus to the process $\log \bar{m}(\bar{x}_t, t)$, taking account of the $\bar{m}$ equation, one can compute the right hand side of (4.6.41). In fact one obtains

$$\frac{\mathrm{d}\tilde{P}^T}{\mathrm{d}\tilde{P}} = \frac{\bar{m}(\bar{x}_0, 0)}{p_0(\bar{x}_T)} \exp\left(\int_0^T \mathrm{div}\, \bar{f}(\bar{x}_t)\mathrm{d}t\right)$$

$$= \frac{m(x(T), T)}{p_0(\xi)} \exp\left(\int_0^T \mathrm{div}\, f(x_t)\mathrm{d}t\right). \tag{4.6.42}$$

If $\mathrm{D}\log m$ has linear growth, the expected value of the random variable (4.6.46) is one. Thus $\tilde{P}^T$ is a probability. For the system $\Omega, \bar{\mathcal{X}}^T \otimes \tilde{\mathcal{Z}}^T, \tilde{P}^T, \bar{\mathcal{X}}^t$ the process

$$w_0(t) = \hat{w}_t + \int_0^t \overline{\sigma^* \mathrm{D}\log m(\bar{x}_s, s)}\mathrm{d}s$$

is a standard Wiener process, and thus, by (4.6.40), the process $\bar{x}_t$ appears as the solution of

$$\mathrm{d}\bar{x} = (\bar{g} - 2\bar{f})(\bar{x}_t, t)\mathrm{d}t + \bar{\sigma}(\bar{x}_t, t)\mathrm{d}w_0$$
$$\bar{x}(0) = x(T) \tag{4.6.43}$$

which is exactly (4.6.28), except for the initial condition. Note that z_t has been unchanged by this new change of probability.

Define ζ_t by (4.6.29) where X_t is replaced by $\bar{x}_t$. Proceeding as above we deduce this time

$$E_0\left[\tilde{q}(\bar{x}_T, T)\zeta_T | \bar{\mathcal{X}}^0\right] = \tilde{q}(\bar{x}_0, 0)$$

where the probability P_0 denotes the restriction of $\tilde{P}^T$ to $\bar{\mathcal{X}}^T$, and $\tilde{z}_t$ is considered as a parameter. Reasoning as for the proof of Theorem 4.6.2, we obtain finally the formula

$$p(x, T) = \tilde{E}^T\left[p_0(\bar{x}_T)\exp\left(-\int_0^T \mathrm{div}\bar{f}(\bar{x}_t)\mathrm{d}t\right)\right.$$

$$\left.\times \exp\left(\int_0^T \bar{h}(\bar{x}_t)\mathrm{d}\tilde{z} - \frac{1}{2}\int_0^T |\bar{h}(\bar{x}_t, t)|^2\mathrm{d}t\right) \middle| \tilde{\mathcal{Z}}^T, \bar{x}_0 = x\right]. \tag{4.6.44}$$

This formula is due to Picard (1987). It can be extended to unbounded h (with linear growth).

4.6.4   A second representation formula

In this paragraph we shall give a representation formula for the random variable $p(T)(\psi)$, where ψ is a test function.

We shall introduce the backward PDE which is dual to (4.6.19). More precisely, define the quantity (analogous to (4.6.18))

$$k_t(h,z) = \frac{\partial h}{\partial t} z_t - A(t)(h_t z_t) - a(t)D(h_t z_t)D(h_t z_t) + \frac{1}{2}|h_t|^2 \qquad (4.6.45)$$

and the PDE

$$-\frac{\partial v}{\partial t} + A(t)v + 2aDvD(hz) + vk_t(h,z) = 0$$
$$v(x,T) = \bar{v}(x). \qquad (4.6.46)$$

It is dual to (4.6.19) in the sense that

$$\int p_0(x)v(x,0)\mathrm{d}x = \int q(x,T)\bar{v}(x)\mathrm{d}x. \qquad (4.6.47)$$

Now take

$$\bar{v}(x) = \psi(x)\exp[h(x,T)z_T] \qquad (4.6.48)$$

then recalling (4.6.16) we deduce

$$\int p_0(x)v(x,0)\mathrm{d}x = p(T)(\psi). \qquad (4.6.49)$$

We then proceed in a way similar to that of § 4.6.4. Let us consider

$$\tilde{v}(x,t) = v(x,t)\exp[-h(x,t)z_T] \qquad (4.6.50)$$

and introduce the process

$$\hat{z}_t = z_t - z_T \qquad (4.6.51)$$

(note the difference from (4.6.24)). Then we can check that $\tilde{v}$ is the solution of

$$-\frac{\partial \tilde{v}}{\partial t} + A(t)\tilde{v} + 2aD\tilde{v}D(h_t\hat{z}_t) + \tilde{v}k_t(h,\hat{z}) = 0$$
$$\tilde{v}(x,T) = \psi(x). \qquad (4.6.52)$$

Consider the filtration $\hat{Z}^t = \sigma(\hat{z}_s,\ t \leq s \leq T)$, which is a decreasing family. We have

Lemma 4.6.3

$$\tilde{E}[\hat{z}_{t_1} - \hat{z}_{t_2}|\hat{Z}^{t_2}] = 0 \quad \forall t_1 \leq t_2$$
$$\tilde{E}\left[(\hat{z}_{t_1} - \hat{z}_{t_2})(\hat{z}_{t_1} - \hat{z}_{t_2})^*|\hat{Z}^{t_2}\right] = (t_2 - t_1)I. \qquad (4.6.53)$$

Proof

$$\hat{z}_{t_1} - \hat{z}_{t_2} = z_{t_1} - z_{t_2}$$

which is independent from $z_s - z_t\ \forall s \geq t_0$, hence $\hat{z}_{t_1} - \hat{z}_{t_2}$ is independent from $\hat{Z}^{t_2}$. The result (4.6.53) follows immediately. ∎

The process $\hat{z}_t$ is a *backward Wiener process*. If ϕ_t is a process adapted to $\hat{Z}^t$, continuous and bounded, one easily defines the *backward Ito integral*

$$\int_0^T \phi_t d\hat{z}_t = P - \lim \sum_i \phi_{t_{i+1}}(\hat{z}_{t_{i+1}} - \hat{z}_{t_i})$$

where $t_0 = 0 < t_1 < t_2 < \ldots$ is a subdivision of the interval $(0, T)$, which becomes thinner and thinner.

One can extend the notion to processes ϕ_t adapted to $\hat{Z}^t$ such that

$$E \int_0^T |\phi_t|^2 dt < \infty.$$

The process $\int_0^T \phi_s d\hat{z}_s$ is a $\hat{Z}^t$ *backward martingale*.

Moreover one can establish a *backward Ito formula*.

Consider the process

$$\xi_t = \xi_T + \int_t^T a(s) ds + \int_t^T b(s) d\hat{z}_s$$

where $a(t)$ and $b(t)$ are $\hat{Z}^t$ adapted and

$$\int_0^T a(s) ds < \infty, \qquad \int_0^T |b(s)|^2 ds < \infty, \quad \text{a.s.}$$

then if $\Phi(x, t) \in C^{2,1}$ one has the formula

$$\Phi(\xi_t, t) = \Phi(\xi_T, T) + \int_t^T \left(-\frac{\partial \Phi}{\partial s} + a(s) D\Phi + \frac{1}{2} \text{tr } D^2 \Phi b b^* \right) ds$$
$$+ \int_t^T D\Phi b d\hat{z}_s. \tag{4.6.54}$$

We have used the same notation for forward and backward stochastic integrals. This should not introduce confusion, since one knows whether the Wiener process is forward or backward. Note the difference between forward and backward calculus, as far as the additional term is concerned. We then consider the process

$$u(x, t) = \tilde{v}(x, t) \exp[-h(x, t)\hat{z}_t] \tag{4.6.55}$$

which is adapted to the family $\hat{Z}^t$. Using the backward Ito's formula we derive the relation

$$- du + A(t) u dt = u h^* d\hat{z}_t$$
$$u(x, t) = \psi(x). \tag{4.6.56}$$

Note also that

$$u(x, 0) = v(x, 0)$$

and thus from (4.6.54) it follows

$$p(T)(\psi) = \int p_0(x) u(x, 0) dx. \tag{4.6.57}$$

Equation (4.6.56) bears a lot of similarity to (4.6.18). It will be studied directly in Section 4.7.

Remark 4.6.4 In formula (4.6.62), the derivatives of h do not appear. They have been used for technical reasons in our derivation. Hence the equality (4.6.57) extends to h bounded (not regular), by an approximation argument. ∎

Remark 4.6.5 Consider the filtering problem, with initial condition playing the role of parameters:

$$dx = g(x_t, t)dt + \sigma(x_t, t)dw, \quad x(s) = x$$
$$dz = h(x_t, t)dt + db, \quad z(s) = 0. \tag{4.6.58}$$

Define next

$$d\eta = \eta h^*(x_t, t)dz, \quad \eta(s) = 1. \tag{4.6.59}$$

The probability $\tilde{P}$ analogous to (4.1.18) depends on x, s, since $\eta(t)$ depends on x, s. Precisely,

$$\left.\frac{d\tilde{P}_{x,s}}{dP}\right|_{\mathcal{F}_s^t} = \rho(t) = \frac{1}{\eta(t)}$$

where $\mathcal{F}_s^t = \sigma(w(\lambda) - w(s), b(\lambda) - b(s), s \leq \lambda \leq t)$. We can then set

$$p_{x,s}(T)(\psi) = \tilde{E}_{x,s}\left[\psi(x(T))\eta(T)|\mathcal{Z}_s^T\right]. \tag{4.6.60}$$

By analogy with (4.6.62) one can check that

$$p_{x,s}(T)(\psi) = u(x, s). \tag{4.6.61}$$

∎

4.7　　　Study of stochastic PDEs

4.7.1　　General remarks

We have encountered two types of stochastic PDE in Section 4.6, namely (4.6.13) and (4.6.56), which are comparable to the Fokker–Planck and to the Kolmogorov equations. In the form (4.6.11), (4.6.56) there is no correlation between the signal noise and the error noise. On the other hand, for this case we have derived the equation for the conditional probability in (4.5.47), in weak form (differentiations taking place on the test function). Our objective in this section is to study the corresponding equations in strong form, like (4.6.13), (4.6.56). The difficulty is that we cannot use the robust form any more and thus the technique of Section 4.6 does not extend.

A direct study of these stochastic PDEs is necessary. This has been done extensively by Pardoux (1982). The general idea is to use the Galerkin approximation, which leads to Ito equations, and proceed analogously to the deterministic case. We shall give another approach, based on fixed point arguments.

4.7.2　　A stochastic PDE set up

We introduce the Hilbert spaces
$$H = L^2(R^n), \quad V = H^1(R^n)\dagger$$
equipped with the scalar products
$$(v_1, v_2) = \int v_1 v_2 \mathrm{d}x$$
$$((v_1, v_2)) = \int v_1 v_2 + \sum_i \frac{\partial v_1}{\partial x_i} \frac{\partial v_2}{\partial x_i} \mathrm{d}x.$$
We note that
$V \subset H$, V is dense in H with continuous injection.

We identify H and its dual, and consider V' to be the dual of V.　　　　(4.7.1)

It is also denoted by $V' = H^{-1}(R^n)$.

---

† We recall that
$$H^1(R^n) = \left\{ z \in L^2(R^n) \,\middle|\, \frac{\partial z}{\partial x_i} \in L^2(R^n) \right\}.$$

Then we have by duality

$$H \subset V', \ H \text{ is dense in } V' \text{ with continuous injection.}$$

We denote by $\langle \, , \, \rangle$ the duality between V and V'. It is easy to check that $A(t) \in \mathcal{L}(V; V')$ and satisfies the coercivity condition

$$\langle A(t)v, v \rangle + \lambda |v|^2 \geq \alpha \|v\|^2, \quad \alpha > 0, \ \lambda \geq 0. \tag{4.7.2}$$

It is convenient to write the operator $A(t)$ in divergence form

$$A(t) = -\sum_{i,j} \frac{\partial}{\partial x_i} \left(a_{ij} \frac{\partial}{\partial x_j} \cdot \right) + \sum_i a_i \frac{\partial}{\partial x_i} \tag{4.7.3}$$

where

$$a_i = -g_i + \sum_j \frac{\partial}{\partial x_j} a_{ij}$$

note that by virtue of the symmetry of the matrix a_{ij},

$$A^*(t) = -\sum_{i,j} \frac{\partial}{\partial x_i} \left(a_{ij} \frac{\partial}{\partial x_j} \cdot \right) - \sum_i \frac{\partial}{\partial x_i} (a_i \cdot). \tag{4.7.4}$$

We recall that (see § 4.1.2) for the system $(\Omega, \mathcal{A}, \tilde{P}, \mathcal{F}^t)$ the processes $w(\cdot)$ and $z(\cdot)$ are $\mathcal{F}^t$ standard Wiener processes, mutually independent, also independent of χ. We define

$$L_Z^2(0, T; V) = \left\{ v \in L^2(\Omega, \mathcal{A}, \tilde{P}; L^2(0, T; V)) \Big| \text{a.e. } T \ v(t) \in L^2(\Omega, Z^t, \tilde{P}; V) \right\}$$

where of course $Z^t = \sigma(z(s), s \leq t)$. One can naturally define spaces similar to (4.7.5), replacing Z^t by another filtration and V by another Hilbert space. We shall make use of the following result, dealing with linear stochastic PDE (cf. Bensoussan 1972, Pardoux 1979)

Lemma 4.7.1 *Let $a(t) \in L_Z^2(0, T; H)$ and $b(t) \in L_Z^2(0, T; H^m)$. Let ξ_o in H be given. Then there exists one and only one $\xi \in L_Z^2(0, T; V) \cap L^2(\Omega, \mathcal{A}, \tilde{P}; C(0, T; H))$ such that*

$$\xi(t) + \int_0^t A(s)\xi(s) ds = \xi_0 + \int_0^t a(s) ds + \int_0^t b(s) dz. \tag{4.7.6}$$

Moreover the following energy equality is satisfied:

$$|\xi(t)|^2 + 2 \int_0^t \langle A(s)\xi(s), \xi(s) \rangle ds = |\xi_0|^2 + 2 \int_0^t (a(s), \xi(s)) ds$$

$$+ 2 \int_0^t (\xi(s), b(s)) dz + \int_0^t |b(s)|_{H^m}^2 ds \quad \text{a.s. } \forall t. \tag{4.7.7}$$

∎

4.7.3   Assumptions

Call (cf. (7.5.47))

$$M_t(x, z) = D_t(z)^{-1}\alpha_t(z)Q_t\sigma_t^*(x, z);$$

we are going to assume that

M_t depends on x only and M_{ij}, $\partial M_{ij}/\partial x_j$ are bounded functions.   (4.7.8)

We shall introduce the family of operators $B(t) \in \mathcal{L}(V; H^m)$ and $B^*(t) \in \mathcal{L}(V; H^m)$ by

$$B(t)v(x) = v(x)D^{-1}(t)h(x, t) + M(x, t)\mathrm{D}v \qquad (4.7.9)$$

$$B^*(t)p(x) = p(x)\tilde{h}(x, t) - M(x, t)\mathrm{D}p \qquad (4.7.10)$$

where we have set

$$\tilde{h}_i(x, t) = (D^{-1}h)_i(x, t) - \sum_j \frac{\partial M_{ij}}{\partial x_j}(x, t). \qquad (4.7.11)$$

We shall consider the stochastic PDE (respectively forward and backward)

$$dp + A^*(t)p \cdot dt = B^*(t)pdz$$
$$p(0) = p_0, \quad p \in L_Z^2(0, T; V) \cap L^2(\Omega, \mathcal{A}, \tilde{P}; C(0, T; H)) \qquad (4.7.12)$$

and

$$-du + A(t)u = B(t)u \cdot d\hat{z}$$
$$u(x, T) = \psi(x) \qquad (4.7.13)$$
$$u \in L_{\hat{Z}}^2(0, T; V) \cap L^2(\Omega, \mathcal{A}, \tilde{P}; C(0, T; H)).$$

These equations must be read in integrated form. For instance for (4.7.12) one has

$$p(t) + \int_0^t A^*(s)p(s)\mathrm{d}s = p_0 + \int_0^t B^*(s)p \cdot dz, \quad \forall t, \qquad (4.7.14)$$

which is an equality between random variables with values in V'. Note that the integrand $B^*(t)p \in L_Z^2(0, T; H^m)$, by virtue of the functional space where p stands.

Remark 4.7.1 If $\alpha = 0$, hence $M = 0$, then $B(t) = B^*(t)$ and (4.7.12), (4.7.13) coincide with (4.6.13), (4.6.56) (recall that $D = I$). We use in the right hand side of (4.7.12), (4.7.13) the scalar product notation $h \cdot dz$ instead of the matrix notation h^*dz, to avoid confusion with the $*$.   ∎

Remark 4.7.2 The solution $*$ for $B^*(t)$ reminds us of duality. This must be understood in the following sense. Let $B_i(t)$, $B_i^*(t)$ be the components of B, B^* which belong to $\mathcal{L}(V; H)$:

$$B_i(t)v(x) = v(x)(D^{-1}(t)h)_i + M_{ij}(x, t)\frac{\partial v}{\partial x_j}$$

$$B_i^*(t)p(x) = p(x)\tilde{h}_i(x, t) - \sum_j M_{ij}(x, t)\frac{\partial p}{\partial x_j}$$

then clearly one has

$$(B_i(t)v, p) = (v, B_i^*(t)p).$$

In other words, if I denotes the injection from $V \to H$ and I^* its dual (from H to V'), then $I^*B \in \mathcal{L}(V; V')$, $I^*B^* \in \mathcal{L}(V; V')$ and I^*B^* is the dual of I^*B.   ∎

4.7.4   Statement of the results

We can state

Theorem 4.7.1 *We assume*

$$2\langle Av, v\rangle + \lambda_0 |v|^2 \geq |D^{1/2} B^* v|^2 + \alpha_0 \|v\|^2$$
$$\forall v \in V, \quad \alpha_0 > 0, \quad \lambda_0 \geq 0$$

(4.7.15)

then there exists one and only one solution of (4.7.12).

Proof

By changing $p(t)$ into $p(t)e^{kt}$, it is sufficient to solve the equation

$$dp + (A^* + k)p \cdot dt = B^*(t)p dz, \quad p(0) = p_0.$$

(7.16)

Consider the map $\eta \to \zeta$ from $L_Z^2(0, T; V)$ into itself (in fact into a subspace of it), defined by

$$d\zeta + (A^* + k)\zeta dt = B^*(t)\eta \cdot dz, \quad \zeta(0) = p_0$$
$$\zeta \in L_Z^2(0, T; V) \cap L^2(\Omega, \mathcal{A}, \tilde{P}; C(0, T; H)).$$

(4.7.17)

We are going to show that it is a contraction, for a convenient choice of k. Take two values of η, say η_1, η_2 and let ζ_1, ζ_2 be the corresponding solutions of (4.7.17). Making use of the energy equality (4.7.7) we obtain

$$\tilde{E}|\zeta_1(t) - \zeta_2(t)|^2 + 2\tilde{E}\int_0^t \langle A(\zeta_1 - \zeta_2), \zeta_1 - \zeta_2\rangle ds + 2k\tilde{E}\int_0^t |\zeta_1(s) - \zeta_2(s)|^2 ds$$

$$= \tilde{E}\int_0^t |D^{1/2}B^*(\eta_1 - \eta_2)|^2 ds$$

$$\leq 2\tilde{E}\int_0^t \langle A(\eta_1 - \eta_2), \eta_1 - \eta_2\rangle ds - \alpha_0 \tilde{E}\int_0^t \|\eta_1 - \eta_2\|^2 ds + \lambda_0 \tilde{E}\int_0^t |\eta_1 - \eta_2|^2 ds,$$

(4.7.18)

using (4.7.15).

We define a norm on V by setting (for any t)

$$\||v\||_t^2 = \langle A(t)v, v\rangle + \gamma |v|^2, \quad \gamma > \lambda$$

for a convenient γ. We easily deduce from (4.7.18) that

$$\tilde{E}\int_0^T \||\zeta_1 - \zeta_2\||_t^2 dt + (k - \gamma)\tilde{E}\int_0^T |\zeta_1 - \zeta_2|^2 dt$$

$$\leq \left(1 - \frac{\alpha_0}{1\|A\|}\right)\tilde{E}\int_0^T \||\eta_1 - \eta_2\||_t^2 dt + \left[\frac{\lambda_0}{2} - \gamma\left(1 - \frac{\alpha_0}{2\|A\|}\right)\right]\tilde{E}\int_0^T |\eta_1 - \eta_2|^2 dt$$

where $\|A\|$ is a constant such that

$$\langle A(t)v, v\rangle \leq \|A\|\|v\|^2, \quad \forall t.$$

It is then clear that picking

$$\gamma > \frac{\lambda_0}{2}\left(1 - \frac{\alpha_0}{2\|A\|}\right), \quad k > \gamma$$

we get the desired contraction property. ∎

A similar result holds for equation (4.7.13), which we state without proof. In fact the second part of the assumption (4.7.8) is not needed. We replace it by

$$M \text{ is bounded.} \tag{4.7.19}$$

Then we can assert

Theorem 4.7.2 *We assume*

$$2\langle Av, v\rangle + \lambda_0 |v|^2 \geq |D^{1/2}Bv|^2 + \alpha_0 \|v\|^2,$$
$$\forall v \in V, \quad \alpha_0 > 0, \quad \lambda_0 \geq 0 \tag{4.7.20}$$

then there exists one and only one solution of (4.7.13).

Remark 4.7.3 When $\alpha = 0$, hence $M = 0$, the assumptions (4.7.15) and (4.7.20) are automatically verified.   ∎

Remark 4.7.4 When the assumption (4.7.8) is not verified (more precisely the second part), then B^* can be interpreted as

$$B^*(t)p(x) = p(x)D^{-1}(t)h(x,t) - \operatorname{div} pM(x,t)$$

and thus maps V into V'^m. In this case the study of equation (4.7.12) is an open problem.   ∎

4.7.5   Interpretation

We start with the interpretation of the solution of (4.7.12). One could derive it by using the equation of the conditional probability in weak form (see Theorem 4.5.1). How ever, this would require the use of the uniqueness theorem, which we have proved only when the coefficients are sufficiently smooth. We thus do it directly.

Theorem 4.7.3 *Under the assumptions of Theorem 4.7.1, the solution of (4.7.12) can be interpreted as follows:*

$$\int p(x,T)\psi(x)dx = p(T)(\psi), \quad \forall \psi \text{ continuous and bounded.} \tag{4.7.21}$$

Proof

Note that the equality (4.7.21) will imply that $p(x,T) \geq 0$ (and thus more generally $p(x,t) \geq 0$) and moreover $p(\cdot,t)$ is in $L^1(R^n)$, $\forall t$. It is also sufficient to prove the relation for $\psi \in C_0^\infty(R^n)$.

We can also assume that the coefficients are slightly smoother. It is indeed clear that the dependence of both sides of (4.7.21) with respect to the coefficients is sufficiently continuous.

Let $\beta(t) \in L^\infty(0, T; R^m)$; we are going to prove that

$$\tilde{E}\theta_T \int p(x,T)\psi(x)dx = \tilde{E}\theta_T p(T)(\psi) \tag{4.7.22}$$

where θ is defined by (4.1.28) and this is sufficient by virtue of Lemma 4.1.4. Note that again we can take β slightly smoother and proceed by approximation. The right hand side of (4.7.22) is equal to $\tilde{E}\theta_T\eta_T\psi(x_T)$ and thus we have to prove that

$$\tilde{E}\theta_T(p(T),\psi) = \tilde{E}\theta_T\eta_T\psi(x_T). \tag{4.7.23}$$

Let us consider the solution of

$$\frac{\partial v}{\partial t} - A(t)v + i\beta^* Bv = 0$$
$$v(x,T) = \psi(x). \tag{4.7.24}$$

This equation has a $C_b^{2,1}$ complex valued solution. Recall that

$$d\eta_t = \eta_t h^*(x_t)D_t^{-1}dz_t, \quad \eta_0 = 0$$
$$d\theta_t = i\theta_t\beta_t^* D_t^{-1}dz_t, \quad \theta_0 = 0$$

and (cf. the calculation on (4.5.19)), $\theta_t\eta_t v(x_t,t)$ is a $\tilde{P}$ martingale, and thus

$$\tilde{E}\theta_T\eta_T\psi(x_T) = (v(0),p_0). \tag{4.7.25}$$

On the other hand v belongs also to $L^2(0,T;V)$. One can compute the Ito differential of $(p(t),v(t))$ and obtain

$$d(p(t),v(t)) = (p(t),B(t)v)dz - i\beta^*(p,Bv)dt.$$

Therefore

$$d\theta_t(p(t),v(t)) = \theta_t\left[(p,Bv) + i(p,v)\beta^* D^{-1}\right]dz$$

and thus

$$\tilde{E}\theta_T(p(T),\psi) = (v(0),p_0)$$

which with (4.7.25) completes the proof of the desired result. ∎

Let us now turn to the interpretation of the solution of (4.7.18). We can assert

Theorem 4.7.4 *We make the assumptions of Theorem 4.7.2; then the solution of (4.7.13) satisfies*

$$(u(0),p_0) = p(T)(\psi). \tag{4.7.26}$$

Proof
Choose $\beta(\cdot)$ as in the proof of Theorem 4.7.3. We consider this time the backward martingale

$$d\hat{\theta}_t = -i\hat{\theta}_t\beta_t^* D_t^{-1}d\hat{z}_t, \quad \hat{\theta}_T = 1. \tag{4.7.27}$$

Let $\zeta \in V$ be fixed; we can write from (4.7.13)

$$-d(u(t),\zeta) + \langle A(t)u(t),\zeta\rangle dt = (\zeta,B(t)u)d\hat{z}$$

and by the backward Ito's formula

$$\begin{aligned}
d(u(t),\zeta)\hat{\theta}_t &= \hat{\theta}_t d(u(t),\zeta) + (u(t),\zeta)d\hat{\theta}_t - d(u(t),\zeta)d\hat{\theta}_t\\
&= \left[\hat{\theta}_t\langle A(t)u(t),\zeta\rangle - i\hat{\theta}_t\beta_t^*(Bu,\zeta)\right]dt\\
&\quad - \hat{\theta}_t\left[(\zeta,Bu) + i(u(t),\zeta)\beta_t^* D_t^{-1}\right]d\hat{z}_t.
\end{aligned}$$

Integrating between t and T, taking the mathematical expectation and noting that

$$v(t) = \tilde{E}u(t)\hat{\theta}_t$$

it follows that

$$(\psi, \zeta) - (v(t), \zeta) = \int_t^T \left[\langle A(s)v(s), \zeta\rangle - i\beta_s^*(Bv, \zeta)\right] ds$$

i.e. $v(t)$ is the solution of

$$\frac{\partial v}{\partial t} - A(t)v + i\beta^* Bv = 0$$

$$v(x, T) = \psi(x)$$

which coincides with the function already introduced in (4.7.24). Hence

$$v(0) = \tilde{E}u(0)\hat{\theta}_0$$

and also

$$(v(0), p_0) = \tilde{E}(u(0), p_0)\hat{\theta}_0. \tag{4.7.28}$$

But

$$\hat{\theta}_t = \exp\left(\int_t^T i\beta_s^* D_s^{-1} d\hat{z}_s + \frac{1}{2}\int_t^T \beta_s^* D_s^{-1}\beta_s ds\right)$$

and thus

$$\hat{\theta}_0 = \exp\left\{\int_t^T i\beta_t^* D_t^{-1} d\hat{z}_t + \frac{1}{2}\int_t^T \beta_t^* D_t^{-1}\beta_t dt\right\}$$

$$= \theta_T$$

since

$$\int_0^T \beta_t^* D_t^{-1} d\hat{z}_t = \int_0^T \beta_t^* D_t^{-1} dz_t$$

by virtue of the fact that the integrand is deterministic. Therefore (4.7.28) implies

$$(v(0), p_0) = \tilde{E}(u(0), p_0)\theta_T$$

and from the previous theorem

$$\tilde{E}(u(0), p_0)\theta_T = \tilde{E}(p(T), \psi)\theta_T = \tilde{E}p(T)(\psi)\theta_T$$

which implies the desired result (4.7.26).   ∎

4.7.6   Duality relation

The relation (4.7.26) can be read as

$$(u(0), p(0)) = (u(T), p(T)).$$

In fact there is a more general property, namely

Theorem 4.7.5 *We make the assumptions of Theorem 4.7.1 and Theorem 4.7.2; then one has*

$$(u(t), p(t)) = (u(T), p(T)), \quad \forall t. \tag{4.7.29}$$

Proof

Note that

$$\hat{\theta}_t = \exp\left(\int_t^T \beta_s^* D_s^{-1} \mathrm{d}z_s + \frac{1}{2} \int_t^T \beta_s^* D_s^{-1} \beta_s \mathrm{d}s\right)$$

$$= \theta_T / \theta_t$$

hence

$$\theta_T = \theta_t \hat{\theta}_t.$$

We thus write

$$\tilde{E}\theta^T(u(t), p(t)) = \tilde{E}(u(t)\hat{\theta}_t, p(t)\theta_t)$$

$$= (\tilde{E}u(t)\hat{\theta}_t, \tilde{E}p(t)\theta_t) \tag{4.7.30}$$

$$= (v(t), \chi(t))$$

where we have denoted

$$\chi(t) = \tilde{E}p(t)\theta_t$$

and of course $v(t)$ is the solution of (4.7.24). We can easily find the equation satisfied by $\chi(t)$. In fact from (4.7.16) and

$$\mathrm{d}\theta_t = \theta_t \mathrm{i}\beta^* D^{-1} \mathrm{d}z$$

it follows easily that

$$\frac{\partial \chi}{\partial t} + A^* \chi = \mathrm{i}\beta^* B^* \chi$$

$$\chi(0) = p_0.$$

We easily check by integration by parts that $(v(t), \chi(t))$ is a constant; hence from (4.7.30)

$$\tilde{E}\theta^T(u(t), p(t)) = \tilde{E}\theta^T(p(T), \psi)$$

which implies the desired result (4.7.29). ∎

4.7.7   Smoothing problem

Let us show in this paragraph that the preceding considerations (namely the stochastic PDE) can also be used in order to solve the smoothing problem (see Pardoux 1982). This problem consists in finding an expression for the quantity $\tilde{E}[\psi(x_s)\eta_T | \mathcal{Z}^T]$. The corresponding conditional probability will follow after dividing by $\tilde{E}[\eta_T | \mathcal{Z}^T]$.

Consider the process $\mu(x, t)$ obtained in a way similar to (4.7.13):

$$-\mathrm{d}\mu + A(t)\mu \mathrm{d}t = B(t)\mu \mathrm{d}\hat{z}$$

$$\mu(x, T) = 1. \tag{4.7.31}$$

Since the initial datum is not in H, we cannot consider the same functional space as in (4.7.13). But the study can be done in a Hilbert space with weights. Consider for instance

$$\rho(x) = (1 + |x|^2)^{-n}$$

and H_ρ is the space of function $v(x)$ such that $v^2\rho$ is integrable. Similarly one defines $H^1_\rho = V_\rho$ and the solution of (4.7.31) can be found in

$$\mu \in L^2_{\hat{Z}}(0, T; V_\rho) \cap L^2(\Omega, \mathcal{A}, \tilde{P}; C(0, T; H_\rho)). \qquad (4.7.32)$$

We then have

Theorem 4.7.6 *We make the assumptions of Theorem 4.7.1 and Theorem 4.7.2; then one has*

$$\tilde{E}\left[\psi(x_s)\eta_T | \mathcal{Z}^T\right] = \int p(x, s)\psi(s)\mu(x, s)dx. \qquad (4.7.33)$$

Proof

It is enough to prove that

$$\tilde{E}\theta_T\eta_T\psi(x_s) = \tilde{E}\theta_T \int p(x, s)\psi(x)\mu(x, s)dx. \qquad (4.7.34)$$

But

$$\tilde{E}\theta_T\eta_T\psi(x_s) = \tilde{E}\left[\theta_s\eta_s\psi(x_s)\tilde{E}(\hat{\theta}_s\eta_s^T|\mathcal{Z}^s, x_s)\right]$$

where

$$\eta_s^T = \exp\left(\int_s^T h^*(x_t)D_t^{-1}dz - \frac{1}{2}\int_s^T h^*(x_t)D_t^{-1}h(x_t)dt\right).$$

But clearly

$$\begin{aligned}
\tilde{E}(\hat{\theta}_s\eta_s^T|\mathcal{Z}^t, x_s) &= \tilde{E}(\hat{\theta}_s\eta_s^T|x_s) \\
&= \tilde{E}\left[\hat{\theta}_s\tilde{E}(\eta_s^T|x_s, \mathcal{Z}_s^T)\Big| x_s\right] \\
&= \tilde{E}\left[\hat{\theta}_s\mu(x_s, s)\Big| x_s\right].
\end{aligned}$$

(cf. Remark (4.6.5)). Therefore using this relation in (4.7.34) yields

$$\begin{aligned}
\tilde{E}\theta_T\eta_T\psi(x_s) &= \tilde{E}\left\{\theta_s\eta_s\psi(x_s)\tilde{E}\left[\hat{\theta}_s\mu(x_s, s)\Big| x_s\right]\right\} \\
&= \tilde{E}\theta_s \int p(x, s)\psi(x)\tilde{E}\left[\hat{\theta}_s\mu(x, s)\right] dx.
\end{aligned}$$

Since $\hat{\theta}_s\mu(x, s)$ and $\theta_sp(x, s)$ are independent, we deduce that

$$\begin{aligned}
\tilde{E}\theta_T\eta_T\psi(x_s) &= \tilde{E}\theta_s \int p(x, s)\psi(x)\hat{\theta}_s\mu(x, s)dx \\
&= \tilde{E}\theta_T \int p(x, s)\psi(x)\mu(x, s)dx
\end{aligned}$$

which is the desired result.   ∎

4.8   Concluding remarks

The stochastic PDE introduced for the conditional density and its dual (see Section 4.7) seem to require h bounded, at least when one does not use the robust form. In this case, it is possible to prove existence results for unbounded h. These results are valid for each part of the observation (see Pardoux 1982; Baras, Blankenship and Hopkins 1983). When the strong formulation is possible, it leads to a uniqueness theorem which requires fewer assumptions than the uniqueness theorem corresponding to the weak formulation (this is not surprising).

We have not considered the stochastic PDE when the coefficients depend on z. The main difficulty arises with the backward equation (4.7.18). Indeed one cannot use $\hat{Z}^t$ any longer. It is nevertheless possible to state a backward equation in a slightly different way. We refer to Pardoux (1982).

There has been in the literature some interest in the logarithmic transformation (cf. Fleming and Mitter 1982). It consists in setting

$$p(x,t) = \exp[-\Sigma(x,t)]$$

which is possible by the positivity of p.

One obtains that Σ is the solution of a stochastic non linear PDE. It can be considered as the value function of a stochastic (in fact doubly stochastic) control problem.

5　Perturbation methods in non linear filtering

Introduction

As we have seen in Chapter 4, the non linear filtering problem is infinite dimensional in nature, although the system we are interested in is finite dimensional. When the system is linear, the filtering problem is also linear and finite dimensional. With the objective of simplifying the calculations as much as possible, it is important to derive cases where the non linear filtering is either finite dimensional (see Chapter 4), or approximately finite dimensional, which is the case studied in this chapter.

The approximation depends on a small parameter ϵ, which will be present either in the signal equation, or the observation equation, or in both. This brings the theory of perturbation into the context of non linear filtering, and a large variety of situations can be considered, including singular perturbations. Note that this approach has been extensively used in practice, with little mathematical justification (extended Kalman filter). A mathematical theory of these approximations has been developed by Picard (1986a,b; 1987) where he shows that alongside the extended Kalman filter a large variety of approximations can be considered. He gives the various rates of approximation. The method he uses is based on the explicit representation formula for the conditional probability density (see Chapter 4, (5.6.49)), and its expansion. In this chapter we use a more analytic approach, based on the stochastic PDE of which q is a solution, and energy estimates.

Briefly speaking, consider the model

$$\mathrm{d}x = g(x_t)\mathrm{d}t + \epsilon^\delta \sigma(x_t)\mathrm{d}w_t$$
$$\mathrm{d}y = h(x_t)\mathrm{d}t + \epsilon^{1-\delta}\mathrm{d}b_t.$$

Two cases will be considered: $0 \le \delta < \frac{1}{2}$, where the noise on the observation is small compared with the noise on the signal; and $\frac{1}{2} \le \delta < 1$, where the reverse is true. Sections 5.1 and 5.2 are denoted to the first case, with a particular treatment for the linear case in Section 5.1. Section 5.3 is devoted to the second case.

In Section 5.4 we consider a singular perturbation problem, where the model is of the following type:

$$\mathrm{d}x_t = f(x_t, y_t)\mathrm{d}t + \sqrt{2}dw_t^1$$
$$\epsilon \mathrm{d}y_t = g(x_t, y_t)\mathrm{d}t + \sqrt{2\epsilon}dw_t^2.$$

Ergodic properties are exploited in this context. This problem has also been considered by Kushner (1986, 1989), with a different approach from ours (purely probabilistic instead of analytic).

5.1   Linear systems with small noise in the observation

5.1.1   The model

The signal is described by the equation

$$dx = Fx dt + dw$$
$$x(0) = \xi$$

$(5.1.1)$

and the observation process is given by

$$dz = Hx dt + \epsilon db$$
$$z(0) = 0.$$

$(5.1.2)$

The matrices F and H belong to $\mathcal{L}(R^n; R^n)$ and $\mathcal{L}(R^n; R^m)$. The processes w, b are independent Wiener processes with covariance Q, R respectively; R is invertible. All these quantities may depend on time. For simplicity we assume that they are constant.

The random variable ξ is gaussian with mean x_0 and covariance matrix P_0. It is independent of w, b. The parameter ϵ is small.

The Kalman filter is given by the relation

$$d\hat{x}_t = F\hat{x}_t dt + \frac{P_\epsilon(t)}{\epsilon^2} H^* R^{-1}(dz - H\hat{x}_t dt)$$
$$\hat{x}_0 = x_0$$

$(5.1.3)$

and $P_\epsilon(t)$ is the solution of the Riccati equation

$$\frac{dP_\epsilon}{dt} = FP_\epsilon + P_\epsilon F^* - \frac{1}{\epsilon^2} P_\epsilon H^* R^{-1} H P_\epsilon + Q$$
$$P_\epsilon(0) = P_0.$$

$(5.1.4)$

Our objective is to study the behaviour of the system $(5.1.3)$, $(5.1.4)$ as $\epsilon \to 0$.

5.1.2   Rescaling

It is convenient to rescale $(5.1.4)$ as follows. Call

$$\frac{P_\epsilon(t)}{\epsilon} = \nu_\epsilon(t)$$

$(5.1.5)$

then we get at once the relation

$$\dot{\nu}_\epsilon(t) = F\nu_\epsilon + \nu_\epsilon F^* - \frac{1}{\epsilon}\nu_\epsilon H^* R^{-1} H\nu_\epsilon + \frac{Q}{\epsilon}$$
$$\nu_\epsilon(0) = \frac{P_0}{\epsilon}.$$

$(5.1.6)$

The form (5.1.6) shows formally that apart from an initial layer term the quantity ν_ϵ should converge as $\epsilon \to 0$ to the solution of

$$\nu H^* R^{-1} H \nu = Q. \tag{5.1.7}$$

From now on we assume that $n = m$ and

$$H \text{ invertible, } Q \text{ invertible, } P_0 \text{ invertible.} \tag{5.1.8}$$

The solution of (5.1.7) is then given by

$$\nu = H^{-1} R^{1/2} (R^{-1/2} H Q H^* R^{-1/2})^{1/2} R^{1/2} H^{*-1}. \tag{5.1.9}$$

Set

$$\gamma = (R^{-1/2} H Q H^* R^{-1/2})^{1/2}$$
$$\theta = \nu H^* R^{-1} = H^{-1} R^{1/2} \gamma R^{-1/2} \tag{5.1.10}$$

and introduce the process (cf. (5.1.3))

$$dm_t = F m_t dt + \frac{\theta}{\epsilon} (dz - H m_t dt)$$
$$m(t) = m_0 \tag{5.1.11}$$

where the initial condition is arbitrary. Clearly we have approximated the gain of the Kalman filter by the quantity θ/ϵ.

5.1.3 Study of the approximation

The conditional probability measure $\Pi(t)(\psi) = \Pi_\epsilon(t)(\psi)$ is defined by (see Chapter 4)

$$\Pi(t)(\psi) = E\left[\psi(x_t)|Z^t\right]$$
$$= \int \psi(\hat{x}_t + \sqrt{\epsilon} \nu_\epsilon^{1/2}(t)\xi) \frac{\exp(-\frac{1}{2}|\xi|^2)}{(2n)^{n/2}} d\xi \tag{5.1.12}$$

which we are going to approximate by

$$\tilde{\Pi}(t)(\psi) = \int \psi(m_t + \sqrt{\epsilon} \nu^{1/2} \xi) \frac{\exp(-\frac{1}{2}|\xi|^2)}{(2n)^{n/2}} d\xi. \tag{5.1.13}$$

We can give some preliminary estimates, which will be improved later.
Lemma 5.1.1 *We have the estimate*

$$E|x_t - m_t|^2 \leq C(\epsilon + e^{-\delta t/\epsilon}|x_0 - m_0|^2) \quad \delta > 0, C > 0, \forall t \tag{5.1.14}$$
$$E|x_t - m_t|^4 \leq C(\epsilon^2 + e^{-\delta t/\epsilon}|x_0 - m_0|^4). \tag{5.1.15}$$

Proof
From (5.1.1), (5.1.2) and (5.1.11) we deduce

$$d(x_t - m_t) = \left[F(x_t - m_t) - \frac{\theta H}{\epsilon}(x - m_t)\right] dt + dw - \theta db$$
$$x(0) - m(0) = \xi - m_0. \tag{5.1.16}$$

Note that from (5.1.10) and (5.1.7)

$$\theta H = \nu H^* R^{-1} H = Q \nu^{-1}.$$

Therefore

$$
\frac{1}{2} \mathrm{d}(x_t - m_t)^* Q^{-1}(x_t - m_t) = \left[(x_t - m_t)^* \left(Q^{-1} F - \frac{\nu^{-1}}{\epsilon} \right) (x_t - m_t) \right.
$$
$$
\left. + \frac{1}{2}(n + \mathrm{tr}\, Q^{-1}) \right] \mathrm{d}t + (x_t - m_t)^* Q^{-1}(\mathrm{d}w - \theta \mathrm{d}b)
$$

$$(5.1.17)$$

where we have made use of

$$R^{1/2} \theta^* Q^{-1} \theta R^{1/2} = \gamma R^{1/2} H^{*-1} Q^{-1} H^{-1} R^{1/2} \gamma$$
$$= \gamma (\gamma^2)^{-1} \gamma = I.$$

Pick

$$\epsilon \le \epsilon_0 \text{ such that } 2\epsilon_0 \|Q^{-1} F\| \|\nu\| \le 1, \qquad (5.1.18)$$

then (5.1.17) implies

$$\frac{1}{2} \frac{\mathrm{d}}{\mathrm{d}t} E(x_t - m_t)^* Q^{-1}(x_t - m_t) \le -\frac{1}{2\epsilon \|\nu\|} E|x_t - m_t|^2 + \frac{1}{2}(n + \mathrm{tr}\, Q^{-1})$$

from which (5.1.14) follows easily.

Computing $\frac{1}{4} \mathrm{d} \left[(x_t - m_t)^* Q^{-1}(x_t - m_t) \right]^2$ yields easily the estimate (5.1.15). ∎

We deduce easily a crude estimate of the approximation of $\Pi(t)$ by $\tilde{\Pi}(t)$, namely

Lemma 5.1.2 *Let ψ be C_b^2; then one has*

$$
E|\Pi(t)(\psi) - \tilde{\Pi}(t)(\psi)| \le C \|\mathrm{D}\psi\| (\sqrt{\epsilon} + \mathrm{e}^{-\delta t/2\epsilon} |x_0 - m_0|)
$$
$$
+ C \|\mathrm{D}^2 \psi\| (\epsilon + \mathrm{e}^{-\delta t/\epsilon} |x_0 - m_0|^2). \qquad (5.1.19)
$$

Proof

In fact we have

$$
E|\Pi(t)(\psi) - \psi(m_t)| \le C \|\mathrm{D}\psi\| (\sqrt{\epsilon} + \mathrm{e}^{-\delta t/2\epsilon} |x_0 - m_0|)
$$
$$
+ C \|\mathrm{D}^2 \psi\| (\epsilon + \mathrm{e}^{-\delta t/\epsilon} |x_0 - m_0|^2) \qquad (5.1.20)
$$

which follows from Taylor's expansion of $\psi(x_t) - \psi(m_t)$ and the use of Lemma 5.1.1, since

$$E|\Pi(t)(\psi) - \psi(m_t)| \le E|\psi(x_t) - \psi(m_t)|.$$

Moreover from (5.1.13)

$$|\tilde{\Pi}(t)(\psi) - \psi(m_t)| \le \sqrt{\epsilon} \|\mathrm{D}\psi\| \nu^{1/2} \sqrt{m}$$

and (5.1.19) follows. ∎

In fact we shall show that the left hand side of (5.1.19) is of order $\epsilon^{3/2}$, instead of $\sqrt{\epsilon}$.

Remark 5.1.1 Since $P_\epsilon(t)$ is of order ϵ, we have

$$E|\hat{x}_t - x_t|^2 = P_\epsilon(t) = O(\epsilon).$$

Therefore, it is impossible to find estimates m_t which improve that degree of approximation. The estimate m_t defined in (5.1.11) has the same degree of approximation as the Kalman filter, but the constant may be badly approximated. By showing that $E|\hat{x}_t - m_t|^2$ is of order ϵ^3, we show that it is very close to the Kalman filter and thus the approximation is very close to that of the Kalman filter.

Remark 5.1.2 The power 4 in (5.1.15) will be useful in the following. Of course we could pick any power and obtain the corresponding approximation. ∎

5.1.4   Statement of the results

Let us assume also that

$$P_0 \text{ is invertible.} \tag{5.1.21}$$

In that case $\nu_\epsilon(t)$ is invertible and $\Pi(t)(\psi)$ has a density that we call $q(x, t)$, given explicitly by

$$q(x, t) = \frac{1}{(2n)^{n/2}\epsilon^{n/2}|\nu_\epsilon(t)|^{1/2}} \exp\left[-\frac{1}{2\epsilon}(x - \hat{x}_t)^*\nu_\epsilon^{-1}(t)(x - \hat{x}_t)\right]. \tag{5.1.22}$$

Note that the numerator is the density of the unnormalized conditional probability $p(t)(\psi)$, which we denote $p(x, t)$, where

$$p(x, t) = \exp\left[-\frac{1}{2\epsilon}(x - \hat{x}_t)^*\nu_\epsilon^{-1}(t)(x - \hat{x}_t)\right]. \tag{5.1.23}$$

Define

$$\begin{aligned}\zeta(x, t) &= \log p(x, t)\\ &= -\frac{1}{2\epsilon}(x - \hat{x}_t)^*\nu_\epsilon^{-1}(t)(x - \hat{x}_t)\end{aligned} \tag{5.1.24}$$

which we wish to approximate by $(-1/2\epsilon)\psi(x, t)$, where

$$\psi(x, t) = (x - m_t)^*\nu^{-1}(x - m_t). \tag{5.1.25}$$

Hence we shall consider

$$u(x, t) = \zeta(x, t) + \frac{1}{2\epsilon}\psi(x, t). \tag{5.1.26}$$

Consider also the gradient

$$Du = -\frac{1}{\epsilon}\nu_\epsilon^{-1}(t)(x - \hat{x}_t) + \frac{1}{\epsilon}\nu^{-1}(x - m_t) \tag{5.1.27}$$

and the 'energy'

$$\rho(t) = \int q(x, t)Du^*QDudx. \tag{5.1.28}$$

Our first task will be to establish estimates concerning the energy. We shall let $\nu_t = \nu_\epsilon(t)$.

Lemma 5.1.3 *We have the estimate*

$$E\rho(t) \leq \rho_0^\epsilon e^{-pt/\epsilon} + C\epsilon \qquad (5.1.29)$$

where

$$\rho_0^\epsilon = E\rho(0) \text{ and } |\rho_0^\epsilon| \leq \frac{C_0}{\epsilon^2}.$$

Proof

We write

$$Du = -\frac{1}{\epsilon}(\nu_t^{-1} - \nu^{-1})(x - \hat{x}_t) + \frac{1}{\epsilon}\nu^{-1}(\hat{x}_t - m_t)$$

hence

$$\rho(t) = \frac{1}{\epsilon^2} \int q(x - \hat{x}_t)^*(\nu_t^{-1} - \nu^{-1})Q(\nu_t^{-1} - \nu^{-1})(x - \hat{x}_t)dx$$
$$+ \frac{1}{\epsilon^2}(\hat{x}_t - m_t)^*\nu^{-1}Q\nu^{-1}(\hat{x}_t - m_t)$$

and from (5.1.22) therefore

$$\rho(t) = \frac{1}{\epsilon}\text{tr } \nu_t(\nu_t^{-1} - \nu^{-1})Q(\nu_t^{-1} - \nu^{-1}) + \frac{1}{\epsilon}(\hat{x}_t - m_t)^*\nu^{-1}Q\nu^{-1}(\hat{x}_t - m_t).$$

We set

$$\beta_t = \hat{x}_t - w_t, \quad \mu_t = \nu_t^{-1}$$

and collecting results we finally obtain

$$\rho(t) = \frac{1}{\epsilon}\text{tr } (Q\mu_t + \nu_t\nu^{-1}Q\nu^{-1} - 2\nu^{-1}Q) + \frac{1}{\epsilon^2}\beta_t^*\nu^{-1}Q\nu^{-1}\beta_t. \qquad (5.1.30)$$

Combining (5.1.3) with (5.1.11) we can write

$$d\beta_t = \left(F - \frac{1}{\epsilon}Q\nu^{-1}\right)\beta_t dt + \frac{1}{\epsilon}(\nu_t - \nu)H^*R^{-1}(dz - H\hat{x}_t dt),$$

hence we can take the Ito differential of (5.1.30) to yield the formula

$$d\rho = \frac{1}{\epsilon}\text{tr } (Q\dot{\mu}_t + \dot{\nu}_t\nu^{-1}Q\nu^{-1})dt$$
$$+ \frac{2}{\epsilon^2}\beta_t^*\nu^{-1}Q\nu^{-1}\left[\left(F - \frac{1}{\epsilon}Q\nu^{-1}\right)\beta_t dt + \frac{1}{\epsilon}(\nu_t - \nu)H^*R^{-1}(dz - H\hat{x}_t dt)\right]$$
$$+ \frac{1}{\epsilon^2}\text{tr } (\nu_t - \nu)\nu^{-1}Q\nu^{-1}(\nu_t - \nu)\nu^{-1}Q\nu^{-1}dt.$$

We replace $\dot{\nu}_t$ by its value taken from (5.1.6) and at the same time we can assert that

$$\dot{\mu}_t = -\mu_t F - F^*\mu_t + \frac{1}{\epsilon}H^*R^{-1}H - \frac{1}{\epsilon}\mu_t Q\mu_t. \qquad (5.1.31)$$

We thus use the expressions of $\dot\nu_t$ and $\dot\mu_t$ in the expression of $d\rho$; it follows that

$$
\begin{aligned}
d\rho =&\frac{2}{\epsilon^2}\beta_t^*\nu^{-1}Q\nu^{-1}\left(F-\frac{1}{\epsilon}Q\nu^{-1}\right)\beta_t dt\\
&+\frac{2}{\epsilon^3}\beta_t^*\nu^{-1}Q\nu^{-1}(\nu_t-\nu)H^*R^{-1}(dz-H\hat{x}_t)dt\\
&+\frac{1}{\epsilon^2}\mathrm{tr}\left(-2\epsilon Q\mu_t F+2\epsilon F\nu_t\nu^{-1}Q\nu^{-1}+3Q\nu^{-1}Q\nu^{-1}\right.\\
&\left.-Q\mu_t Q\mu_t-2\nu_t\nu^{-1}Q\nu^{-1}Q\nu^{-1}\right)dt.
\end{aligned}
\tag{5.1.32}
$$

We want to estimate the right hand side of (5.1.32) in terms of ρ. For that purpose we first compute the quantity

$$
\eta_t=\int q\mathrm{D}u^*Q\left(\nu^{-1}F\nu-\frac{1}{\epsilon}\nu^{-1}Q\right)\mathrm{D}u\,dx.
$$

Using the definition of $\mathrm{D}u$, it is easily seen that

$$
\begin{aligned}
\eta_t=&\frac{1}{\epsilon^2}\beta_t^*\nu^{-1}Q\nu^{-1}\left(F-\frac{1}{\epsilon}Q\nu^{-1}\right)\beta_t\\
&+\frac{1}{\epsilon^2}\mathrm{tr}\left(\epsilon Q\nu^{-1}F\nu\mu_t-2\epsilon Q\nu^{-1}F+\epsilon\nu_t\nu^{-1}Q\nu^{-1}F\right.\\
&\left.-Q\nu^{-1}Q\mu_t+2\nu^{-1}Q\nu^{-1}Q-\nu_t\nu^{-1}Q\nu^{-1}Q\nu^{-1}\right).
\end{aligned}
$$

Then we use this expression to compute the first term and insert it in (5.1.32). We deduce

$$
\begin{aligned}
d\rho =&2\eta_t dt+\frac{2}{\epsilon^3}\beta_t^*\nu^{-1}Q\nu^{-1}(\nu_t-\nu)H^*R^{-1}(dz-Hx_t dt)\\
&+\frac{1}{\epsilon^2}\mathrm{tr}\left(-Q\mu_t Q\mu_t-Q\nu^{-1}Q\nu^{-1}+2Q\nu^{-1}Q\mu_t\right.\\
&\left.+4\epsilon Q\nu^{-1}F-2\epsilon Q\mu_t F-2\epsilon Q\nu^{-1}F\nu\mu_t\right).
\end{aligned}
$$

We integrate between 0 and t and take the mathematical expectation. The stochastic integral involving the innovation $dz-H\hat{x}_t dt$ cancels out. Differentiating again in t, we deduce

$$
\begin{aligned}
\frac{d}{dt}E\rho(t)=&2E\eta_t+\frac{1}{\epsilon^2}\mathrm{tr}\left\{-Q\mu_t Q\mu_t-Q\nu^{-1}Q\nu^{-1}\right.\\
&\left.+2Q\nu^{-1}Q\mu_t+4\epsilon Q\nu^{-1}F-2\epsilon Q\mu_t F-2\epsilon Q\nu^{-1}F\nu\mu_t\right\}\\
=&2E\eta_t-\frac{1}{\epsilon^2}\mathrm{tr}\,Q(\mu_t-\nu^{-1}+\epsilon F^*Q^{-1}+\epsilon Q^{-1}\nu F^*\nu^{-1})Q\\
&(\mu_t-\nu^{-1}+\epsilon Q^{-1}F+\epsilon\nu^{-1}F\nu Q^{-1})\\
&+\mathrm{tr}\left(QF^*Q^{-1}F+2\nu F^*\nu^{-1}F+\nu F^*\nu^{-1}Q\nu^{-1}F\nu Q^{-1}\right).
\end{aligned}
\tag{5.1.33}
$$

Since

$$
Q^{1/2}\nu^{-1}Q^{1/2}\ge pI,\quad p>0
\tag{5.1.34}
$$

we can define ϵ_0 such that

$$
\chi\epsilon_0<p,\quad\text{where }\chi=\|Q^{1/2}\nu^{-1}F\nu Q^{-1/2}\|
\tag{5.1.35}
$$

and for $\epsilon \leq \epsilon_0$ we have

$$E\eta_t \leq \left(\chi - \frac{1}{\epsilon}p\right)E\rho_t.$$

Therefore using (5.1.33) we deduce

$$\frac{\mathrm{d}}{\mathrm{d}t}E\rho(t) \leq 2\left(\chi - \frac{1}{\epsilon}p\right)E\rho_t + C$$

where C is a constant. Choosing

$$\epsilon \leq \epsilon_1 = \frac{p}{2\chi}$$

we can write

$$\frac{\mathrm{d}}{\mathrm{d}t}E\rho(t) \leq -\frac{p}{\epsilon}E\rho(t) + C$$

hence the desired result.   ■

Remark 5.1.3 One has

$$\rho_0^\epsilon = \frac{1}{\epsilon^2}(x_0 - m_0)^* \nu^{-1} Q \nu^{-1}(x_0 - m_0) + \frac{1}{\epsilon}\mathrm{tr}\left(\epsilon Q P_0^{-1} + \frac{1}{\epsilon}P_0 \nu^{-1} Q \nu^{-1} - 2\nu^{-1}Q\right)$$

hence the estimate on ρ_0^ϵ.   ■

We can now state the main result of this section.

Theorem 5.1.1 *Assume (5.1.8), then one has the estimate*

$$E|\hat{x}_t - m_t|^2 \leq C_0 e^{(-p/\epsilon)t} + C\epsilon^3. \tag{5.1.36}$$

Proof
From (5.1.27) one has

$$\int q(x,t)\mathrm{D}u\mathrm{d}x = \frac{1}{\epsilon}\nu^{-1}(\hat{x}_t - m_t)$$

hence

$$E|\hat{x}_t - m_t|^2 \leq \epsilon^2 \|\nu\|^2 \|Q^{-1}\| E\rho(t)$$

and from Lemma 5.1.3 the desired result follows.   ■

We next prove a result concerning the approximation of ν_t by ν. We can state

Theorem 5.1.2 *Assume (5.1.8), then one has the estimate*

$$\|\nu_t - \nu\|^2 \leq C\left[\epsilon^2 + \frac{1}{\epsilon}e^{(-p/\epsilon)t} + \frac{1}{\epsilon^2}e^{(-2p/\epsilon)t}\right]. \tag{5.1.37}$$

Proof
We compute

$$E\int q\mathrm{D}u^*\nu_t\mathrm{D}u\mathrm{d}x = \frac{1}{\epsilon}\mathrm{tr}\,\nu_t(\nu_t^{-1} - \nu^{-1})\nu_t(\nu_t^{-1} - \nu^{-1})$$

$$+ \frac{1}{\epsilon^2}E(\hat{x}_t - m_t)^* \nu^{-1}\nu_t\nu^{-1}(\hat{x}_t - m_t) \tag{5.1.38}$$

$$\geq \frac{1}{\epsilon}\mathrm{tr}\,(\nu - \nu_t)\nu^{-1}(\nu - \nu_t)\nu^{-1}.$$

We define

$$\|\nu_t - \nu\|^2 = \text{tr } (\nu_t - \nu)(\nu_t - \nu)$$
$$= \sum |(\nu_t - \nu)e_i|^2$$

$\forall \, e_i$ where $e_1, \ldots, e_n$ is an orthonormal system.

We have

$$\|\nu_t - \nu\|^2 \leq \alpha \text{tr } (\nu - \nu_t)\nu^{-1}(\nu - \nu_t)\nu^{-1}, \quad \alpha > 0.$$

Indeed take $e_1, \ldots, e_n$ to be the eigenvectors of ν, i.e.

$$\nu e_i = \lambda_i e_i, \quad e_i^* e_j = \delta_{ij}, \quad \lambda_i > 0$$

then

$$\begin{aligned}
\text{tr } (\nu - \nu_t)\nu^{-1}(\nu - \nu_t)\nu^{-1} &= \text{tr } \nu^{-1/2}(\nu - \nu_t)\nu^{-1}(\nu - \nu_t)\nu^{-1/2} \\
&= \sum e_i^* \nu^{-1/2}(\nu - \nu_t)\nu^{-1}(\nu - \nu_t)\nu^{-1/2} e_i \\
&= \sum \frac{1}{\lambda_i} e_i^*(\nu - \nu_t)\nu^{-1}(\nu - \nu_t)e_i \\
&\geq c_0 \text{tr } (\nu - \nu_t)\nu^{-1}(\nu - \nu_t) \\
&= c_0 \text{tr } \nu^{-1/2}(\nu - \nu_t)^2 \nu^{-1/2}
\end{aligned}$$

and, by the same reasoning

$$\text{tr } (\nu - \nu_t)\nu^{-1}(\nu - \nu_t)\nu^{-1} \geq c_0^2 \text{tr } (\nu - \nu_t)^2$$

where

$$c_0 = \frac{1}{\max_i \lambda_i}.$$

Therefore from (5.1.38) it follows that

$$\begin{aligned}
\|\nu_t - \nu\|^2 &\leq \alpha \epsilon E \int q \text{D}u^* \nu_t \text{D}u dx \\
&\leq \alpha \epsilon \|\nu_t\| E \int q |\text{D}u|^2 \mathrm{d}x \\
&\leq \alpha \epsilon \|\nu_t\| \|Q^{-1}\| E \rho(t) \\
&\leq \alpha \epsilon \|Q^{-1}\| \|\nu_t - \nu\| E \rho(t) + \alpha \epsilon \|\nu\| \|Q^{-1}\| E \rho(t)
\end{aligned}$$

and the desired result follows easily. ∎

We can finally give the following improvement of Lemma 5.1.2, namely

Theorem 5.1.3 *Assume (5.1.8), then for ψ Lipschitz one has*

$$E|\Pi(t)(\psi) - \tilde{\Pi}(t)(\psi)| \leq C\|\text{D}\psi\| \left(\epsilon^{3/2} + e^{(-p/2\epsilon)t} + \frac{1}{\sqrt{\epsilon}} e^{(-p/\epsilon)t} \right). \qquad (5.1.39)$$

where we have used the notation of (5.1.12), (5.1.13).

Proof

From formulas (5.1.12), (5.1.13) we deduce

$$E|\Pi(t)(\psi) - \tilde{\Pi}(t)(\psi)| \leq C\|\text{D}\psi\|(E|\hat{x}_t - m_t| + \sqrt{\epsilon}\|\nu_t^{1/2} - \nu^{1/2}\|)$$

and the result follows from the estimates of Theorem 5.1.1 and 5.1.2. ∎

5.2   Non linear systems with small noise in the observation

5.2.1   Setting of the problem

Let g, σ, h be such that:

$$g : R^n \to R^n, \text{ is } C^2 \text{ with bounded derivatives.} \tag{5.2.1}$$

$$\sigma : R^n \to L(R^n; R^m) \text{ is } C^3; \ \sigma \text{ and its derivatives are bounded;}$$
$$\text{the function } a = \tfrac{1}{2}\sigma\sigma^* \text{ is uniformly positive definite.} \tag{5.2.2}$$

$$h : R^n \to R^n \text{ is } C^3, \text{ with bounded derivatives.} \tag{5.2.3}$$

Moreover

$$g_i \partial/\partial x_i a_{jk}, \ g_i \partial^2 h_k/\partial x_i \partial x_j \text{ are bounded, } \forall j, k. \tag{5.2.3'}$$

Let also p_0 be such that:

$$p_0 : R^n \to R^+ - \{0\}; \ \int p_0(x)\mathrm{d}x = 1; \ p_0 \text{ is a smooth function,}$$
$$\int |Dp_0|^2/p_0 \mathrm{d}x < \infty, \ p_0 \exp -(1 + |x|^2)^{1/2} \in H^1(R^n), \ \int |x|^4 p_0(x)\mathrm{d}x < \infty. \tag{5.2.4}$$

Let $(\Omega, \mathcal{A}, P)$ be a probability space equipped with a filtration $\mathcal{F}^t$ and $w(t), b(t)$ are two independent standard $\mathcal{F}^t$ Wiener processes with values in R^n; ξ is a $\mathcal{F}^0$ measurable random variable with values in R^n. The variable ξ has a probability distribution Π_0 which has a density with respect to Lebesgue measure given by p_0. \hfill (5.2.5)

We consider the signal and the observation given by

$$\mathrm{d}x = g(x_t)\mathrm{d}t + \sigma(x_t)\mathrm{d}w$$
$$x(0) = \xi \tag{5.2.6}$$

$$\mathrm{d}z = h(x_t)\mathrm{d}t + \epsilon \mathrm{d}b_t$$
$$z(0) = 0. \tag{5.2.7}$$

Let A be the second order differential operator

$$A = -g_i \frac{\partial}{\partial x_i} - a_{ij} \frac{\partial^2}{\partial x_i \partial x_h} \tag{5.2.8}$$

and its formal adjoint

$$A^* = \frac{\partial}{\partial x_i}(g_i \cdot) - \frac{\partial^2}{\partial x_i \partial x_j}(a_{ij} \cdot). \tag{5.2.9}$$

Let $\tilde{P}$ be the probability on $(\Omega, \mathcal{A})$ where the Radon–Nikodym derivative with respect to P is given by

$$\left.\frac{\mathrm{d}\tilde{P}}{\mathrm{d}P}\right|_{\mathcal{F}^t} = \rho_t \tag{5.2.10}$$

where

$$\rho_t = \exp\left[\int_0^t -\frac{h^*(x_s)\mathrm{d}b}{\epsilon} - \frac{1}{2\epsilon^2}\int_0^t |h(x_s)|^2 \mathrm{d}s\right]. \tag{5.2.11}$$

We have omitted to indicate explicitly that $\tilde{P}$ and ρ_t depend on ϵ.

We know that for the system $(\Omega, \mathcal{A}, \tilde{P}, \mathcal{F}^t)$ the processes $w(\cdot)$ and $z(\cdot)$ are $\mathcal{F}^t$ Wiener processes. Note that $w(\cdot)$ is standard and that $z(\cdot)$ has a covariance operator given by $\epsilon^2 I$.

We recall that $\mathcal{Z}^t = \sigma(z(s), s \leq t)$ and

$$\eta_t = \frac{1}{\rho_t} = \exp\left[\int_0^t \frac{h^*(x_s)}{\epsilon^2}\mathrm{d}z_s - \frac{1}{2\epsilon^2}\int_0^t |h(x_s)|^2 \mathrm{d}s\right]. \tag{5.2.12}$$

We recall the definitions

$$\begin{aligned}
\Pi(t)(\psi) &= E\left[\psi(x_t)|\mathcal{Z}^t\right] \\
p(t)(\psi) &= \tilde{E}\left[\psi(x_t)\eta_t|\mathcal{Z}^t\right] \\
\Pi(t)(\psi) &= \frac{p(t)(\psi)}{p(t)(1)}.
\end{aligned} \tag{5.2.13}$$

5.2.2 Formal calculations: the approximation

We postulate at his stage the existence of densities

$$\Pi(t)(\psi) = \int q(x, t)\psi(x)\mathrm{d}x \tag{5.2.14}$$

$$p(t)(\psi) = \int p(x, t)\psi(x)\mathrm{d}x. \tag{5.2.15}$$

Necessarily

$$q(x, t) = p(x, t)\mu_t \tag{5.2.16}$$

where

$$\mu_t = \frac{1}{\zeta_t}; \quad \zeta_t = p(t)(1) \quad \text{(cf. Chapter 4)}.$$

The functions $p(x, t)$, $q(x, t)$ will be solutions of the stochastic PDE

$$\begin{aligned}
\mathrm{d}p + A^*p\mathrm{d}t &= \frac{p}{\epsilon^2}h^*\mathrm{d}z \\
p(x, 0) &= p_0(x)
\end{aligned} \tag{5.2.17}$$

$$dq + A^*q dt = \frac{q}{\epsilon^2}(h^* - \hat{h}^*)(dz - \hat{h}dt)$$

$$q(x,0) = p_0(x)$$

(5.2.18)

where we make use of the notation:

$$\hat{\psi} = \hat{\psi}_t = \Pi(t)(\psi).$$

In the following we shall perform formal manipulations on p and q. In particular we shall differentiate on x formally. Our objective is to obtain a priori estimates.

We shall introduce some notation

$$H(x) = h'(x)$$

$$H_{ij} = h_{i,j} = \frac{\partial h_i}{\partial x_j}$$

(5.2.19)

$$\gamma = (H\sigma^* H^*)^{1/2} = \sqrt{2}(HaH^*)^{1/2}$$

(5.2.20)

$$\phi = \gamma^{-1}, \quad \tau = \phi H$$

(5.2.21)

$$\theta = \tau^{-1} = H^{-1}\gamma, \quad \nu = \theta H^{*-1} = H^{-1}\gamma H^{*-1}.$$

(5.2.22)

We easily deduce the algebraic relations

$$\theta\theta^* = 2a$$

(5.2.23)

$$\tau a \tau^* = \frac{1}{2}I$$

(5.2.24)

$$H^*\tau = \tau^* H.$$

(5.2.25)

Consider now the process m_t to be the solution of

$$dm_t = g(m_t)dt + \frac{1}{\epsilon}\theta(m_t)(dz - h(m_t)dt)$$

$$m(0) = m_0.$$

(5.2.26)

This process will be, in a sense to be made precise, an approximation of $\hat{x}_t$.

We then introduce

$$\psi(x,t) = (h(x) - h(m_t))^*\phi(m_t)(h(x) - h(m_t))$$

(5.2.27)

$$\zeta(x,t) = \log p(x,t)$$

(5.2.28)

$$u(x,t) = \zeta(x,t) + \frac{1}{2\epsilon}\psi(x,t).$$

(5.2.29)

We shall use the following notation for a given function $\Lambda(x)$:

$$\Lambda_t = \Lambda(m_t) \text{ (or } \Lambda^t \text{ to avoid confusion with other lower indices)}$$

and

$$\Lambda_i = \frac{\partial\Lambda}{\partial x_i}; \quad \Lambda_{ij} = \frac{\partial^2\Lambda}{\partial x_{ij}}.$$

If χ_i is a function already indexed by i, we write

$$\chi_{i,j} = \frac{\partial \chi_i}{\partial x_j}.$$

We also use the summation convention.

It is easy to check the Ito differential

$$
d\zeta = \left[a_{\lambda\mu}\zeta_{\lambda\mu} + a_{\lambda\mu}\zeta_\lambda\zeta_\mu - \zeta_\lambda(g_\lambda - 2a_{\lambda\mu,\mu}) \right.
$$
$$
\left. - g_{\lambda,\lambda} + a_{\lambda\mu,\lambda\mu} - \frac{1}{2\epsilon^2}|h|^2 \right] dt + \frac{1}{\epsilon^2} h^* dz. \tag{5.2.30}
$$

Now using a shortened notation for (5.2.26) written in scalar form, we have successively

$$
dm_i^t = g_i^t dt + \frac{1}{\epsilon}\theta_{ij}^t(dz_j - h_j^t dt) \tag{5.2.31}
$$

$$
dh_i^t = (h_{i,j}^t g_j^t + h_{i,jk}^t a_{jk}^t)dt + \frac{1}{\epsilon}\gamma_{ij}^t(dz_j - h_j^t dt). \tag{5.2.32}
$$

To obtain (5.2.32) we have made use of (5.2.23) and (5.2.29). We next derive

$$
\begin{aligned}
d\psi = & \left\{ \left[-2h_{i,k}^t\phi_{ij}^t(h_j - h_j^t) + (h_i - h_i^t)\phi_{ij,k}^t(h_j - h_j^t) \right] g_k^t \right. \\
& + \left[(h_i - h_i^t)\phi_{ij,kl}^t a_{kl}^t(h_j - h_j^t) - 2h_{i,kl}^t a_{kl}^t\phi_{ij}^t(h_j - h_j^t) \right. \\
& \left. -4h_{i,k}^t a_{kl}^t\phi_{ij,l}^t(h_j - h_j^t) + \operatorname{tr}\gamma^t \right] \right\} dt \\
& + \frac{1}{\epsilon}\left[(h_i - h_i^t)\phi_{ij,k}^t(h_j - h_j^t)\theta_{kl}^t - 2(h_l - h_l^t) \right] (dz_l - h_l^t dt)
\end{aligned} \tag{5.2.33}
$$

where we have made use of the relations (5.2.23), (5.2.20), (5.2.21), (5.2.22). Note also

$$
\zeta_\lambda = u_\lambda - \frac{1}{\epsilon}h_{i,\lambda}\phi_{ij}^t(h_j - h_j^t)
$$
$$
\zeta_{\lambda\mu} = u_{\lambda\mu} - \frac{1}{\epsilon}h_{i,\lambda}\phi_{ij}^t h_{j,\mu} - \frac{1}{\epsilon}h_{i,\lambda\mu}\phi_{ij}^t(h_j - h_j^t).
$$

Collecting results we arrive at

$$
\begin{aligned}
du = & a_{\lambda\mu}u_{\lambda\mu} + a_{\lambda\mu}u_\lambda u_\mu - u_\lambda\left[g_\lambda - 2a_{\lambda\mu,\mu} + \frac{2}{\epsilon}a_{\lambda\mu}h_{i,\mu}\phi_{ij}^t(h_j - h_j^t) \right] \\
& + \frac{1}{\epsilon^2}\left\{ \left[\frac{1}{2}(\phi^t(\gamma^2 - (\gamma^t)^2)\phi^t)_{ij}(h_i - h_i^t)(h_j - h_j^t) + \frac{1}{2}|h^t|^2 \right] \right. \\
& \left. + \left[\frac{1}{2}(h_i - h_i^t)\phi_{ij,k}^t(h_j - h_j^t)\theta_{kl}^t + h_l^t \right](dz_l - h_l^t dt) \right\} \\
& + \frac{1}{\epsilon}\left\{ \frac{1}{2}(h_i - h_i^t)\phi_{ij,k}^t(h_j - h_j^t)g_k^t + (h_{i,k}g_k - h_{i,k}^t g_k^t)\phi_{ij}^t(h_j - h_j^t) \right. \\
& + \frac{1}{2}\phi_{ij,kl}^t a_{kl}^t(h_i - h_i^t)(h_j - h_j^t) - \frac{1}{2}\operatorname{tr}(\gamma^2 - (\gamma^t)^2)\phi^t \\
& \left. + (-2h_{i,\lambda}a_{\lambda\mu,\mu}\phi_{ij}^t - h_{i,kl}^t a_{kl}^t\phi_{ij}^t - a_{\lambda\mu}h_{i,\lambda\mu}\phi_{ij}^t - 2h_{i,k}^t a_{kl}^t\phi_{ij,l}^t)(h_j - h_j^t) \right\} \\
& - g_{\lambda,\lambda} + a_{\lambda\mu,\lambda\mu}.
\end{aligned} \tag{5.2.34}
$$

The next step is to differentiate with respect to x_i, and obtain

$$
\begin{aligned}
du_i = \Bigg\{ & a_{\lambda\mu}u_{\lambda\mu,i} + a_{\lambda\mu,i}u_{\lambda\mu} + a_{\lambda\mu,i}u_{\lambda}u_{\mu} + 2a_{\lambda\mu}u_{\lambda i}u_{\mu} \\
& - u_{\lambda i}\left[g_{\lambda} - 2a_{\lambda\mu,\mu} + \frac{2}{\epsilon}a_{\lambda\mu}h_{k,\mu}\phi^t_{kj}(h_j - h^t_j) \right] \\
& - u_{\lambda}\left[g_{\lambda,i} - 2a_{\lambda\mu,\mu i} + \frac{2}{\epsilon}a_{\lambda\mu}h_{k,\mu}\phi^t_{kj}h_{j,i} \right. \\
& \left. \qquad + \frac{2}{\epsilon}(h_j - h^t_j)\phi^t_{kj}(a_{\lambda\mu,i}h_{k,\mu} + a_{\lambda\mu}h_{k,\mu i}) \right] \\
& + \frac{1}{\epsilon^2}(h_j - h^t_j)\left[(\phi^t(\gamma^2 - (\gamma^t)^2)\phi^t)_{jk}h_{k,i} + \frac{1}{2}(\phi^t\gamma^2\phi^t)_{jk,i}(h_k - h^t_k) \right] \\
& + \frac{1}{\epsilon}\left[(h_j - h^t_j)\Big[h_{m,i}(\phi^t_{mj,kl}a^t_{kl} + \phi^t_{mj,k}g^t_k) \right. \\
& \qquad + \phi^t_{mj}(h_{m,ki}g_k + h_{m,k}g_{k,i} - 2h_{m,\lambda i}a_{\lambda\mu,\mu} \\
& \qquad - 2h_{m,\lambda}a_{\lambda\mu,\mu i} - a_{\lambda\mu,i}h_{m,\lambda\mu} - a_{\lambda\mu}h_{m,\lambda\mu i}) \Big] \\
& \qquad + (h_{m,k}g_k - h^t_{m,k}g^t_k)\phi^t_{mj}h_{j,i} \\
& \qquad - \phi^t_{mj}\left[a_{\lambda\mu}(h_{m,\lambda\mu}h_{j,i} + 2h_{m,\mu}h_{j,\lambda i}) \right. \\
& \qquad \left. + h_{m,\lambda}(2a_{\lambda\mu,\mu}h_{j,i} + h_{j,\mu}a_{\lambda\mu,i}) + a^t_{\lambda\mu}h^t_{m,\lambda\mu}h_{j,i} \right] - 2h^t_{m,\lambda}a^t_{\lambda\mu}\phi^t_{mj,\mu}h_{j,i} \Big] \\
& - g_{\lambda,\lambda i} + a_{\lambda\mu,\lambda\mu i} \Bigg\} dt \\
& + \frac{1}{\epsilon^2}(h_j - h^t_j)h_{\lambda,i}\phi^t_{\lambda j,k}\theta^t_{kl}(dz_l - h^t_l dt).
\end{aligned}
\tag{5.2.35}
$$

Let us now define the 'energy' by setting

$$
\rho(t) = \int qu_i a_{ir}u_r dx. \tag{5.2.36}
$$

We shall express the Ito differential of $\rho(t)$. We use the formula

$$
\begin{aligned}
d\rho = & \int dq u_i a_{ir}u_r dx + \int q(2a_{ir}u_r du_i + du_i a_{ir}du_r)dx \\
& + 2\int a_{ir}u_r dq du_i dx
\end{aligned}
\tag{5.2.37}
$$

where

$$
du_i du_r = \frac{2}{\epsilon^2}(h_j - h^t_j)(h_k - h^t_k)h_{\lambda,i}h_{\mu,r}\phi^t_{\lambda j,l}\phi^t_{\mu k,m}a^t_{lm}dt \tag{5.2.38}
$$

using (5.2.23) and

$$
dq du_i = \frac{q}{\epsilon^2}(h_j - h^t_j)(h_k - h^t_k)h_{\lambda,i}\phi^t_{\lambda j,l}\theta^t_{lk}dt. \tag{5.2.39}
$$

On the right hand side of (5.2.37) we first collect the terms in u of order greater than

1 in u. We consider the partial sum

$$
\begin{aligned}
\mathrm{d}\rho_1 = &\int \mathrm{d}q u_i a_{ir} u_r \mathrm{d}x \\
&+ 2\int q a_{ir} u_r \Big\{ a_{\lambda\mu} u_{\lambda\mu i} + a_{\lambda\mu,i} u_{\lambda\mu} + a_{\lambda\mu,i} u_\lambda u_\mu + 2a_{\lambda\mu} u_{\lambda i} u_\mu \\
&\qquad - u_{\lambda i}\Big[g_\lambda - 2a_{\lambda\mu,\mu} + \frac{2}{\epsilon} a_{\lambda\mu} h_{k,\mu} \phi_{kj}^t (h_j - h_j^t)\Big] \\
&\qquad - u_\lambda \Big[g_{\lambda i} - 2a_{\lambda\mu,\mu i} + \frac{2}{\epsilon} a_{\lambda\mu} h_{k,\mu} \phi_{kj}^t h_{j,i} \\
&\qquad\quad + \frac{2}{\epsilon}(h_j - h_j^t)\phi_{kj}^t (a_{\lambda\mu,i} h_{k,\mu} + a_{\lambda\mu} h_{k,\mu i})\Big]\Big\} \mathrm{d}x \mathrm{d}t
\end{aligned} \tag{5.2.40}
$$

and make use of (5.2.18), which yields

$$
\begin{aligned}
&\int u_i a_{ir} u_r \left[-(qg_\lambda)_{,\lambda} + (a_{\lambda\mu} q)_{,\lambda\mu} \right] \mathrm{d}x \mathrm{d}t + 2\int q a_{ir} u_r a_{\lambda\mu} u_{\lambda\mu i}\, \mathrm{d}x \mathrm{d}t \\
&= \int q g_\lambda u_r (a_{ir,\lambda} u_i + 2u_{i\lambda} a_{ir}) \mathrm{d}x \mathrm{d}t \\
&\quad + \int q a_{\lambda\mu}\left[4(u_{i\lambda\mu} a_{ir} u_r + u_{i\lambda} a_{ir,\mu} u_r) + 2u_{i\lambda} a_{ir} u_{r\mu} + u_i u_r a_{ir,\lambda\mu} \right] \mathrm{d}x.
\end{aligned} \tag{5.2.41}
$$

We also use the integration by parts formula

$$
\begin{aligned}
4\int q a_{\lambda\mu}(u_{i\lambda\mu} a_{ir} u_r + u_{i\lambda} a_{ir,\mu} u_r) \mathrm{d}x = - 4\int u_{i\lambda} a_{ir}(& a_{\lambda\mu,\mu} u_r q \\
& + a_{\lambda\mu} u_{r\mu} q + a_{\lambda\mu} u_r q_\mu) \mathrm{d}x
\end{aligned}
$$

and (see (5.2.28))

$$
q_\mu = q\zeta_\mu = q\left(u_\mu - \frac{1}{\epsilon} h_{k,\mu} \phi_{kj}^t (h_j - h_j^t) \right).
$$

Collecting these results and (5.2.41) in (5.2.40) yields

$$
\begin{aligned}
\mathrm{d}\rho_1 = &\int q \Big\{ g_\lambda a_{ir,\lambda} u_i u_r - 2a_{\lambda\mu} a_{ir} u_{i\lambda} u_{r\mu} + a_{\lambda\mu} a_{ir,\lambda\mu} u_i u_r. \\
&\qquad + 2a_{ir} a_{\lambda\mu,i} u_r u_{\lambda\mu} + 2a_{ir} a_{\lambda\mu,i} u_r u_\lambda u_\mu \\
&\qquad - 2a_{ir} u_r u_\lambda \Big[g_{\lambda,i} - 2a_{\lambda\mu,\mu i} + \frac{2}{\epsilon} a_{\lambda\mu} h_{k,\mu} \phi_{kj}^t h_{j,i} \\
&\qquad\quad + \frac{2}{\epsilon}(a_{\lambda\mu,i} h_{k,\mu} + a_{\lambda\mu} h_{k,\mu i})\phi_{kj}^t(h_j - h_j^t)\Big]\Big\} \mathrm{d}x \mathrm{d}t \\
&+ \int \frac{q}{\epsilon^2} u_i a_{ir} u_r (h_j - \hat{h}_j)(\mathrm{d}z_j - \hat{h}_j \mathrm{d}t).
\end{aligned}
$$

But also

$$
\begin{aligned}
\int 2q a_{ir} a_{\lambda\mu,i} u_r u_\lambda u_\mu \mathrm{d}x = -2\int \Big[& q_\mu a_{ir} a_{\lambda\mu,i} u_r u_\lambda \\
& + q u_\lambda \left(a_{ir,\mu} a_{\lambda\mu,i} u_r + a_{ir} a_{\lambda\mu,i} u_{r\mu} + a_{ir} a_{\lambda\mu,i\mu} u_r \right)\Big] \mathrm{d}x.
\end{aligned}
$$

We deduce

$$
\begin{aligned}
\mathrm{d}\rho_1 = -\,2\int & q(a_{\lambda\mu}a_{ir}u_{i\lambda}u_{r\mu} + a_{ir}a_{\lambda\mu,i}u_{r\mu}u_\lambda)\mathrm{d}x\mathrm{d}t \\
+\,2\int & qu_\lambda u_r \left\{ \frac{1}{2}(g_i a_{\lambda r,i} + a_{i\mu}a_{\lambda r,i\mu}) - a_{ir,\mu}a_{\lambda\mu i} \right. \\
& - a_{ir}\left[g_{\lambda,i} - a_{\lambda\mu,\mu i} + \frac{2}{\epsilon}a_{\lambda\mu}h_{k,\mu}\phi^t_{kj}h_{j,i} \right. \\
& \left. + \frac{1}{\epsilon}(a_{\lambda\mu,i}h_{k,\mu} + 2a_{\lambda\mu}h_{k,\mu i})\phi^t_{kj}(h_j - h^t_j) \right] \Big\} \,\mathrm{d}x\mathrm{d}t \\
+\int & \frac{q}{\epsilon^2}u_i a_{ir}u_r(h_j - \hat{h}_j)(\mathrm{d}z_j - \hat{h}_j\mathrm{d}t).
\end{aligned}
\tag{5.2.42}
$$

In addition we use the property

$$
\begin{aligned}
-\frac{2}{\epsilon}\int & qa_{ir}u_r u_\lambda(a_{\lambda\mu,i}h_{k,\mu} + 2a_{\lambda\mu}h_{k,\mu i})\phi^t_{kj}(h_j - h^t_j)\mathrm{d}x \\
= -\frac{2}{\epsilon}\int & qa_{ir}u_r \left(\frac{q_\lambda}{q} + \frac{1}{\epsilon}h_{m,\lambda}\phi^t_{ml}(h_l - h^t_l) \right) \\
& (a_{\lambda\mu,i}h_{k,\mu} + 2a_{\lambda\mu}h_{k,\mu i})\phi^t_{kj}(h_j - h^t_j)\mathrm{d}x \\
= \frac{2}{\epsilon}\int & qu_r[a_{ir,\lambda}(a_{\lambda\mu,i}h_{k,\mu} + 2a_{\lambda\mu}h_{k,\mu i}) \\
& + a_{ir}(a_{\lambda\mu,i\lambda}h_{k,\mu} + a_{\lambda\mu,i}h_{k,\mu\lambda} + 2a_{\lambda\mu,\lambda}h_{k,\mu i} + 2a_{\lambda\mu}h_{k,\mu i\lambda})]\,\phi^t_{kj}(h_j - h^t_j)\mathrm{d}x \\
+ \frac{2}{\epsilon}\int & qa_{ir}u_{r\lambda}(a_{\lambda\mu,i}h_{k,\mu} + 2a_{\lambda\mu}h_{k,\mu i})\phi^t_{kj}(h_j - h^t_j)\mathrm{d}x \\
+ \frac{2}{\epsilon}\int & qa_{ir}u_r h_{j,\lambda}(a_{\lambda\mu,i}h_{k,\mu} + 2a_{\lambda\mu}h_{k,\mu i})\phi^t_{kj}\mathrm{d}x \\
- \frac{2}{\epsilon^2}\int & qa_{ir}u_r(a_{\lambda\mu,i}h_{k,\mu} + 2a_{\lambda\mu}h_{k,\mu i})\phi^t_{kj}\phi^t_{ml}h_{m,\lambda}(h_j - h^t_j)(h_l - h^t_l)\mathrm{d}x.
\end{aligned}
\tag{5.2.43}
$$

Note that using (5.2.20)

$$
\begin{aligned}
\frac{1}{\epsilon^2}\int & qa_{ir}u_r(\phi^t\gamma^2\phi^t)_{jl,i}(h_j - h^t_j)(h_l - h^t_l)\mathrm{d}x \\
= \frac{2}{\epsilon^2}\int & qa_{ir}u_r h_{m,\lambda}(h_{k,\mu}a_{\lambda\mu,i} + 2h_{k,\mu i}a_{\lambda\mu})\phi^t_{jk}\phi^t_{ml}(h_j - h^t_j)(h_l - h^t_l)\mathrm{d}x.
\end{aligned}
\tag{5.2.44}
$$

We can finally express $\mathrm{d}\rho$ using the preceding calculations. There are some cancellations. The last term of (5.2.43) cancels out with (5.2.44). There are other reductions between the terms of (5.2.43) involving u_r/ϵ and corresponding terms arising in the computation of $2\int qa_{ir}u_r\mathrm{d}u_i$.

We finally obtain the expression

$$
\begin{aligned}
\mathrm{d}\rho = &- 2 \int q(a_{\lambda\mu}a_{ir}u_{i\lambda}u_{r\mu} + a_{ir}a_{\lambda\mu,i}u_{r\mu}u_\lambda)\mathrm{d}x\mathrm{d}t \\
&+ 2 \int qu_\lambda u_r \left[\frac{1}{2}(g_i a_{\lambda r,i} + a_{i\mu}a_{\lambda r,i\mu}) - a_{ir,\mu}a_{\lambda\mu,i}\right. \\
&\left. -a_{ir}(g_{\lambda,i} - a_{\lambda\mu,\mu i}) - \frac{2}{\epsilon}a_{ir}a_{\lambda\mu}h_{k,\mu}\phi^t_{kj}h_{j,i}\right]\mathrm{d}x\mathrm{d}t \\
&+ \frac{2}{\epsilon}\int qa_{ir}u_{r\lambda}(a_{\lambda\mu,i}h_{k,\mu} + 2a_{\lambda\mu}h_{k,\mu i})\phi^t_{kj}(h_j - h^t_j)\mathrm{d}x\mathrm{d}t \\
&+ \frac{2}{\epsilon^2}\int qa_{ir}u_r(h_j - h^t_j)\Big[(\phi^t(\gamma^2 - (\gamma^t)^2)\phi^t)_{jk}h_{k,i} \\
&+ h_{\lambda,i}\phi^t_{\lambda j,k}\theta^t_{kl}(h_l - h^t_l) + (h_k - \hat{h}_k)h_{\lambda,i}\phi^t_{\lambda j,l}\theta^t_{lk}\Big]\mathrm{d}x\mathrm{d}t \\
&+ \frac{2}{\epsilon}\int qu_r\left\{a_{ir,\lambda}\phi^t_{kj}(h_j - h^t_j)(a_{\lambda\mu,i}h_{k,\mu} + 2a_{\lambda\mu}h_{k,\mu i})\right. \\
&+ a_{ir}\Big[-\phi^t_{mj}(a_{\lambda\mu}h_{m,\lambda\mu}h_{j,i} + 2a_{\lambda\mu,\mu}h_{m,\lambda}h_{j,i} + a^t_{\lambda\mu}h^t_{m,\lambda\mu}h_{j,i}) \\
&- 2h_{j,i}h^t_{m,\lambda}a^t_{\lambda\mu}\phi^t_{mj,\mu} + (h_j - h^t_j)[h_{m,i}(\phi^t_{mj,kl}a^t_{kl} + \phi^t_{mj,k}g^t_k) \\
&+ \phi^t_{mj}(h_{m,ki}g_k + h_{m,k}g_{k,i} - h_{m,\lambda}a_{\lambda\mu,\mu i} + a_{\lambda\mu}h_{m,\lambda\mu i})) \\
&\left.+(h_{m,k}g_k - h^t_{m,k}g^t_k)\phi^t_{mj}h_{j,i}\Big]\right\}\mathrm{d}x\mathrm{d}t \\
&+ 2\int qa_{ir}u_r(-g_{\lambda,\lambda i} + a_{\lambda\mu,\lambda\mu i})\mathrm{d}x\mathrm{d}t \\
&+ \frac{2}{\epsilon^2}\int qa_{ir}(h_j - h^t_j)(h_k - h^t_k)h_{\lambda,i}h_{\mu,r}\phi^t_{\lambda j,l}\phi^t_{\mu k,m}a^t_{lm}\mathrm{d}x\mathrm{d}t \\
&+ \frac{1}{\epsilon^2}\int qa_{ir}u_i u_r(h_j - \hat{h}_j)(\mathrm{d}z_j - \hat{h}_j\mathrm{d}t) + \frac{2}{\epsilon}\int qa_{ir}u_r(h_j - h^t_j)h_{\lambda,i}\phi^t_{\lambda j,k}\theta^t_{kl}\mathrm{d}b_l.
\end{aligned}
$$

$$(5.2.45)$$

We integrate between 0 and t and then take the mathematical expectation. We can differentiate again in t and obtain

$$\frac{\mathrm{d}}{\mathrm{d}t}E\rho(t) = -2E\int q(a_{\lambda\mu}a_{ir}u_{i\lambda}u_{r\mu} + a_{ir}a_{\lambda\mu,i}u_{r\mu}u_{\lambda})dx$$

$$+2E\int qu_{\lambda}u_r\left[\frac{1}{2}(g_i a_{\lambda r,i} + a_{\lambda\mu}a_{\lambda r,i\mu}) - a_{ir,\mu}a_{\lambda\mu,i}\right.$$

$$\left.-a_{ir}(g_{\lambda,i} - a_{\lambda\mu,\mu i}) - \frac{2}{\epsilon}a_{ir}a_{\lambda\mu}h_{k,\mu}\phi^t_{kj}h_{j,i}\right]dx$$

$$+\frac{2}{\epsilon}E\int qa_{ir}u_{r\lambda}(a_{\lambda\mu,i}h_{k,\mu} + 2a_{\lambda\mu}h_{k,\mu i})\phi^t_{kj}(h_j - h^t_j)dx$$

$$+\frac{2}{\epsilon^2}E\int qa_{ir}u_r(h_j - h^t_j)\left[(\phi^t(\gamma^2 - (\gamma^t)^2)\phi^t)_{jk}h_{k,i}\right.$$

$$\left.+ h_{\lambda,i}\phi^t_{\lambda j,k}\theta^t_{kl}(h_l - h^t_l) + (h_k - \hat{h}_k)h_{\lambda,i}\phi^t_{\lambda j,l}\theta^t_{lk}\right]dx \qquad (5.2.46)$$

$$+\frac{2}{\epsilon}E\int qu_r\Big(a_{ir,\lambda}\phi^t_{kj}(h_j - h^t_j)(a_{\lambda\mu,i}h_{k,\mu} + 2a_{\lambda\mu}h_{k,\mu i})$$

$$+ a_{ir}\{-\phi^t_{mj}(a_{\lambda\mu}h_{m,\lambda\mu}h_{j,i} + 2a_{\lambda\mu,\mu}h_{m,\lambda}h_{j,i} + a^t_{\lambda\mu}h^t_{m,\lambda\mu}h_{j,i})$$

$$- 2h_{j,i}h^t_{m,\lambda}a^t_{\lambda\mu}\phi^t_{mj,\mu} + (h_j - h^t_j)[h_{m,i}(\phi^t_{mj,kl}a^t_{kl} + \phi^t_{mj,k}g^t_k)$$

$$+ \phi^t_{mj}(h_{m,ki}g_k + h_{m,k}g_{k,i} - h_{m,\lambda}a_{\lambda\mu,\mu i} + a_{\lambda\mu}h_{m,\lambda\mu i})]$$

$$+(h_{m,k}g_k - h^t_{m,k}g^t_k)\phi^t_{mj}h_{j,i}\}\Big)dx$$

$$+2E\int qa_{ir}u_r(-g_{\lambda,\lambda i} + a_{\lambda\mu,\lambda\mu i})dx$$

$$+\frac{2}{\epsilon^2}E\int qa_{ir}(h_j - h^t_j)(h_k - h^t_k)h_{\lambda,i}h_{\mu,r}\phi^t_{\lambda j,l}\phi^t_{\mu k,m}a^t_{lm}dx.$$

5.2.3 Justification of the formal calculations

We begin by justifying the calculations made in § 5.2.2. For that we make precise the equations (5.2.17), (5.2.18). We cannot use directly the results on stochastic PDEs described in Section 4.7, since the coefficients are unbounded, in particular h.

The easiest way is to use the robust form introduced in § 4.6.3. Define

$$\alpha(x,t) = p(x,t)\exp[-h^*(x)z(x)]\dagger \qquad (5.2.47)$$

then α appears as the solution of

$$\frac{\partial\alpha}{\partial t} + A^*\alpha - 2a_{ij}h_{k,i}z_k\frac{\partial\alpha}{\partial x_j}$$

$$+ \alpha\left(\frac{1}{2}|h|^2 - a_{ij}h_{k,ij}z_k + h_{k,i}z_k g_i - 2h_{k,i}z_k a_{ij,j}\right) = 0 \qquad (5.2.48)$$

$$\alpha(x,0) = p_0(x).$$

---

† In Chapter 4, the function α was denoted by q. We change the notation to avoid confusion with the solution of (5.2.18).

We can study (5.2.48) directly, for each ω. The function $z(t)$ is then bounded on every finite time interval. However the coefficients of the PDE (5.2.48) are not bounded. To recover a PDE with bounded coefficients, we follow an idea suggested by W. Fleming, used in Pardoux (1982).

Let

$$\gamma(x,t) = \alpha(x,t) \exp\left[-(1 + |x|^2)^{1/2} \exp Kt\right] \tag{5.2.49}$$

where K is a constant to be chosen. We can deduce a PDE for γ, namely

$$\frac{\partial \gamma}{\partial t} + A^*\gamma - 2a_{ij}(h_{k,i}z_k + G_i)\frac{\partial \gamma}{\partial x_j}$$

$$+ \gamma \left(\frac{1}{2}|h|^2 - a_{ij}h_{k,ij}z_k + h_{k,i}z_k g_i - 2h_{k,i}z_k a_{ij,j} + G_t + g_i G_i\right. \tag{5.2.50}$$

$$\left. - a_{ij}G_{ij} - a_{ij}G_i G_j - 2a_{ij,j}G_i - 2a_{ij}h_{k,i}z_k G_j\right) = 0$$

$$\gamma(x,0) = p_0(x) \exp\left[-(1 + |x|^2)^{1/2}\right]$$

where we have set

$$G(x,t) = (\exp Kt)(1 + |x|^2)^{1/2}$$

hence

$$G_t = K(\exp Kt)(1 + |x|^2)^{1/2}$$
$$G_i = (\exp Kt)(1 + |x|^2)^{-1/2}x_i$$
$$G_{ij} = (\exp Kt)\left[\delta_{ij} - x_i x_j(1 + |x|^2)^{-1}\right](1 + |x|^2)^{-1/2}.$$

If we fix the interval of time $(0, T)$, it is possible to find K depending on T and $\sup_{t \in [0,T]} |z(t)|$, such that

$$e_0(x,t) = h_{k,i}z_k g_i + G_t + g_i G_i \geq 0.$$

Let

$$e_1(x,t) = \frac{1}{2}g_{i,i} - \frac{1}{2}a_{ij,ij} - a_{ij,j}(h_{k,i}z_k + G_i) - a_{ij}G_i G_j - 2a_{ij}h_{k,i}z_k G_j$$

which is bounded. From (5.2.50) we can deduce the energy equality

$$\frac{1}{2}\frac{d}{dt}\int |\gamma|^2 dx + \int a_{ij}\frac{\partial \gamma}{\partial x_j}\frac{\partial \gamma}{\partial x_i}dx + \int \gamma^2 \left(\frac{1}{2}|h|^2 + e_0 + e_1\right) dx = 0 \tag{5.2.51}$$

Since $e_0 \geq 0$ and e_1 is bounded, (5.2.51) leads to an energy estimate proving that

$$\gamma \in L^2(0,T;H^1(R^n)), \quad \gamma \in L^\infty(0,T;L^2(R^n)) \tag{5.2.52}$$

where we have made use of the assumption on the initial condition p_0.

Therefore we can solve (5.2.50) in the functional space (5.2.52) and through (5.2.49) we can define α; then by (5.2.47) we can define p.

Let $\gamma_\lambda = \partial\gamma/\partial x_\lambda$ and differentiate (5.2.50) with respect to x_λ. We obtain

$$
\frac{\partial\gamma_\lambda}{\partial t} + A^*\gamma_\lambda - 2a_{ij}(h_{k,i}z_k + G_i)\frac{\partial\gamma_l}{\partial x_j}
$$

$$
+ \gamma_\lambda\left\{\frac{1}{2}|h|^2 + e_0 - a_{ij}h_{k,ij}z_k - 2k_{k,i}z_ka_{ij,j} - a_{ij}G_{ij}\right.
$$

$$
\left. - a_{ij}G_iG_j - 2a_{ij,j}G_i - 2a_{ij}h_{k,i}z_kG_j\right\} \tag{5.2.53}
$$

$$
- \frac{\partial^2}{\partial x_i\partial x_j}(a_{ij,\lambda}\gamma) + \frac{\partial}{\partial x_i}(g_{i,\lambda}\gamma)
$$

$$
- 2\left[a_{ij,\lambda}(h_{k,i}z_k + G_i) + a_{ij}(h_{k,i\lambda}z_k + G_{i\lambda})\right]\gamma_j
$$

$$
+ \gamma\left(h_ih_{i,\lambda} + h_{k,i\lambda}z_kg_i + g_iG_{i\lambda} + e_\lambda\right) = 0
$$

where

$$
e_\lambda = \frac{\partial}{\partial x_\lambda}(-a_{ij}h_{k,ij})z_k + h_{k,i}z_kg_{i,\lambda} - 2\frac{\partial}{\partial x_\lambda}(h_{k,i}a_{ij,j})z_k + G_{t\lambda} + g_{i,\lambda}G_i
$$

$$
- \frac{\partial}{\partial x_\lambda}(a_{ij}G_{ij} + a_{ij}G_iG_j + 2a_{ij,j}G_i + 2a_{ij}h_{k,i}z_kG_j) \tag{5.2.54}
$$

which is a bounded quantity.

We then deduce from (5.2.53) the following energy equality

$$
\frac{1}{2}\frac{d}{dt}\int|\gamma_\lambda|^2dx + \int a_{ij}\frac{\partial\gamma_\lambda}{\partial x_j}\frac{\partial\gamma_\lambda}{\partial x_i}dx + \int \gamma_\lambda^2\left(\frac{1}{2}|h|^2 + e_0 + e_1\right)dx
$$

$$
+ \int\frac{\partial\gamma_\lambda}{\partial x_i}(a_{ij,\lambda j}\gamma + a_{ij,\lambda}\gamma_j - g_{i,\lambda}\gamma)dx \tag{5.2.55}
$$

$$
- 2\int\gamma_\lambda\gamma_j\left[a_{ij,\lambda}(h_{k,i}z_k + G_i) + a_{ij}(h_{k,i\lambda}z_k + G_{i\lambda})\right]dx
$$

$$
+ \int\gamma_\lambda\gamma\left[h_ih_{i,\lambda} + h_{k,i\lambda}z_kg_i + g_iG_{i\lambda} + e_\lambda\right]dx = 0.
$$

Making use of the fact that γ, γ_j, γh_i are L^2 and the assumption (5.2.3′), we deduce that

$$
\gamma_\lambda \in L^2(0,T;H^1(R^n)), \quad \gamma_\lambda \in L^\infty(0,T;L^2(R^n)). \tag{5.2.56}
$$

In fact $\gamma(x,t)$ (hence α and p) is even more regular in x, t. Let us check this. Let $\phi \in C_0^\infty(R^n)$ and set $\tilde\gamma = \gamma_\phi$. From (5.2.50) we deduce

$$
\frac{\partial\tilde\gamma}{\partial t} + A^*\tilde\gamma + a_{ij}\gamma\frac{\partial^2\phi}{\partial x_i\partial x_j}
$$

$$
+ \frac{\partial\phi}{\partial x_i}\left[-g_i\gamma + 2\frac{\partial}{\partial x_j}(a_{ij}\gamma) + 2a_{ij}(h_{k,j}z_k + G_j)\gamma\right] \tag{5.2.57}
$$

$$
- 2a_{ij}(h_{k,j}z_k + G_i)\frac{\partial\tilde\gamma}{\partial x_j} + \tilde\gamma b = 0
$$

where b is the coefficient of γ in (5.2.50).

Since $\tilde{\gamma}$ vanishes outside the support of ϕ, which is a smooth bounded domain denoted by O_ϕ, we deduce by standard regularity results on the parabolic Dirichlet problem

$$\tilde{\gamma} \in L^2(0,T; H^2(O_\phi)), \quad \frac{\partial \tilde{\gamma}}{\partial t} \in L^2(0,T; L^2(O_\phi)) \qquad (5.2.58)$$

and by a similar argument for the derivative $\tilde{\gamma}_\lambda$

$$\tilde{\gamma}_\lambda \in L^2(0,T; H^2(O_\phi)), \quad \frac{\partial \tilde{\gamma}_\lambda}{\partial t} \in L^2(0,T; L^2(O_\phi)). \qquad (5.2.59)$$

The properties (5.2.58), (5.2.59) imply that there exists $2 < r_0 < \infty$, such that $\tilde{\gamma}, \tilde{\gamma}_\lambda \in L^{r_0}(0,T; L^{r_0}(O_\phi))$. Take $\phi^1 \in C_0^\infty(R^n)$ such that the closure of its support is included in O_ϕ. Then $\gamma, \gamma_\lambda \in L^{r_0}(0,T; L^{r_0}(O_{\phi^1}))$. Set $\tilde{\gamma}^1 = \gamma\phi^1$. Reasoning as for (5.2.58) one deduces by standard regularity theorems on PDEs that

$$\tilde{\gamma}^1 \in L^{r_0}(0,T; W^{2,r_0}(O_{\phi^1})), \quad \frac{\partial \tilde{\gamma}^1}{\partial t} \in L^{r_0}(0,T; L^{r_0}(O_{\phi_1})).$$

Using a bootstrap argument we deduce

$$\gamma, \gamma_\lambda \in L^r(0,T; W^{2,r_0}_{\text{loc}}(R^n)), \quad \frac{\partial \gamma}{\partial t}, \frac{\partial \gamma_\lambda}{\partial t} \in L^r(0,T; L^r_{\text{loc}}(R^n))$$
$$\forall r, \ 2 \le r < \infty. \qquad (5.2.60)$$

In particular $\gamma \in C^{2,1}(R^n \times [0,T])$, and therefore the formula (5.2.34) is verified for any x (a regularization of $\log p$ can be used as an intermediary step).

This does not justify (5.2.35) for any x, since in particular the right hand side of (5.2.35) involves the third derivative $u_{\lambda\mu i}$. But we may assume *temporarily more smoothness* on the coefficients, in which case more regularity will be available on the function γ and the relation (5.2.35) will be verified.

From § 4.6.1 we know that

$$p(t)(\psi) = \int p(x,t)\psi(x)\mathrm{d}x.$$

We can then set

$$q(x,t) = \frac{p(x,t)}{p(t)(1)}$$

and see (4.3.20) that

$$q(x,t) = p(x,t)\mu_t$$

where

$$\mu_t = \exp\left(\int_0^t \frac{-\hat{h}^*\mathrm{d}z}{\epsilon^2} + \frac{1}{2\epsilon^2}\int_0^t |\hat{h}|^2\mathrm{d}s\right).$$

When ω is frozen, $q(x,t)$ has the same regularity as $p(x,t)$. Moreover q is the solution of (5.2.18) (unique since the solution of (5.2.17) is unique).

Let us next show that the 'energy' process is well defined and that (5.2.46) holds.

Let us introduce a cut off function. Let

$$\beta(x) = \begin{cases} 1 & \text{if } |x| \le 1 \\ 0 & \text{if } |x| \ge 2 \end{cases}$$
$$|\beta(x)| \le 1, \quad \beta \in C_0^\infty(R^n)$$

and set

$$\beta_R(x) = \beta\left(\frac{x}{R}\right).$$

Define

$$\rho_R(t) = \int \beta_R q u_i a_{ir} u_r dx.$$

This process is well defined since the integral is over a bounded domain. We can then perform all the calculations leading to (5.2.45) on ρ_R. The stochastic differential $d\rho_R$ is then the sum of the right hand side of (5.2.45), in which β_R is inserted in all integrals, and additional terms involving derivatives of β_R; those terms are

$$\left[\int q u_r u_\lambda \left(-2a_{ir}a_{\lambda\mu,i}\beta_{R,i} + g_i a_{\lambda r}\beta_{R,i} + a_{i\mu}a_{\lambda r}\beta_{R,i\mu}\right)dx \right.$$
$$\left. +\frac{2}{\epsilon}\int q a_{ir} u_r \beta_{R,\lambda}(a_{\lambda\mu,i}h_{k,\mu} + 2a_{\lambda\mu}h_{k,\mu i})\phi_{kj}^t(h_j - h_j^t)dx\right]dt = Y_{\epsilon,R}(t)dt \tag{5.2.61}$$

Consider also an increasing sequence of stopping times τ_k defined as follows

$$\tau_k = \inf\left\{t \left| \int q|\mathrm{D}u|^2 dx \ge k\right.\right\}.$$

For k sufficiently large we have $\int q|\mathrm{D}u|^2 dx < k$, at $t = 0$, by virtue of the assumptions on p_0. Therefore $\tau_k > 0$, a.s. The sequence τ_k is an increasing sequence of $\mathcal{F}^t$ stopping times (in fact $\mathcal{Z}^t$ stopping times). For $t \le \tau_k$, $\rho(t)$ is well defined and

$$\rho_R(t) \uparrow \rho(t) \text{ as } R \uparrow +\infty.$$

We then establish an estimate which is not uniform in ϵ, but at this stage ϵ is fixed.

By the ellipticity of the matrix a $(a \ge \alpha_0 I)$ we have

$$a_{\lambda\mu}a_{ir}u_{i\lambda}u_{r\mu} \ge \alpha_0^2 \sum_{\lambda,i}(u_{i\lambda})^2. \tag{5.2.62}$$

Using this property, we deduce from (5.2.45) written, for ρ_R,

$$d\rho_R \le (C_0\rho_R(t) + X_\epsilon(t) + Y_{\epsilon,R}(t))dt$$
$$+ \frac{1}{\epsilon^2}\int \beta_R q a_{ir} u_i u_r(h_j - \hat{h}_j)(dz_j - \hat{h}_j dt)$$
$$+ \frac{2}{\epsilon}\int \beta_R q a_{ir} u_r(h_j - h_j^t)h_{\lambda,i}\phi_{\lambda j,k}^t\theta_{kl}^t db_l$$

where

$$X_\epsilon(t) = \frac{C}{\epsilon^2} + \frac{1}{\epsilon^4}\int q\left(|h - h^t|^4 + |h - \hat{h}|^4 + \|\gamma^2 - (\gamma^t)^2\|^4\right)dx$$
$$+ \frac{1}{\epsilon^2}\int q|Hg - (Hg)^t|^2 dx.$$

Since $E|x_t|^4$, $E|m_t|^4$ are finite, $EX_\epsilon(t)$ is finite and bounded in t on $(0, T)$ (the bound depends on ϵ). Note that $X_\epsilon(t)$ does not depend on R.

It follows that

$$\rho_R(t)e^{-C_0 t} \leq \rho_R(0) + \int_0^t e^{-C_0 s}(X_\epsilon(s) + Y_{\epsilon,R}(s))ds$$

$$+ \frac{1}{\epsilon^2}\int_0^t e^{-C_0 s}\int \beta_R q a_{ir} u_i u_r (h_j - \hat{h}_j)(dz_j - \hat{h}_j dt)$$

$$+ \frac{2}{\epsilon}\int_0^t e^{-C_0 s}\int \beta_R q a_{ir} u_r (h_j - \hat{h}_j)h_{\lambda,i}\phi^t_{\lambda j,k}\theta^t_{kl}db_l.$$

Applying this inequality with t replaced by $t \wedge \tau_k$, we can take the mathematical expectation. The contribution of the stochastic integrals is then 0. It follows that

$$E\rho_R(t \wedge \tau_k)e^{-C_0 t \wedge \tau_k} \leq \rho_R(0) + E\int_0^{t \wedge \tau_k} e^{-C_0 s}(X_\epsilon(s) + Y_{\epsilon,R}(s))ds$$

$$\leq C_\epsilon + E\int_0^{t \wedge \tau_k} e^{-C_0 s}Y_{\epsilon,R}(s)ds. \tag{5.2.63}$$

For $s \leq \tau_k$, $Y_{\epsilon,R}(s) \to 0$ as $R \to \infty$. Indeed, if we look at (5.2.61), the integrand tends to 0 with R, for ω, s, x fixed, and is bounded above by

$$C\left(q|Du|^2 + q/\epsilon|Du|\right),$$

whose integral over R^n is bounded by $c(k + 1/\epsilon)$. We have used the fact that $g_i\beta_{R,i}$, $\beta_{R,\lambda}h_j$ are bounded by a constant independent of R.

Letting R tend to 0 in (5.2.63), we deduce

$$E\rho(t \wedge \tau_k)e^{-C_0 t \wedge \tau_k} \leq C_\epsilon, \quad \text{independent of } k.$$

Let $\tau = \lim \uparrow \tau_k$; we deduce by the continuity of $\rho(t)$ and Fatou's lemma

$$E\rho(t \wedge \tau)e^{-C_0 t \wedge \tau} \leq C_\epsilon.$$

Necessarily $\tau = +\infty$ a.s. Otherwise let

$$\Omega_0 = \{\omega|\tau < \infty\}$$

then $\rho(\tau) = +\infty$ on Ω_0. Since, letting $t \uparrow \infty$,

$$E\rho(\tau)e^{-C_0\tau}\big|_{\tau < \infty} \leq C$$

we get a contradiction when Ω_0 is not negligible. But then we can assert that $E\rho(t)$ is finite and bounded on $(0, T)$. A more precise use of (5.2.45) alongside the previous estimates yields also

$$\sum_{i,\lambda} E\int_0^T \int u^2_{i\lambda}dxdt < \infty.$$

We can then justify (5.2.46), by first establishing this relation for ρ_R, and letting R tend to $+\infty$.

Note that we have assumed additional regularity on the coefficients. But now considering (5.2.46) written in integrated form, we see that it can be extended to the case of the original assumptions stated in § 5.2.1. This is because the estimates derived from (5.2.46) require only the assumptions of § 5.2.1. Hence (5.2.46) has been established.

5.2.4   Statement of results

We assume now that

$$|h(x) - h(z)| \geq \delta|x - z|, \quad \delta > 0. \tag{5.2.64}$$

This implies that the matrix $H(x)$ defined in (5.2.19) is uniformly invertible and $|H^{-1}(x)| \leq 1/\delta$.

We begin with

Lemma 5.2.1 *We have for $\epsilon \leq \epsilon_0$*

$$E|x_t - m_t|^2 \leq CE|\xi - m_0|^2 e^{-\gamma_0 t/\epsilon} + C\epsilon, \quad \gamma_0 > 0 \tag{5.2.65}$$

$$E|x_t - m_t|^4 \leq CE|\xi - m_0|^4 e^{-2\gamma_0 t/\epsilon} + C\epsilon e^{-\gamma_0 t/\epsilon} + C\epsilon^2. \tag{5.2.66}$$

Proof

Using Ito's calculus and (5.2.32) we have

$$d|h(x_t) - h(m_t)|^2 = \left\{ -\frac{2}{\epsilon}(h(x_t) - h(m_t))^*\gamma(m_t)(h(x_t) - h(m_t)) \right.$$
$$+ 2(h_i(x_t) - h_i(m_t))\left[(h_{i,j}g_j + h_{i,jk}a_{jk})(x_t) \right.$$
$$\left. -(h_{i,j}g_j + h_{i,jk}a_{jk})(m_t) \right] + \text{tr}\left(\gamma^2(x_t) + \gamma^2(m_t) \right) \right\}dt$$
$$+ (h(x_t) - h(m_t))^*(H\sigma(x_t)dw_t - \gamma(m_t)db_t).$$

It follows that

$$\frac{d}{dt}E|h(x_t) - h(m_t)|^2 \leq -\frac{C_0}{\epsilon}E|h(x_t) - h(m_t)|^2$$
$$+ C_1|x_t - m_t||h(x_t) - h(m_t)| + C_2$$

therefore, using (5.2.64), for $\epsilon \leq \epsilon_0$

$$\frac{d}{dt}E|h(x_t) - h(m_t)|^2 \leq -\frac{\gamma_0}{\epsilon}E|h(x_t) - h(m_t)|^2 + C_3.$$

Making use of (5.2.64), we deduce easily (5.2.65).

Similarly we have

$$d|h(x_t) - h(m_t)|^4$$
$$= 2|h(x_t) - h(m_t)|^2 \left\{ -\frac{1}{\epsilon}(h(x_t) - h(m_t))^*\gamma(m_t)\, h(x_t) - h(m_t)) \right.$$
$$+ 2(h_i(x_t) - h_i(m_t))\left[(h_{i,j}g_j + h_{i,jk}a_{jk})(x_t) - (h_{i,j}g_j + h_{i,jk}a_{jk})(m_t) \right]$$
$$\left. + \text{tr}\left(\gamma^2(x_t) + \gamma^2(m_t) \right) \right\}dt$$
$$+ (h(x_t) - h(m_t))^*(\gamma^2(x_t) + \gamma^2(m_t))(h(x_t) - h(m_t))dt$$
$$+ 2|h(x_t) - h(m_t)|^2(h(x_t) - h(m_t))^*(H\sigma(x_t)dw_t - \gamma(m_t)db_t)$$

hence for $\epsilon \le \epsilon_0$

$$\frac{\mathrm{d}}{\mathrm{d}t}E|h(x_t) - h(m_t)|^4 \le -\frac{2\gamma_0}{\epsilon}E|h(x_t) - h(m_t)|^4 + CE|h(x_t) - h(m_t)|^2$$

$$\le -\frac{2\gamma_0}{\epsilon}E|h(x_t) - h(m_t)|^4 + C\epsilon + Ce^{-\gamma_0 t/\epsilon}$$

hence (5.2.66).∎

We deduce the following estimate on the energy.

Lemma 5.2.2 *We have the estimate*

$$E\rho(t) \le C + \frac{C_1}{\epsilon}e^{-\gamma_0 t/\epsilon} + \frac{C_2}{\epsilon^2}e^{-\gamma_1 t/\epsilon} \tag{5.2.67}$$

where γ_0 is the constant of (5.2.65) and $\gamma_1 > 0$, $\gamma_1 \ne \gamma_0$.

Proof

We consider (5.2.46). Note that

$$u_r u_\lambda a_{ir} a_{\lambda\mu} h_{k,\mu} \phi_{kl}^t h_{l,i} = u_r u_\lambda (aH^* \phi^t Ha)_{\lambda r}$$

$$\ge \delta_0 \sum_r |u_r|^2 \ge \delta_1 a_{ir} u_r u_i$$

hence

$$\int q u_r u_\lambda a_{ir} a_{\lambda\mu} h_{k,\mu} \phi_{kl}^t h_{l,i} \mathrm{d}x \ge \delta_1 \rho.$$

We also make use of (5.2.62). We then deduce from (5.2.46) that

$$\frac{\mathrm{d}}{\mathrm{d}t}E\rho(t) \le -C_0 E \sum_{i,\lambda} \int q(u_{\lambda i})^2 \mathrm{d}x - \frac{C_1}{\epsilon}E\rho(t)$$

$$+ \frac{C}{\epsilon^2}E\int q|h - h^t|^2 \mathrm{d}x + \frac{C}{\epsilon^3}E\int q|h - h^t|^4 \mathrm{d}x$$

$$+ \frac{C}{\epsilon^3}E\int q\|\gamma^2 - (\gamma^2)^t\|^4 \mathrm{d}x + \frac{C}{\epsilon}E\int q|Hg - (Hg)^t|^2 \mathrm{d}x + \frac{C}{\epsilon}$$

$$\le -\frac{C_1}{\epsilon}E\rho(t) + \frac{C}{\epsilon^2}E|x_t - m_t|^2 + \frac{C}{\epsilon^3}E|x_t - m_t|^4 + \frac{C}{\epsilon}$$

and from Lemma 5.2.1

$$\frac{\mathrm{d}}{\mathrm{d}t}E\rho(t) \le -\frac{C_1}{\epsilon}E\rho(t) + \frac{C}{\epsilon} + \frac{C}{\epsilon^2}e^{-\gamma_0 t/\epsilon} + \frac{C}{\epsilon^3}e^{-2\gamma_0 t/\epsilon}.$$

It follows that

$$E\rho(t) \le E\rho(0)e^{-C_1 t/\epsilon} + C + \frac{C}{\epsilon}e^{-\gamma t/\epsilon} + \frac{C}{\epsilon^2}e^{-2\gamma_0 t/\epsilon}.$$

Since $E\rho(0) \le C/\epsilon^2$, we deduce (5.2.67), with $\gamma_1 = \min(2\gamma_0, C_1)$.∎

Remark 5.2.1. In the proof of Lemma 5.2.2, we have used the property

$$E\int q|h - \hat{h}|^4 \mathrm{d}x \le CE\int q|h - h_t|^4 \mathrm{d}x$$

which enters in the coefficient of $1/\epsilon^3$.∎

We can then prove

Theorem 5.2.1 *We make the assumptions (5.2.1)–(5.2.4) and (5.2.64). Then one has*

$$E|\hat{x}_t - m_t|^2 \leq C(\epsilon^2 + \epsilon e^{-\gamma_0 t/\epsilon} + e^{-\gamma_1 t/\epsilon}). \tag{5.2.68}$$

Proof

From the definition of u we have

$$Du = \frac{Dq}{q} + \frac{1}{\epsilon}H^*\phi^t(h - h^t)$$

hence

$$q\frac{h - h^t}{\epsilon} = q\gamma^t(H^*)^{-1}Du - \gamma^t(H^*)^{-1}Dq.$$

Integrating in x, we obtain

$$\frac{1}{\epsilon}(\hat{h}^t - h^t) = \int q\gamma^t(H^*)^{-1}Du\,dx + \int \gamma_{ij}^t(H^*)_{jk,k}^{-1}q\,dx.$$

Therefore

$$\frac{1}{\epsilon^2}E|\hat{h}^t - h^t|^2 \leq c(E\rho(t) + 1)$$

hence from (5.2.67)

$$E|\hat{h}^t - h^t|^2 \leq C\epsilon^2 + C_1\epsilon e^{-\gamma_0 t/\epsilon} + C_2 e^{-\gamma_1 t/\epsilon}. \tag{5.2.69}$$

On the other hand we have

$$|h(x) - h(\hat{x}_t) - H(\hat{x}_t)(x - \hat{x}_t)| \leq c|x - \hat{x}_t|^2$$

hence

$$\left|\int q(h(x) - h(\hat{x}_t))dx\right| = \left|\int q(h(x) - h(\hat{x}_t) - H(\hat{x}_t)(x - \hat{x}_t))dx\right|$$

$$\leq C\int q|x - \hat{x}_t|^2\,dx$$

i.e.

$$|\hat{h}^t - h(\hat{x}_t)| \leq C\int q|x - \hat{x}_t|^2\,dx.$$

Therefore

$$\begin{aligned}
E|\hat{h}^t - h(\hat{x}_t)|^2 &\leq CE\int q|x - \hat{x}_t|^4\,dx \\
&= CE|x_t - \hat{x}_t|^4 \\
&\leq CE|x_t - m_t|^4 \\
&\leq C\epsilon^2 + C\epsilon e^{-\gamma_0 t/\epsilon} + CE|\xi - m_0|^4 e^{-2\gamma_0 t/\epsilon}.
\end{aligned}$$

Using (5.2.69) we deduce

$$E|h(\hat{x}_t) - h(m_t)|^2 \leq C(\epsilon^2 + \epsilon e^{-\gamma_0 t/\epsilon} + e^{-\gamma_1 t/\epsilon})$$

and from (5.2.64) the desired result follows.   ∎

Let $\psi \in C_b^2(R^n)$; it is proposed to approximate

$$\hat{\psi}^t = E\left[\psi(x_t)|\mathcal{Z}^t\right]$$

by $\psi(\hat{x}_t)$. We then have

Corollary 5.2.1 *The following estimate holds:*

$$E|\hat{\psi}^t - \psi(m_t)| \leq C\|D\psi\|(\epsilon + \sqrt{\epsilon}e^{-\gamma_0 t/2\epsilon} + e^{-\gamma_1 t/2\epsilon})$$
$$+ C\|D^2\psi\|(\epsilon + e^{-\gamma_0 t/2\epsilon}).$$

Proof

We have

$$\hat{\psi}^t - \psi(m_t) = E\left[\psi(x_t) - \psi(m_t)|\mathcal{Z}^t\right]$$
$$= \psi(\hat{x}_t) - \psi(m_t)$$
$$+ E\left\{\int_0^1 \int_0^1 D^2\psi\left[\hat{x}_t + \lambda\mu(x_t - \hat{x}_t)\right]\lambda d\lambda d\mu(x_t - \hat{x}_t)^2 \,\bigg|\, \mathcal{Z}^t\right\}$$

hence

$$E|\hat{\psi}^t - \psi(m_t)| \leq \|D\psi\|E|\hat{x}_t - m_t| + \frac{1}{2}\|D^2\psi\|E|x_t - \hat{x}_t|^2$$

and from (5.2.68) the desired result follows. ■

Remark 5.2.2 If we specialize (5.2.46) to the linear case, i.e.

$$H(x) = Hx, \quad g(x) = Fx, \quad a = \frac{I}{2}$$

we obtain

$$\frac{d}{dt}E\rho(t) = -E\int q\sum_{i,\lambda}(u_{i\lambda})^2 dx + E\int qDu^*\left[-F - \frac{2}{\epsilon}H^*(HH^*)^{-1/2}H\right]Dudx$$
$$+ \frac{1}{\epsilon}E\int q(x - m_t)^*\left[H^*(HH^*)^{-1/2}HF + F^*H^*(HH^*)^{-1/2}H\right]Dudx$$
$$\leq -\frac{C_0}{\epsilon}E\rho(t) + C_1 + \frac{C_2}{\epsilon}e^{-\gamma_0 t/\epsilon}$$

and we recover the estimate of Lemma 5.1.3, which is better than the general one, since many terms disappear. In Section 5.1, the proof was a direct one, not relying on the Zakai and Kushner equations.

Remark 5.2.3 The estimate obtained in Theorem 5.1.3 is more precise than the one desired from Corollary 5.2.1. We used there a real approximation of the conditional probability law, which is not the case in Corollary 5.2.1.

5.2.5   Another approximation

From this approach developed previously, it is clear that the role of the function $\psi(x,t)$ introduced in (5.2.27) has been instrumental. This is in some sense an approximation to $-2\epsilon \log p$. But the approximation has been used only in a very specific context, namely to majorize the integral $\int q(x_t - m_t)^2 dx$.

Let us check here that we could have used a different one to get the same result, namely

$$\psi(x,t) = (x - m_t)^* \beta(m_t)(x - m_t) \tag{5.2.70}$$

where

$$\beta(x) = \nu^{-1}(x) = H^* \phi H(x).$$

In some sense, we consider here a linearized version of (5.2.27). Again let u be defined by

$$u(x,t) = \zeta(x,t) + \frac{1}{2\epsilon}\psi(x,t). \tag{5.2.71}$$

Then we have

$$
\begin{aligned}
du = \Big\{ & a_{\lambda\mu}u_{\lambda\mu} + a_{\lambda\mu}u_\lambda u_\mu - u_\lambda\Big[g_\lambda - 2a_{\lambda\mu,\mu} + \frac{2}{\epsilon}a_{\lambda\mu}\beta^t_{\mu j}(x_j - m^t_j)\Big] \\
& + \frac{1}{\epsilon^2}\Big[a_{\lambda\mu}\beta^t_{\lambda j}\beta^t_{\mu k}(x_j - m^t_j)(x_k - m^t_k) - \frac{1}{2}|h|^2 + h_j h^t_j\Big] \\
& + \frac{1}{\epsilon}\Big[\frac{1}{2}\beta^t_{ij,k}g^t_k(x_i - m^t_i)(x_j - m^t_j) + (g_k - g^t_k)\beta^t_{kj}(x_j - m^t_j) \\
& + \frac{1}{2}\beta^t_{ij,kl}a^t_{kl}(x_i - m^t_i)(x_j - m^t_j) - 2a_{\lambda\mu,\mu}\beta^t_{\lambda j}(x_j - m^t_j) - a_{\lambda\mu}\beta^t_{\lambda\mu} + \frac{1}{2}\gamma^t_{\lambda\lambda}\Big] \\
& - g_{\lambda,\lambda} + a_{\lambda\mu,\lambda\mu}\Big\}dt \\
& + \Big[\frac{1}{2}\beta^t_{ij,k}\theta^t_{kl}(x_i - m^t_i)(x_j - m^t_j) - \beta^t_{ij}\theta^t_{il}(x_j - m^t_j) + h_l\Big](dz_l - h^t_l dt).
\end{aligned}
\tag{5.2.72}
$$

We obtain from this

$$
\begin{aligned}
du_i = \Big(& a_{\lambda\mu}u_{\lambda\mu i} + a_{\lambda\mu,i}a_{\lambda\mu} + a_{\lambda\mu}u_\lambda u_\mu + 2a_{\lambda\mu}u_{\lambda i}u_\mu \\
& - u_{\lambda i}\Big[g_\lambda - 2a_{\lambda\mu,\mu} + \frac{2}{\epsilon}a_{\lambda\mu}\beta^t_{\mu j}(x_j - m^t_j)\Big] \\
& - u_\lambda\Big[g_{\lambda,i} - 2a_{\lambda\mu,\mu i} + \frac{2}{\epsilon}a_{\lambda\mu,i}\beta^t_{\mu j}(x_j - m^t_j) + \frac{2}{\epsilon}a_{\lambda\mu}\beta^t_{\mu i}\Big] \\
& + \frac{1}{\epsilon^2}\Big\{a_{\lambda\mu,i}\beta^t_{\lambda j}\beta^t_{\mu k}(x_j - m^t_j)(x_k - m^t_k) + 2(a_{\lambda\mu} - a^t_{\lambda\mu})\beta^t_{\lambda j}\beta^t_{\mu i}(x_j - m^t_j) \\
& - (h_{j,i} - h^t_{j,i})[h_j - h^t_j) - h^t_{j,i}(h_j - h^t_j - h^t_{j,k}(x_k - x^t_k)]\Big\} \\
& + \frac{1}{\epsilon}\Big[\beta^t_{ij,k}g^t_k(x_j - m^t_j) + (g_{k,i}\beta^t_{kj} + \beta^t_{ij,kl})(x_j - m^t_j) \\
& + (g_k - g^t_k)\beta^t_{ki} - a_{\lambda\mu,i}\beta^t_{\lambda\mu} - 2a_{\lambda\mu,\mu}\beta^t_{\lambda i}\Big] - g_{\lambda,\lambda i} + a_{\lambda\mu,\mu i}\Big)dt \\
& + \frac{1}{\epsilon^2}\Big[\beta^t_{ij,k}\theta^t_{kl}(x_j - m^t_j) + h_{l,i} - h^t_{l,i}\Big](dz_l - h^t_l dt)
\end{aligned}
\tag{5.2.73}
$$

and for the energy

$$\rho(t) = \int q a_{ir} u_i u_r dx$$

the Ito differential

$$
\begin{aligned}
d\rho = &- 2 \int q(a_{\lambda\mu} a_{ir} u_{i\lambda} u_{r\mu} + a_{ir} a_{\lambda\mu,i} u_\lambda u_{r\mu}) dx dt \\
&+ 2 \int q u_r u_\lambda \left[\frac{1}{2} (g_i a_{\lambda r,i} + a_{i\mu} a_{\lambda r,i\mu}) - a_{ir,\mu} a_{\lambda\mu,i} \right. \\
&\left. - a_{ir} (g_{\lambda,i} - a_{\lambda\mu,\mu i}) - \frac{2}{\epsilon} a_{ir} a_{\lambda\mu} \beta^t_{\mu i} \right] dx dt \\
&+ \frac{2}{\epsilon} \int q a_{ir} u_{r\lambda} a_{\lambda\mu,i} \beta^t_{\mu j} (x_j - m^t_j) dx dt \\
&+ \frac{2}{\epsilon^2} \int q a_{ir} u_r \left\{ 2(a_{\lambda\mu} - a^t_{\lambda\mu}) \beta^t_{\lambda j} \beta^t_{\mu i} (x_j - m^t_j) - (h_{j,i} - h^t_{j,i})(h_j - h^t_j) \right. \\
&\quad - h^t_{j,i} [h_j - h^t_j - h^t_{j,k} (x_k - m^t_k)] \\
&\quad \left. + (h_l - h^t_l) \left[\beta^t_{ij,k} \theta^t_{kl} (x_j - m^t_j) + h_{l,i} - h^t_{l,i} \right] \right\} dx dt \\
&+ \frac{2}{\epsilon} \int q u_r \left\{ \left[(a_{ir,\lambda} a_{\lambda\mu,i} + a_{ir} a_{\lambda\mu,i\lambda}) \beta^t_{\mu j} (x_j - m^t_j) + a_{ir} a_{\lambda\mu,i} \beta^t_{\mu\lambda} \right] \right. \\
&\quad + a_{ir} \left[(\beta^t_{ij,k} g^t_k + g_{k,i} \beta^t_{kj} + \beta^t_{ij,kl})(x_j - m^t_j) + (g_k - g^t_k) \beta^t_{ki} - a_{\lambda\mu,i} \beta^t_{\lambda\mu} \right] \\
&\quad \left. - 2 a_{\lambda\mu,\mu} \beta^t_{\lambda i} \right] \right\} dx dt \\
&+ 2 \int q a_{ir} u_r (-g_{\lambda,\lambda i} + a_{\lambda\mu,\lambda\mu i}) dx dt \\
&+ \frac{1}{\epsilon^2} \int q a_{ir} \left[\beta^t_{ij,k} \theta^t_{kl} (x_j - m^t_j) + h_{l,i} - h^t_{l,i} \right] \\
&\quad \times \left[\beta^t_{r\lambda,\mu} \theta^t_{\mu l} (x_l - m^t_l) + h_{l,r} - h^t_{l,r} \right] dx dt \\
&+ \frac{1}{\epsilon^2} \int q \left\{ a_{ir} u_i u_r (h_l - \hat{h}_l) + 2 a_{ir} u_r \left[\beta^t_{ij,k} \theta^t_{kl} (x_j - m^t_j) + h_{l,i} - h^t_{l,i} \right] \right\} \\
&\quad (dz_l - \hat{h}_l dt)
\end{aligned}
$$

$$(5.2.74)$$

and we can proceed as in the proof of Theorem 5.2.1, to prove again the estimate (5.2.68).

5.2.6   Generalization

Consider the following model

$$
\begin{aligned}
dx &= g(x_t) dt + \epsilon^\delta \sigma(x_t) dw, \quad x(0) = \xi \\
dy &= h(x_t) dt + \epsilon^{1-\delta} db
\end{aligned}
$$

$$(5.2.75)$$

with the same notation as in § 5.2.1 (see (5.2.6), (5.2.7)), except for the exponent of ϵ. Here we assume that $0 \le \delta < \frac{1}{2}$. The case $\frac{1}{2} \le \delta \le 1$ will be considered in the next section. The model (5.2.75) will behave very similarly to the case considered previously, which corresponds to $\delta = 0$.

The approximation to $\hat{x}_t$ will be defined by

$$dm_t = g(m_t) + \frac{\theta(m_t)}{\epsilon^{1-2\delta}}(dz - h(m_t)dt). \qquad (5.2.76)$$

One can study the accuracy of the approximation m_t to $\hat{x}_t$, by the same methods as before. We shall have

$$E\rho(t) \leq c\max(1, \epsilon^{1-4\delta}) \text{ up to initial layer terms}$$

and

$$E|\hat{x}_t - m_t|^2 \leq C\max(\epsilon^2, \epsilon^{3-4\delta}) \text{ up to initial layer terms.}$$

Therefore for $0 \leq \delta \leq \frac{1}{4}$ the error $E|\hat{x}_t - m_t|^2$ is of order ϵ^2, and of order $\epsilon^{3-4\delta}$ for $\frac{1}{4} \leq \delta < \frac{1}{2}$. It deteriorates in this interval as δ increases.

5.3 Dynamic systems with small noise and small signal to noise ratio

5.3.1 Setting of the problem

Consider here the following signal and noise processes

$$
\begin{aligned}
\mathrm{d}x_t &= g(x_t)\mathrm{d}t + \epsilon^\delta \sigma(x_t)\mathrm{d}w \\
\mathrm{d}z_t &= h(x_t)\mathrm{d}t + \epsilon^{1-\delta}\mathrm{d}b \\
x(0) &= x_0 + \epsilon^{1/2}\xi.
\end{aligned}
\tag{5.3.1}
$$

We assume that

$$
g : R^n \to R^n \text{ is } C^2 \text{ with bounded derivatives.} \tag{5.3.2}
$$

$$
\sigma : R^n \to L(R^n; R^n) \text{ is } C^3; \ \sigma \text{ and its derivatives are bounded.}
$$
$$
a = \tfrac{1}{2}\sigma\sigma^* \text{ is uniformly positive definite.} \tag{5.3.3}
$$

$$
h : R^n \to R^n \text{ is } C^2, \text{ with bounded derivatives.} \tag{5.3.4}
$$

There exists a probability space $(\Omega, \mathcal{A}, P)$ equipped with a filtration $\mathcal{F}^t$ and

$$
g_i \partial/\partial x_i a_{jk}, \ g_i \partial^2 h_k/\partial x_i \partial x_j \text{ are bounded } \forall\, j, k. \tag{5.3.4'}
$$

$w(t)$, $b(t)$ are two independent standard $\mathcal{F}^t$ Wiener processes

with values in R^n; ξ is a $\mathcal{F}^0$ measurable random variable with values in R^n;

the variable ξ has a probability distribution Π_0 which has a a density

with respect to Lebesgue measure given by p_0; the mean of ξ is 0.

$$\tag{5.3.5}$$

We assume that p_0 satisfies

$p_0 : R^n \to R^+ - \{0\}; \quad \int p_0(x)\mathrm{d}x = 1;$

p_0 is a smooth function; $\int |Dp_0|^2/p_0 \mathrm{d}x < \infty;$

$\int |x|^2 p_0 \mathrm{d}x < \infty;$ the variance $P_0 = E\xi\xi^* = \int xx^* p_0 \mathrm{d}x$ is an invertible matrix.

$$\tag{5.3.6}$$

The number δ satisfies

$$
\frac{1}{2} \leq \delta \leq 1 \tag{5.3.7}
$$

Hence conversely to what was considered in Section 5.2, the noise in the signal is somewhat smaller than the observation noise. When $\delta = 1$, the observation noise is not even small in itself, whereas the signal noise remains small.

We write the generator corresponding to the diffusion representing the signal as

$$A^2 = -g_i \frac{\partial}{\partial x_i} - \epsilon^{2\delta} a_{ij} \frac{\partial^2}{\partial x_i \partial x_j}$$

$$= A_0 - \epsilon^{2\delta} A_1$$

and the formal adjoint

$$(A^2)^* = A_0^* - \epsilon^{2\delta} A_1^*$$

with

$$A_0^* = \frac{\partial}{\partial x_i}(g_i \cdot), \quad A_1^* = -\frac{\partial^2}{\partial x_i \partial x_j}(a_{ij} \cdot).$$

We consider the probability $\tilde{P}$ on $(\Omega, \mathcal{A})$ such that

$$\left. \frac{\mathrm{d}\tilde{P}}{\mathrm{d}P} \right|_{\mathcal{F}^t} = \rho_t \tag{5.3.8}$$

with

$$\rho_t = \exp\left[\int_0^t -\frac{h^*(x_s)\mathrm{d}b}{\epsilon^{1-\delta}} - \frac{1}{2\epsilon^{2-2\delta}} \int_0^t |h(x_s)|^2 \mathrm{d}s \right]. \tag{5.3.9}$$

We define

$$\mathcal{Z}^t = \sigma(z(s), s \leq t).$$

On $(\Omega, \mathcal{A}, \tilde{P}, \mathcal{F}^t)$ the processes $w(\cdot)$ and $z(\cdot)$ are independent Wiener processes ($w(\cdot)$ is standard and $z(\cdot)$ has covariance $\epsilon^{2-2\delta} I$). Setting

$$\eta_t = \frac{1}{\rho_t}$$

we have

$$\eta_t = \exp\left[\int_0^t \frac{h^*(x_s)}{\epsilon^{2-2\delta}} \mathrm{d}z - \frac{1}{2\epsilon^{2-2\delta}} \int_0^t |h(x_s)|^2 \mathrm{d}s \right]. \tag{5.3.10}$$

The non linear filter is defined by the quantities

$$\Pi(t)(\psi) = E\left[\psi(x_t) | \mathcal{Z}^t \right]$$
$$p(t)(\psi) = \tilde{E}\left[\psi(x_t) \eta_t | \mathcal{Z}^t \right]. \tag{5.3.11}$$

5.3.2   The linear case

Consider the linear case corresponding to

$$g(x) = F(x), \quad h(x) = Hx, \quad \sigma(x) = \sigma$$

and set

$$Q = \sigma\sigma^* = 2a.$$

The Kalman filter is defined by the equations

$$\mathrm{d}\hat{x}_t = F\hat{x}_t \mathrm{d}t + \frac{P_\epsilon}{\epsilon^{2-2\delta}} H^*(\mathrm{d}z - H\hat{x}_t \mathrm{d}t) \tag{5.3.12}$$

$$\hat{x}_0 = x_0$$

where $P_\epsilon(t)$ is the solution of the Riccati equation

$$\frac{\mathrm{d}P_\epsilon}{\mathrm{d}t} = FP_\epsilon + P_\epsilon F^* - \frac{1}{\epsilon^{2-2\delta}} P_\epsilon H^* H P_\epsilon + \epsilon^{2\delta} Q$$
$$P_\epsilon(0) = \epsilon P_0. \tag{5.3.13}$$

If we consider the scaling

$$\frac{P_\epsilon(t)}{\epsilon} = \nu_\epsilon(t)$$

then we get for ν_ϵ the equation

$$\dot\nu_\epsilon = F\nu_\epsilon + \nu_\epsilon F^* - \epsilon^{2\delta-1}\nu_\epsilon H^* H \nu_\epsilon + \epsilon^{2\delta-1} Q$$
$$\gamma_\epsilon(0) = P_0 \tag{5.3.14}$$

and $\hat x_t$ is the solution of

$$\mathrm{d}\hat x_t = F\hat x_t \mathrm{d}t + \epsilon^{2\delta-1}\nu_\epsilon H^* (\mathrm{d}z - H\hat x_t \mathrm{d}t)$$
$$\hat x_0 = x. \tag{5.3.15}$$

The differences between this and the situation examined in Section 5.1 are clear. The system (5.3.14), (5.3.15) contains only *regular perturbation* terms, whereas in Section 5.1 we encountered singular perturbations. On the other hand, the memory is now important. That is why the assumption on the initial condition is necessary. Otherwise we would have kept a singularity as time evolves. If there is stability built into the system, this singularity could be removed for very large times only.

If we reason on an interval of time $[0, T]$ bounded, it is clear from (5.3.14) that

$$\|\nu_\epsilon(t)\| \leq C$$

therefore at any rate

$$\sup_{0 \leq t \leq T} E|x_t - \hat x_t|^2 \leq C\epsilon. \tag{5.3.16}$$

In the linear case, there is little which can be done. If $\delta = 1/2$, for instance, $\nu_\epsilon = \nu_1$ and no approximation whatsoever is available. If $\delta > 1/2$ we could think of approximating (5.3.14) by

$$\dot\nu = F\nu + \nu F^*, \quad \nu(0) = P_0$$

and approximate (5.3.15) replacing ν_ϵ by ν.

At any rate, to get a *significant result*, we must build a process m_t which should satisfy the following type of estimate

$$\frac{1}{\epsilon} \sup_{0 \leq t \leq T} E|\hat x_t - m_t|^2 \to 0 \text{ as } \epsilon \to 0. \tag{5.3.17}$$

It would be meaningless to build a m_t such that

$$\sup_{0 \leq t \leq T} E|\hat x_t - m_t|^2 \leq C\epsilon \tag{5.3.18}$$

since the average $\bar x_t$ defined by

$$\frac{\mathrm{d}}{\mathrm{d}t}\bar x_t = F\bar x_t, \quad \bar x_0 = x_0$$

is such that
$$\sup_{0 \le t \le T} E|x_t - \bar{x}_t|^2 \le C\epsilon$$
and thus, by (5.3.16), the average $\bar{x}_t$ already satisfies (5.3.18).

5.3.3   Formal calculations: the approximation

We proceed as in §5.2.2 and postulate that

$$\Pi(t)(\psi) = \int q(x,t)\psi(x)\mathrm{d}x \tag{5.3.19}$$

$$p(t)(\psi) = \int p(x,t)\psi(x)\mathrm{d}x \tag{5.3.20}$$

and the functions $p(x,t)$, $q(x,t)$ are solutions of the stochastic PDEs

$$\mathrm{d}p + A_0^*p + \epsilon^{2\delta}A_1^*p = \frac{p}{\epsilon^{2-2\delta}}h^*\mathrm{d}z \tag{5.3.21}$$
$$p(x,0) = p_0^\epsilon(x)$$

$$\mathrm{d}q + A_0^*q + \epsilon^{2\delta}A_1^*q = \frac{q}{\epsilon^{2-2\delta}}(h^* - \hat{h}^*)(\mathrm{d}z - \hat{h}\mathrm{d}t) \tag{5.3.22}$$
$$q(x,0) = p_0^\epsilon(x)$$

where

$$p_0^\epsilon(x) = p_0\left(\frac{x - x_0}{\epsilon^{1/2}}\right)\frac{1}{\epsilon^{n/2}}. \tag{5.3.23}$$

In order to guess what could be a good candidate to approximate $\hat{x}_t$, we compute the Ito differential of $\hat{x}_t$, by
$$\mathrm{d}\hat{x}_t = \int x\mathrm{d}q.$$

It is easy to check that

$$\mathrm{d}\hat{x}_t = \hat{g}\mathrm{d}t + \epsilon^{2\delta-2}p\left[\widehat{(x - \hat{x})(h - \hat{h})^*}\right](\mathrm{d}z - \hat{h}\mathrm{d}t), \quad \hat{x}_0 = x_0 \tag{5.3.24}$$

noting that

$$\left[\widehat{(x - \hat{x})(h - \hat{h})^*}\right] = \left[\widehat{(x - \hat{x})h^*}\right] = \left[\widehat{x(h - \hat{h})^*}\right].\dagger$$

Consider the second moment, or more precisely

$$\Gamma_t = \frac{1}{\epsilon}E\left[(x_t - \hat{x}_t)(x_t - \hat{x}_t)^*|\mathcal{Z}^t\right]$$
$$= \frac{1}{\epsilon}\int q(x,t)(x - \hat{x}_t)x^*\mathrm{d}x. \tag{5.3.25}$$

After some calculations we arrive at the Ito differential

$$\mathrm{d}\Gamma_t = \left\{\frac{1}{\epsilon}\left[\widehat{[g(x - \hat{x})^*]} + \widehat{[(x - \hat{x})g^*]}\right] + 2\epsilon^{2\delta-1}\hat{a} - \epsilon^{2\delta-3}\left[\widehat{(x - \hat{x})(h - \hat{h})^*}\right]\right.$$
$$\left.\left[\widehat{(h - \hat{h})(x - \hat{x})^*}\right]\right\}\mathrm{d}t + \epsilon^{2\delta-3}\left[\widehat{(x - \hat{x})(x - \hat{x})^*(h - \hat{h})^*}\right](\mathrm{d}z - \hat{h}\mathrm{d}t) \tag{5.3.26}$$

$$\Gamma_0 = P_0.$$

---

† We recall that $\hat{\psi} = \hat{\psi}_t = \Pi(t)(\psi)$.

Let us consider

$$F(x) = g'(x) \quad H(x) = h'(x).$$

We shall approximate the pair $\hat{x}_t$, Γ_t by m_t, ν_t defined by

$$dm_t = g(m_t)dt + \epsilon^{2\delta-1}\theta_t(dz - h(m_t)dt)$$
$$m_t = x_0$$
(5.3.27)

where

$$\theta_t = \nu_t H^*(m_t)$$
(5.3.28)

and ν_t is the solution of

$$\dot{\nu}_t = F(m_t)\nu_t + \nu_t F^*(m_t) - \epsilon^{2\delta-1}\nu_t H^*(m_t)H(m_t)\nu_t + 2\epsilon^{2\delta-1}a(m_t)$$
$$\nu_0 = P_0.$$
(5.3.29)

We consider $\beta_t = \nu_t^{-1}$ which is the solution of

$$\dot{\beta}_t = -\beta_t F(m_t) - F^*(m_t)\beta_t - 2\epsilon^{2\delta-1}\beta_t a(m_t)\beta_t + \epsilon^{2\delta-1}H^*(m_t)H(m_t)$$
$$\beta_0 = P_0^{-1}.$$
(5.3.30)

We introduce

$$\psi(x,t) = (x - m_t)^*\beta_t(x - m_t)$$
(5.3.31)

and recalling that

$$\zeta(x,t) = \log p(x,t)$$

we set

$$u(x,t) = \zeta(x,t) + \frac{1}{2\epsilon}\psi(x,t).$$
(5.3.32)

We compute several Ito differentials. We have first

$$d\zeta = \left[\epsilon^{2\delta}(a_{\lambda\mu}\zeta_{\lambda\mu} + a_{\lambda\mu}\zeta_\lambda\zeta_\mu + 2\zeta_\lambda a_{\lambda\mu,\mu} + a_{\lambda\mu,\lambda\mu}) - \zeta_\lambda g_\lambda - g_{\lambda,\lambda} - \frac{1}{2\epsilon^{2-2\delta}}|h|^2\right]dt$$
$$+ \frac{1}{\epsilon^{2-2\delta}}h^*dz.$$
(5.3.33)

We write (5.3.27) in scalar form to yield, with the usual notation,

$$dm_i^t = g_i^t dt + \epsilon^{2\delta-1}\theta_{ij}^t(dz_j - h_j^t dt).$$
(5.3.34)

Similarly we write (5.3.30) by components, which yields

$$\dot{\beta}_{ij}^t = \beta_{ik}^t g_{k,j}^t - \beta_{jk}^t g_{k,i}^t - 2\epsilon^{2\delta-1}\beta_{ik}^t a_{kl}^t\beta_{lj}^t + \epsilon^{2\delta-1}h_{k,i}^t h_{k,j}^t.$$
(5.3.35)

We can then write the differential of ψ as follows:

$$d\psi = \left[-2g_i^t\beta_{ij}^t(x_j - m_j^t) - (\beta_{ik}^t g_{k,j}^t + \beta_{jk}^t g_{k,i}^t + 2\epsilon^{2\delta-1}\beta_{ik}^t a_{kl}^t\beta_{lj}^t)\right.$$
$$\left. -\epsilon^{2\delta-1}h_{k,i}^t h_{k,j}^t(x_i - m_i^t)(x_j - m_j^t) + \epsilon^{2\delta}\nu_{ij}^t h_{k,i}^t h_{k,j}^t\right]dt$$
$$- 2\epsilon^{2\delta-1}h_{i,j}^t(x_j - m_j^t)(dz_i - h_i^t dt).$$
(5.3.36)

We note also that

$$u_\lambda = \zeta_\lambda + \frac{1}{\epsilon}\beta^t_{\lambda j}(x_j - m^t_j)$$

$$u_{\lambda\mu} = \zeta_{\lambda\mu} + \frac{1}{\epsilon}\beta^t_{\lambda\mu}.$$

The Ito differential of u is then given by

$$
\begin{aligned}
du =& \epsilon^{2\delta}(a_{\lambda\mu}u_{\lambda\mu} + a_{\lambda\mu}u_\lambda u_\mu) - u_\lambda\left[g_\lambda - 2\epsilon^{2\delta}a_{\lambda\mu,\mu} + 2\epsilon^{2\delta-1}a_{\lambda\mu}\beta^t_{\mu j}(x_j - m^t_j)\right] \\
&+ \frac{1}{\epsilon}\left[(g_i - g^t_i)\beta^t_{ij}(x_j - m^t_j) - \beta^t_{ik}g^t_{k,j}(x_i - m^t_i)(x_j - m^t_j)\right] \\
&+ \frac{1}{\epsilon^{2-2\delta}}\left[(a_{\lambda\mu} - a^t_{\lambda\mu})\beta^t_{\lambda j}\beta^t_{\mu k}(x_j - m^t_j)(x_k - m^t_k) - \frac{1}{2}|h|^2 + h^t_i h^t_{i,j}(x_j - m^t_j)\right. \\
&\left.+ \frac{1}{2}h^t_{k,i}h^t_{k,j}(x_i - m^t_i)(x_j - m^t_j) + (h_i - h^t_{i,j}(x_j - m^t_j))dz_i\right] \\
&- g_{\lambda,\lambda} + \epsilon^{2\delta-1}\left[-a_{\lambda\mu}\beta^t_{\lambda\mu} - 2a_{\lambda\mu,\mu}\beta^t_{\lambda j}(x_j - m^t_j) + \frac{1}{2}\nu^t_{ij}h^t_{k,i}h^t_{k,j}\right] + \epsilon^{2\delta}a_{\lambda\mu,\lambda\mu}.
\end{aligned}
$$

$$(5.3.37)$$

We now differentiate with respect to x_i and obtain

$$
\begin{aligned}
du_i =& \epsilon^{2\delta}(a_{\lambda\mu,i}u_{\lambda\mu} + a_{\lambda\mu}u_{\lambda\mu i} + a_{\lambda\mu,i}u_\lambda u_\mu + 2a_{\lambda\mu}u_{\lambda i}u_\mu) \\
&- u_{\lambda i}\left[g_\lambda - 2\epsilon^{2\delta}a_{\lambda\mu,\mu} + 2\epsilon^{2\delta-1}a_{\lambda\mu}\beta^t_{\mu j}(x_j - m^t_j)\right] \\
&- u_\lambda\left[g_{\lambda,i} - 2\epsilon^{2\delta}a_{\lambda\mu,\mu i} + 2\epsilon^{2\delta-1}a_{\lambda\mu,i}\beta^t_{\mu j}(x_j - m^t_j) + 2\epsilon^{2\delta-1}a_{\lambda\mu}\beta^t_{\mu i}\right] \\
&+ \frac{1}{\epsilon}\left[(g_{k,i} - g^t_{k,i})\beta^t_{kj}(x_j - m^t_j) + \beta^t_{ji}(g_j - g^t_j - g^t_{j,k}(x_k - m^t_k))\right] \\
&+ \frac{1}{\epsilon^{2-2\delta}}\left[a_{\lambda\mu,i}\beta^t_{\lambda j}\beta^t_{\mu k}(x_j - m^t_j)(x_k - m^t_k) + 2(a_{\lambda\mu} - a^t_{\lambda\mu})\beta^t_{\lambda j}\beta^t_{\mu i}(x_j - m^t_j)\right. \\
&- h_{j,i}(h_j - h^t_j - h^t_{j,k}(x_k - m^t_k)) - (h_{j,i} - h^t_{j,i})h^t_{j,k}(x_k - m^t_k) \\
&\left.+ (h_{j,i} - h^t_{j,i})(dz_j - h^t_j dt)\right] - g_{\lambda,\lambda i} \\
&+ \epsilon^{2\delta-1}\left[-a_{\lambda\mu,i}\beta^t_{\lambda\mu} - 2a_{\lambda\mu,\mu i}\beta^t_{\lambda j}(x_j - m^t_j) - 2a_{\lambda\mu,\mu}\beta^t_{\lambda i}\right] + \epsilon^{2\delta}a_{\lambda\mu,\lambda\mu i}.
\end{aligned}
$$

$$(5.3.38)$$

We define the 'energy' by setting

$$\rho(t) = \int qu_i a_{ir}u_r dx. \tag{5.3.39}$$

We express the Ito differential of $\rho(t)$. We proceed as in § 5.2.2. We use

$$
d\rho = \int dqu_i a_{ir}u_r dx + \int q(2a_{ir}u_r du_i + du_i a_{ir}du_r)dx
$$
$$
+ 2\int a_{ir}u_r dqdu_i dx
$$

and

$$
du_i du_r = \frac{1}{\epsilon^{2-2\delta}}(h_{j,i} - h^t_{j,i})(h_{j,r} - h^t_{j,r})
$$

$$
dqdu_i = \frac{1}{\epsilon^{2-2\delta}}q(h_{j,i} - h^t_{j,i})(h_j - \hat{h}_j).
$$

Consider

$$
\mathrm{d}\rho_1 = \int \mathrm{d}q u_i a_{ir} u_r \mathrm{d}x
$$

$$
+ 2 \int q a_{ir} u_r \Big\{ \epsilon^{2\delta} (a_{\lambda\mu,i} u_{\lambda\mu} + a_{\lambda\mu} u_{\lambda\mu i} + a_{\lambda\mu,i} u_\lambda u_\mu + 2 a_{\lambda\mu} u_{\lambda i} u_\mu)
$$

$$
- u_{\lambda i} \big[g_\lambda - 2\epsilon^{2\delta} a_{\lambda\mu,\mu} + 2 a_{\lambda\mu} \beta^t_{\mu j}(x_j - m^t_j) \big]
$$

$$
- u_\lambda \big[g_{\lambda,i} - 2\epsilon^{2\delta} a_{\lambda\mu,\mu i} + 2\epsilon^{2\delta-1} a_{\lambda\mu,i} \beta^t_{\mu j}(x_j - m^t_j) + 2\epsilon^{2\delta-1} a_{\lambda\mu} \beta^t_{\mu i} \big] \Big\} \mathrm{d}x \mathrm{d}t.
$$

Next

$$
\int \Big\{ u_i a_{ir} u_r \big[-(q g_\lambda)_{,\lambda} + \epsilon^{2\delta}(q a_{\lambda\mu})_{,\lambda\mu} \big] + 2\epsilon^{2\delta} q a_{ir} u_r a_{\lambda\mu} u_{\lambda\mu i} \Big\} \mathrm{d}x \mathrm{d}t
$$

$$
= \int q g_\lambda u_r (a_{ir,\lambda} u_i + 2 u_{i\lambda} a_{ir}) \mathrm{d}x \mathrm{d}t
$$

$$
+ \epsilon^{2\delta} \int q a_{\lambda\mu} \big[4(u_{i\lambda\mu} a_{ir} u_r + u_{i\lambda} a_{ir,\mu} u_r) + 2 u_{i\lambda} a_{ir} u_{ru} + u_i u_r a_{ir,\lambda\mu} \big] \mathrm{d}x \mathrm{d}t.
$$

We use

$$
q_\mu = q \left[u_\mu - \frac{1}{\epsilon} \beta^t_{\mu j}(x_j - m^t_j) \right]
$$

and deduce, as in § 5.2.2,

$$
\mathrm{d}\rho_1 = \int q \Big\{ g_\lambda a_{ir,\lambda} u_i u_r
$$

$$
+ \epsilon^{2\delta} (-2 a_{\lambda\mu} a_{ir} u_{i\lambda} u_{r\mu} + a_{\lambda\mu} a_{ir,\lambda\mu} u_i u_r
$$

$$
+ 2 a_{ir} a_{\lambda\mu,i} u_r u_{\lambda\mu} + 2 a_{ir} a_{\lambda\mu,i} u_r u_\lambda u_\mu)
$$

$$
- 2 a_{ir} u_r u_\lambda \big[g_{\lambda,i} - 2\epsilon^{2\delta} a_{\lambda\mu,\mu i}
$$

$$
+ 2\epsilon^{2\delta-1} a_{\lambda\mu} \beta^t_{\mu i} + 2\epsilon^{2\delta-1} a_{\lambda\mu,i} \beta^t_{\mu j}(x_j - m^t_j) \big] \Big\} \mathrm{d}x \mathrm{d}t
$$

$$
+ \int \frac{q}{\epsilon^{2-2\delta}} u_i a_{ir} u_r (h_j - \hat{h}_j)(\mathrm{d}z_j - \hat{h}_j \mathrm{d}t)
$$

and

$$d\rho = -2\epsilon^{2\delta} \int q(a_{\lambda\mu}a_{ir}u_{i\lambda}u_{r\mu} + a_{ir}a_{\lambda\mu,i}u_\lambda u_{r\mu})dxdt$$

$$+ 2\int qu_\lambda u_r \left[\frac{1}{2}(g_i a_{\lambda r,i} + \epsilon^{2\delta}a_{i\mu}a_{\lambda r,i\mu}) - \epsilon^{2\delta}a_{ir,\mu}a_{\lambda\mu,i}\right.$$

$$\left. - a_{ir}(g_{\lambda,i} - \epsilon^{2\delta}a_{\lambda\mu,\mu i}) - 2\epsilon^{2\delta-1}a_{ir}a_{\lambda\mu}\beta^t_{\mu i}\right]dxdt$$

$$+ 2\epsilon^{2\delta-1}\int qu_{r\lambda}a_{ir}a_{\lambda\mu,i}\beta^t_{\mu j}(x_j - m^t_j)dxdt$$

$$+ \frac{2}{\epsilon}\int qa_{ir}u_r\left[\beta^t_{ji}(g_j - g^t_j - g^t_{j,k}(x_k - m^t_k)) + \beta^t_{kj}(g_{k,i} - g^t_{k,i})(x_j - m^t_j)\right]dxdt$$

$$+ \frac{2}{\epsilon^{2-2\delta}}\int qa_{ir}u_r\left[2(a_{\lambda\mu} - a^t_{\lambda\mu})\beta^t_{\lambda j}\beta^t_{\mu i}(x_j - m^t_j)\right.$$

$$- h_{j,i}(h_j - h^t_j - h^t_{j,k}(x_k - m^t_k))$$

$$\left. - (h_{j,i} - h^t_{j,i})h^t_{j,k}(x_k - m^t_k) + (h_{j,i} - h^t_{j,i})(h_j - h^t_j)\right]dxdt$$

$$+ 2\int qa_{ir}u_r g_{\lambda,\lambda i}dxdt$$

$$+ 2\epsilon^{2\delta-1}\int qu_r\left\{(a_{ir,\lambda}a_{\lambda\mu,i} + a_{ir}a_{\lambda\mu,i\lambda}) \times \beta^t_{\mu j}(x_j - m^t_j)\right.$$

$$\left. + a_{ir}\left[-2a_{\lambda\mu,\mu}\beta^t_{\lambda i} - 2a_{\lambda\mu,\mu i}\beta^t_{\lambda j}(x_j - m^t_j)\right]\right\}dxdt$$

$$+ 2\epsilon^{2\delta}\int qa_{ir}u_r a_{\lambda\mu,\lambda\mu i}dxdt + \frac{1}{\epsilon^{2-2\delta}}\int qa_{ir}(h_{j,i} - h^t_{j,i})(h_{j,r} - h^t_{j,r})dxdt$$

$$+ \frac{1}{\epsilon^{2-2\delta}}\int q\left[u_i a_{ir}u_r(h_j - \hat{h}_j) + 2a_{ir}u_r(h_{j,i} - h^t_{j,i})\right](dz_j - \hat{h}_j dt).$$

$$(5.3.40)$$

Integrating between 0 and t and taking the mathematical expectation, then differentiating in t again, yields the relation

$$\frac{\mathrm{d}}{\mathrm{d}t}E\rho(t)$$

$$= -2\epsilon^{2\delta}E\int q(a_{\lambda\mu}a_{ir}u_{i\lambda}u_{r\mu} + a_{ir}a_{\lambda\mu,i}u_{\lambda}u_{r\mu})\mathrm{d}x$$

$$+ 2E\int qu_{\lambda}u_r\left[\frac{1}{2}(g_ia_{\lambda r,i} + \epsilon^{2\delta}a_{i\mu}a_{\lambda r,i\mu}) - \epsilon^{2\delta}a_{ir,\mu}a_{\lambda\mu,i}\right.$$

$$\left. - a_{ir}(g_{\lambda,i} - \epsilon^{2\delta}a_{\lambda\mu,\mu i}) - 2\epsilon^{2\delta-1}a_{ir}a_{\lambda\mu}\beta_{\mu i}^t\right]\mathrm{d}x$$

$$+ 2\epsilon^{2\delta-1}E\int qu_{r\lambda}a_{ir}a_{\lambda\mu,i}\beta_{\mu j}^t(x_j - m_j^t)\mathrm{d}x$$

$$+ \frac{2}{\epsilon}E\int qa_{ir}u_r\left[\beta_{ji}^t(g_j - g_j^t - g_{j,k}^t(x_k - m_k^t)) + \beta_{kj}^t(g_{k,i} - g_{k,i}^t)(x_j - m_j^t)\right]\mathrm{d}x$$

$$+ \frac{2}{\epsilon^{2-2\delta}}E\int qa_{ir}u_r\left[2(a_{\lambda\mu} - a_{\lambda\mu}^t)\beta_{\lambda j}^t\beta_{\mu i}^t(x_j - m_j^t)\right.$$

$$- h_{j,i}(h_j - h_j^t - h_{j,k}^t(x_k - m_k^t))$$

$$\left. -(h_{j,i} - h_{j,i}^t)h_{j,k}^t(x_k - m_k^t) + (h_{j,i} - h_{j,i}^t)(h_j - h_j^t)\right]\mathrm{d}x$$

$$+ 2E\int qa_{ir}u_rg_{\lambda,\lambda i}\mathrm{d}x + 2\epsilon^{2\delta-1}E\int qu_r\left\{(a_{ir,\lambda}a_{\lambda\mu,i} + a_{ir}a_{\lambda\mu,i\lambda})\beta_{\mu j}^t(x_j - m_j^t)\right.$$

$$\left. + a_{ir}\left[-2a_{\lambda\mu,\mu}\beta_{\lambda i}^t - 2a_{\lambda\mu,\mu i}\beta_{\lambda j}^t(x_j - m_j^t)\right]\right\}\mathrm{d}x$$

$$+ 2\epsilon^{2\delta}E\int qa_{ir}u_ra_{\lambda\mu,\lambda\mu i}\mathrm{d}x + \frac{1}{\epsilon^{2-2\delta}}E\int qa_{ir}(h_{j,i} - h_{j,i}^t)(h_{j,r} - h_{j,r}^t)\mathrm{d}x.$$

$$(5.3.41)$$

5.3.4 Statement of results

The justification of the calculations which yield (5.3.41) are made as in § 5.2.3 and will not be repeated here. Note in particular that the technical assumption (5.3.4′) is used at this stage.

We begin with

Lemma 5.3.1 *One has the estimate*

$$\sup_{0\leq t\leq T}E|x_t - m_t|^2 \leq C_T\epsilon \qquad (5.3.42)$$

$$\sup_{0\leq t\leq T}E|x_t - m_t|^4 \leq C_T\epsilon^2. \qquad (5.3.43)$$

Proof

We have from (5.3.1) and (5.3.27)

$$\mathrm{d}(x_t - m_t) = \left[g(x_t) - g(m_t) + \epsilon^{2\delta-1}\theta_t(h(x_t) - h(m_t))\right]\mathrm{d}t + \epsilon^{\delta}\sigma(x_t)\mathrm{d}w_t - \epsilon^{\delta}\theta\mathrm{d}b_t$$

hence

$$\frac{\mathrm{d}}{\mathrm{d}t}E|x_t - m_t|^2 = 2(x_t - m_t)^*[g(x_t) - g(m_t) + \epsilon^{2\delta-1}\theta_t(h(x_t) - h(m_t))]$$

$$+ \epsilon^{2\delta}\mathrm{tr}\,(\sigma\sigma^*(x_t) + \theta_t\theta_t^*).$$

Since $\delta \geq \frac{1}{2}$ it is easy to check that the solutions of the Riccati equations (5.3.29), (5.3.30) are locally bounded in time. The result (5.3.42) follows easily. The proof of (5.3.43) is similar.   ∎

We deduce the following estimate on the energy.

Lemma 5.3.2 *We have the estimate*

$$\sup_{0 \leq t \leq T} E\rho(t) \leq C_T \tag{5.3.44}$$

where C_T does not depend on ϵ.

Proof

We recall that (see (5.2.62))

$$a_{\lambda\mu} a_{ir} u_{i\lambda} r_{r\mu} \geq \alpha_0^2 \sum_{i,\lambda} (u_{i\lambda})^2$$

hence from (5.3.41) we deduce

$$\frac{\mathrm{d}}{\mathrm{d}t} E\rho(t) \leq - C_0 \epsilon^{2\delta} E \sum_{i,\lambda} \int q(u_{\lambda i})^2 \mathrm{d}x + C_1 E\rho(t)$$

$$+ C \epsilon^{2\delta - 2} E|x_t - m_t|^2 + \frac{C}{\epsilon^2} E|x_t - m_t|^4$$

$$+ \frac{C}{\epsilon^{4-4\delta}} E|x_t - m_t|^4 + C$$

and by virtue of Lemma 5.3.1

$$\frac{\mathrm{d}}{\mathrm{d}t} E\rho(t) \leq C_T(E\rho(t) + 1), \quad t \in [0, T].$$

Since $E\rho(0)$ is bounded in ϵ, the desired result obtains.   ∎

We can then deduce

Theorem 5.3.1 *We make the assumptions (5.3.2), (5.3.3), (5.3.4), (5.3.4'), (5.3.5), (5.3.6), (5.3.7); then one has*

$$\sup_{0 \leq t \leq T} E|\hat{x}_t - m_t|^2 \leq C\epsilon^2. \tag{5.3.45}$$

Proof

We have

$$Du = \frac{Dq}{q} + \frac{1}{\epsilon} \beta_t (x - m_t)$$

hence

$$q\nu_t Du = \nu_t Dq + \frac{1}{\epsilon} q(x - m_t).$$

Integrating in x yields

$$\frac{1}{\epsilon} (\hat{x}_t - m_t) = \int q\nu_t Du \, dx$$

and

$$\frac{1}{\epsilon^2} |\hat{x}_t - m_t|^2 \leq C_T \rho(t)$$

and (5.3.45) obtains.   ∎

5.4   Non linear filtering for dynamic systems with singular perturbations

5.4.1   Setting of the problem

We begin by stating the assumptions. Let

$$
\begin{aligned}
f(x,y) &: R^n \times R^p \to R^n \\
g(x,y) &: R^n \times R^p \to R^p \\
h(x,y) &: R^n \times R^p \to R^p
\end{aligned}
\tag{5.4.1}
$$

be C_2 functions, periodic in y (with period 1 in all components).†

> g and its derivatives (first and second) are bounded.
> The derivatives of f, h (first and second) are bounded.
> $\hspace{10cm}$ (5.4.2)

Let $(\Omega, \mathcal{A}, P)$ be a probability space equipped with a filtration $\mathcal{F}^t$ and let w_t^1, w_t^2, z_t be standard $\mathcal{F}^t$ Wiener processes independent one from each other. Let also

> ξ, y_0 be random variables $\mathcal{F}^0$ measurable with values in R^n, Y respectively.
> The probability law of the pair (ξ, y_0) is denoted by q_0 and the marginal probability law of ξ is Π_0.
> $\hspace{10cm}$ (5.4.3)

We define the processes x_t^ϵ, y_t^ϵ by solving the equations

$$
\begin{aligned}
\mathrm{d}x_t &= f(x_t, t_t)\mathrm{d}t + \sqrt{2}\mathrm{d}w_t^1 \\
\epsilon \mathrm{d}y_t &= g(x_t, y_t)\mathrm{d}t + \sqrt{2}\epsilon \mathrm{d}w_t^2 \\
x(0) &= \xi, \quad y(0) = y_0.
\end{aligned}
\tag{5.4.4}
$$

Next let η_t^ϵ be defined by

$$
\eta_t^\epsilon = \exp\left[\int_0^t h(x_s^\epsilon, y_s^\epsilon)dz_s - \frac{1}{2}\int_0^t |h(x_s^\epsilon, y_s^\epsilon)|^2 \mathrm{d}s \right]
\tag{5.4.5}
$$

i.e. η_t^ϵ has the Ito differential

$$
\begin{aligned}
\mathrm{d}\eta_t^\epsilon &= \eta_t^\epsilon h(x_t^\epsilon, y_t^\epsilon)dz_t \\
\eta_0^\epsilon &= 1.
\end{aligned}
\tag{5.4.6}
$$

---

† To fix the ideas, we denote by Y the torus $(0,1)^p$ identifying the opposite sides of the cube $(0,1)^p$.

As in Lemma 4.1.1 one can check that

$$E\eta_t^\epsilon = 1,$$

and thus we may define a new probability P^ϵ on $(\Omega, \mathcal{A})$ by setting

$$\left.\frac{\mathrm{d}P^\epsilon}{\mathrm{d}P}\right|_{\mathcal{F}^t} = \eta_t^\epsilon. \tag{5.4.7}$$

Consider next the process

$$b_t^\epsilon = z_t - \int_0^t h(x_s^\epsilon, y_s^\epsilon)\mathrm{d}s \tag{5.4.8}$$

then, on the setup $(\Omega, \mathcal{A}, P^\epsilon, \mathcal{F}^t), w_t^1, w_t^2, b_t^\epsilon$ become independent standard $\mathcal{F}^t$ Wiener processes and z_t can be written as

$$\begin{aligned}\mathrm{d}z_t &= h(x_t^\epsilon, y_t^\epsilon)\mathrm{d}t + \mathrm{d}b_t^\epsilon \\ z(0) &= 0\end{aligned} \tag{5.4.9}$$

and appears as the observation process corresponding to the state $(x_t^\epsilon, y_t^\epsilon)$.

We shall be interested in the unnormalized conditional probability $p^\epsilon(t)$, given by

$$p^\epsilon(t)(\psi) = E[\psi(x_t^\epsilon)\eta_t^\epsilon | \mathcal{Z}^t] \tag{5.4.10}$$

for any test function ψ (Borel bounded), and the normalized conditional probability $\Pi^\epsilon(t)$

$$\begin{aligned}\Pi^\epsilon(t)(\psi) &= \frac{p^\epsilon(t)(\psi)}{p^\epsilon(t)(1)} \\ &= E^\epsilon[\psi(x_t^\epsilon)|\mathcal{Z}^t].\end{aligned} \tag{5.4.11}$$

The question of interest is the behaviour of $p^\epsilon(t)(\psi)$ and $\Pi^\epsilon(t)(\psi)$ as ϵ tends to 0.

5.4.2   A duality argument

We shall use the results of Section 4.2. Let $\beta(t) \in L^\infty(0, T; R^p)$ and $\phi(x) \in \mathcal{C}_0^\infty(R^n)$. Consider the PDE

$$\begin{aligned}&-\frac{\partial v}{\partial t} - f(x, y) \cdot \mathrm{D}_x v - \frac{1}{\epsilon}g(x, y) \cdot \mathrm{D}_y v - \Delta_x v - \frac{1}{\epsilon}\Delta_y v \\ &+ ivh(x, y) \cdot \beta(t) = 0 \\ &v(x, y; T) = \phi(x).\end{aligned} \tag{5.4.12}$$

In fact (5.4.12) is a system; if $v = v_1 + iv_2$, then one has

$$\begin{aligned}&-\frac{\partial v_1}{\partial t} - f \cdot \mathrm{D}_x v_1 - \frac{1}{\epsilon}g \cdot \mathrm{D}_y v_1 - \Delta_x v_1 - \frac{1}{\epsilon}\Delta_y v_1 - v_2 h \cdot \beta = 0 \\ &-\frac{\partial v_2}{\partial t} - f \cdot \mathrm{D}_x v_2 - \frac{1}{\epsilon}g \cdot \mathrm{D}_y v_2 - \Delta_x v_2 - \frac{1}{\epsilon}\Delta_y v_2 + v_1 h \cdot \beta = 0 \\ &v_1(x, y, T) = \phi(x), \quad v_2(x, y, T) = 0.\end{aligned} \tag{5.4.13}$$

By the results of Section 4.2 (see Proposition 4.2.1) there exists a unique solution v_1^ϵ, v_2^ϵ of (5.4.13) such that

v_1^ϵ, v_2^ϵ and their first and second derivatives in x, y are bounded.

$$|v_{1,t}^\epsilon|, \ |v_{2,t}^\epsilon| \le C(1+|x|). \tag{5.4.14}$$

The constants entering in (5.4.14) depend on ϵ.

Consider the process associated to $\beta(t)$,

$$\theta(t) = \exp\left[i\int_0^t \beta(s)\mathrm{d}z + \frac{1}{2}\int_0^t |\beta|^2 \mathrm{d}s\right] \tag{5.4.15}$$

then we have proven in Theorem 4.2.1 that

$$E\theta(T)p^\epsilon(T)(\phi) = q_0(v_1^\epsilon(0)), \tag{5.4.16}$$

where $v_1^\epsilon(0)$ stands for $v_1(x,y,0)$, and we have emphasized the dependence on ϵ.

Therefore a major step in studying the convergence as ϵ tends to 0 is to study the limit of $v_1^\epsilon(x,y,0)$.

This is a classical problem of singular perturbations for a PDE, the solution of which is given in the next paragraph.

5.4.3   The limit of v_1^ϵ, v_2^ϵ

Let us consider the invariant measure $m(x,y)$ which is the solution of

$$-\Delta_y m + \mathrm{div}_y(mg) = 0, \quad m \in H^1(Y), \quad \text{m periodic} \tag{5.4.17}$$

where $Y = (0,1)^p$.

Since g is bounded, there exists for any x a unique solution of (5.4.17) satisfying

$$\int_Y m(x,y)\mathrm{d}y = 1, \quad \forall x \tag{5.4.18}$$

and moreover one has the estimates

$$0 < \delta_0 \le m(x,y) \le \delta_1 \tag{5.4.19}$$

where δ_0, δ_1 depend only on the bound of g. (For details see for instance Bensoussan (1988).) Let us prove some regularity results on $m(x,y)$.

Lemma 5.4.1 *The function m satisfies*

$$\begin{aligned}
\|m(x,\cdot)\|_{W^{2,p}(Y)} &\le C_p \ \forall x, \ 2 \le p < \infty \\
\|m_\lambda(x,\cdot)\|_{W^{2,p}(Y)} &\le C_p \ \forall x
\end{aligned} \tag{5.4.20}$$

where $m_\lambda = \partial m/\partial x_\lambda$.

Proof

We can write (5.4.17) as

$$-\Delta_y m + D_y m \cdot g + (\gamma + \mathrm{div}g)m = \gamma m \tag{5.4.21}$$

where γ is sufficiently large that the second order elliptic operator on the left hand side is strongly elliptic. Since the right hand side belongs to $L^p(Y)$, $\forall p$, the regularity results on elliptic equations yield that $m(x, \cdot)$ belongs to $W^{2,p}(Y)$ for any x. Since the constants δ_0, δ_1 in (5.4.19) do not depend on x, the first result in (5.4.20) obtains.

Let now

$$m_\lambda^h(x, y) = \frac{m(x + he_\lambda, y) - m(x, y)}{h}$$

where e_λ is the λ unit vector corresponding to the x variable. We deduce from (5.4.17) that

$$-\Delta_y m_\lambda^h + \mathrm{div}_y(m_\lambda^h g) + h\mathrm{div}_y(m_\lambda^h g_\lambda^h) + \mathrm{div}_y(m g_\lambda^h) = 0 \qquad (5.4.22)$$

where

$$g_\lambda^h = \frac{g(x + he_\lambda, y) - g(x, y)}{h}$$

and from (5.4.18)

$$\int_Y m_\lambda^h(x, y)\mathrm{d}y = 0, \quad \forall x. \qquad (5.4.23)$$

Let us first check that

$$\|m_\lambda^h(x)\|_{L^p(Y)} \leq C_p, \quad \forall x, \ 2 \leq p < \infty, \qquad (5.4.24)$$

where $m_\lambda^h(x)$ is the function $y \to m_\lambda^h(x; y)$.

Indeed let $\phi(y) \in L^q(Y)$, $1/p + 1/q = 1$ and let $\bar{\phi}(x) = \int_Y \phi(y) m(x, y)\mathrm{d}y$. Consider the equation

$$-\Delta_y \zeta - g \cdot D_y \zeta = \phi - \bar{\phi}. \qquad (5.4.25)$$

By Fredholm's alternative there exists a unique solution of (5.4.25) in $H^1(Y)$, such that $\int_Y \zeta(x, y)\mathrm{d}y = 0$, $\forall x$.

By regularity results for elliptic equations, the solution ζ is in fact in $W^{2,p}(Y)$ and since the map $\phi \to \zeta$ has a closed graph it is continuous. Therefore

$$\|\zeta(x)\|_{W^{2,p}(Y)} \leq C_p \|\phi\|_{L^p(Y)}, \quad \forall p, \qquad (5.4.26)$$

the constant C_p being independent of x. Now comparing (5.4.22) and (5.4.25) and making use of (5.4.23) we have

$$\int_Y m_\lambda^h(x, y)\phi(y)\mathrm{d}y = \int_Y D_y\zeta \left(h g_\lambda^h m_\lambda^h + g_\lambda^h m\right) \mathrm{d}y. \qquad (5.4.27)$$

Note that g_λ^h is bounded. By making use of (5.4.26) it is easy to deduce from (5.4.27) that

$$\left| \int_Y m_\lambda^h(x, y)\phi(y)\mathrm{d}y \right| \leq C_p \|\phi\|_{L^q(Y)} (h\|m_\lambda^h(x)\|_{L^p(Y)} + 1).$$

Since ϕ is arbitrary we deduce

$$\|m_\lambda^h(x)\|_{L^p(Y)} \leq C_p (h\|m_\lambda^h(x)\|_{L^p(Y)} + 1).$$

Since h is small the result (5.4.24) obtains.

Making use of (5.4.24) in (5.4.22) and operating as for (5.4.21), we deduce the second estimate, (5.4.20). ∎

We turn now to (5.4.13) and first obtain a *priori* estimates. In order to cope with the non boundedness of f, we introduce the function

$$G(x,t) = e^{Kt}(1 + |x|^2)^{1/2}$$

and change v_1, v_2 into

$$u_1 = v_1 e^{-G}, \quad u_2 = v_2 e^{-G}.$$

Note that

$$\frac{\partial v_1}{\partial t} = \left(\frac{\partial u_1}{\partial t} + u_1 KG\right) e^G$$

$$D_x v_1 = (D_x u_1 + u_1 D_x G)e^G$$

$$\Delta_x v_1 = \left[\Delta_x u_1 + 2D_x u_1 \cdot D_x G + u_1(\Delta_x G + |D_x G|^2)\right] e^G$$

hence (5.4.13) yields

$$-\frac{\partial u_1}{\partial t} - (f + 2D_x G) \cdot D_x u_1 - \frac{1}{\epsilon}g \cdot D_y u_1 - \Delta_x u_1 - \frac{1}{\epsilon}\Delta_y u_1$$

$$+ u_1 \gamma + u_2 h \cdot \beta = 0$$

$$-\frac{\partial u_2}{\partial t} - (f + 2D_x G) \cdot D_x u_2 - \frac{1}{\epsilon}g \cdot D_y u_2 - \Delta_x u_2 - \frac{1}{\epsilon}\Delta_y u_2 \qquad (5.4.28)$$

$$+ u_2 \gamma - u_1 h \cdot \beta = 0$$

$$u_1(x,y,T) = \phi(x)e^{-G(x,T)}, \quad u_2(x,y,T) = 0$$

where

$$\gamma(x,y,t) = KG - f \cdot D_x G - \Delta_x G - |D_x G|^2. \qquad (5.4.29)$$

Lemma 5.4.2 *One has the estimates*

$$\|v_1^\epsilon\|_{L^\infty}, \quad \|v_2^\epsilon\|_{L^\infty} \leq \|\phi\|_{L^\infty} \qquad (5.4.30)$$

$$\|u_k^\epsilon\|_{L^2}, \quad \|D_x u_k^\epsilon\|_{L^2} \leq C$$

$$\|D_y u_k^\epsilon\|_{L^2} \leq C\epsilon, \quad k = 1,2. \qquad (5.4.31)$$

Proof

The estimates (5.4.30) follow from the calculations made already in the proof of Proposition 4.2.1.

Let us proceed with the proof of (5.4.31). We multiply the u_1 equation by $u_1 m$ and the u_2 equation by $u_2 m$. We integrate in x,y to obtain, taking account of the equation for m,

$$-\frac{d}{dt}\int\int \frac{1}{2}m(u_1^2 + u_2^2)dxdy + \int\int m(|D_x u_1|^2 + |D_x u_2|^2)dxdy$$

$$+ \frac{1}{\epsilon}\int\int m(|D_y u_1|^2 + |D_y u_2|^2)dxdy + \int\int D_x m(u_1 D_x u_1 + u_2 D_x u_2)dxdy$$

$$+ \int\int (u_1^2 + u_2^2)\left[m\gamma + \frac{1}{2}m(\text{div}_x f + 2\Delta_x G) + \frac{1}{2}(f + 2D_x G) \cdot D_x m\right]dxdy = 0.$$

$$(5.4.32)$$

Note that

$$
\begin{aligned}
E =\,& m\gamma + \frac{1}{2}m(\operatorname{div}_x f + 2\Delta_x G) + \frac{1}{2}(f + 2D_x G) \cdot D_x m \\
=\,& m\left[KG + \frac{1}{2}\operatorname{div} f - f \cdot DG - |D_x G|^2 \right] + \left(\frac{1}{2}f + D_x G \right) \cdot D_x m \\
=\,& m\left[Ke^{Kt}(1 + |x|^2)^{1/2} + \frac{1}{2}\operatorname{div} f - e^{Kt} f \cdot x(1 + |x|^2)^{-1/2} \right. \\
& \left. - e^{2Kt}|x|^2(1 + |x|^2)^{-1} \right] + \left[\frac{1}{2}f + e^{Kt} x(1 + |x|^2)^{-1/2} \right] \cdot D_x m.
\end{aligned}
$$

By Lemma 5.4.1 $|D_x m| \leq C$, and $0 < \delta_0 \leq m \leq \delta_1$, hence for K sufficiently large

$$
\begin{aligned}
& m[Ke^{Kt}(1 + |x|^2)^{1/2} - e^{Kt} f \cdot x(1 + |x|^2)^{-1/2}] + \frac{1}{2}f \cdot D_x m \\
& \geq \delta_0 Ke^{Kt}(1 + |x|^2)^{1/2} - Ce^{Kt}(1 + |x|^2)^{1/2} \geq \delta_0 K - C
\end{aligned}
$$

therefore

$$
\begin{aligned}
c \geq\,& m\left[\frac{1}{2}\operatorname{div} f - e^{2Kt}|x|^2(1 + |x|^2)^{-1} \right] + e^{Kt} x \cdot D_x m(1 + |x|^2)^{-1/2} \\
\geq\,& - C_0 + \delta_0 K - C.
\end{aligned}
$$

Use also the fact that

$$
\begin{aligned}
\int\int D_x m(u_1 D_x u_1 + u_2 D_x u_2)\,dx\,dy \geq\,& - c_1 \int\int m(|D_x u_1|^2 + |D_x u_2|^2)\,dx\,dy \\
& - \frac{1}{4c_1} \int\int (u_1^2 + u_2^2)\frac{|D_x m|^2}{m}\,dx\,dy
\end{aligned}
$$

with c_1 arbitrary. Collecting results, one deduces from (5.4.32) the estimates (5.4.31).

∎

Lemma 5.4.3 We have for $k = 1, 2$

$$
u_k^\epsilon \to u_k \quad \text{in } L^2(0, T; H^1(R^n \times Y)) \text{ weakly} \tag{5.4.33}
$$

where u_1, u_2 are the solutions of

$$
\begin{aligned}
& -\frac{\partial u_1}{\partial t} - (\bar{f} + 2D_x G) \cdot D_x u_1 - \Delta_x u_1 + u_1 \bar{\gamma} + u_2 \bar{h} \cdot \beta = 0 \\
& -\frac{\partial u_2}{\partial t} - (\bar{f} + 2D_x G) \cdot D_x u_2 - \Delta_x u_2 + u_2 \bar{\gamma} - u_1 \bar{h} \cdot \beta = 0 \\
& u_1(x, T) = \phi(x)e^{-G(x,T)} \\
& u_2(x, T) = 0,
\end{aligned} \tag{5.4.34}
$$

u_1, u_2 do not depend on y, and where $\bar{f}$, $\bar{\gamma}$, $\bar{h}$ are defined by

$$
\bar{f}(x) = \int_Y f(x, y)m(x, y)\,dy
$$

and similarly for $\bar{\gamma}, \bar{h}$.

Proof

Choose a subsequence u_k^ϵ converging weakly in $L^2(0, T; H^1(R^n \times Y))$, to u_k, $k = 1, 2$. Let $\mu(x) \in C_0^\infty(R^n)$, and $\theta(t) \in C^1[0, T]$, with $\theta(0) = 0$. Multiply (5.4.28) by $m\mu\theta$ and integrate in x, y, t. We deduce

$$\int_0^T \mathrm{d}t\theta' \int \int u_1^\epsilon m\mu \mathrm{d}x \mathrm{d}y - \int_0^T \mathrm{d}t\theta \int \int (f + 2\mathrm{D}_x G) \cdot \mathrm{D}_x u_1^\epsilon m\mu \mathrm{d}x \mathrm{d}y$$

$$- \int_0^T \mathrm{d}t\theta \int \int \mathrm{D}_x u_1^\epsilon (\mathrm{D}_x m\mu + m\mathrm{D}_x \mu) \mathrm{d}x \mathrm{d}y$$

$$+ \int_0^T \mathrm{d}t\theta \int \int (u_1^\epsilon \gamma + u_2^\epsilon h \cdot \beta) m\mu \mathrm{d}x \mathrm{d}y - \int \phi(x) \mathrm{e}^{-G(x,T)} \mu \mathrm{d}x = 0$$

and a similar relation for u_2^ϵ. It is possible to pass to the limit in (5.4.35). Noting that the limit u_1, u_2 does not depend on y because of the last estimate (5.4.31) and taking account of (5.4.18), we obtain the relation

$$\int_0^T \mathrm{d}t\theta' \int u_1 \mu \mathrm{d}x - \int_0^T \mathrm{d}t\theta \int (\bar{f} + 2\mathrm{D}_x G) \cdot \mathrm{D}_x u_1 \mathrm{d}x$$

$$- \int_0^T \mathrm{d}t\theta \int \mathrm{D}_x u_1 \mathrm{D}_x \mu \mathrm{d}x + \int_0^T \mathrm{d}t\theta \int (u_1 \bar{\gamma} + u_2 \bar{h} \cdot \beta) \mu \mathrm{d}x \qquad (5.4.36)$$

$$- \int \phi \mathrm{e}^{-G(x,T)} \mu \mathrm{d}x = 0.$$

It follows that u_1 is the solution of the first equation of (5.4.34), and similarly u_2 is the solution of the second equation of (5.4.34). Note that

$$-\partial u_k / \partial t - \Delta_x u_k \in L^2(0, T; H^1(R^n)), \quad k = 1, 2.$$

Moreover $u_k \in L^2(0, T; H^1(R^n))$, hence $\Delta u_k \in L^2(0, T; H^{-1}(R^n))$. Therefore $\frac{\partial u_k}{\partial t} \in L^2(0, T; H^{-1}(R^n))$. This implies that $u_k(x, t)$ has a meaning for any t as an element of $L^2(R^n)$. Multiplying the first equation of (5.4.34) by θ_μ and integrating over x, t, we deduce after comparison with (5.4.36) that

$$u_1(x, T) = \phi(x) \mathrm{e}^{-G(x,T)}.$$

Similarly

$$u_2(x, y, T) = 0.$$

Since the solution of (5.4.34) is unique, the whole sequence converges.

Lemma 5.4.4 *We have for* $k = 1, 2$

$$u_k^\epsilon \to u_k \ \ in \ \ L^2(0, T; H^1(R^n \times Y)) \ \ strongly$$
$$u_k^\epsilon(t) \to u_k(t) \ \ in \ \ L^2(R^n \times Y) \ \ strongly, \ \ \forall t \in [0, T]. \qquad (5.4.37)$$

Proof

Let α be a positive number sufficiently large to ensure

$$E + \alpha m \geq -c_0 + \alpha \delta_0 = \alpha_1 > 0.$$

From the energy equality (5.4.32) we first deduce that

$$-\frac{d}{dt}e^{\alpha t}\int\int\frac{1}{2}m((u_1^\epsilon)^2+(u_2^\epsilon)^2)dxdy+e^{\alpha t}\int\int m(|D_xu_1^\epsilon|^2+|D_xu_2^\epsilon|^2)dxdy$$

$$+\frac{e^{\alpha t}}{\epsilon}\int\int m(|D_yu_1^\epsilon|^2+|D_yu_2^\epsilon|^2)dxdy+e^{\alpha t}\int\int((u_1^\epsilon)^2+(u_2^\epsilon)^2)(E+m\alpha)dxdy$$

$$+e^{\alpha t}\int\int D_xm(u_1^\epsilon D_xu_1^\epsilon+u_2^\epsilon D_xu_2^\epsilon)dxdy=0.$$

Therefore, integrating between t and T yields

$$e^{\alpha t}\int\int\frac{1}{2}m((u_1^\epsilon)^2+(u_2^\epsilon)^2)(x,y,t)dxdy+\int_t^T e^{\alpha s}\int\int m(|D_xu_1^\epsilon|^2+|D_xu_2^\epsilon|^2)dxdyds$$

$$+\frac{1}{\epsilon}\int_t^T e^{\alpha s}\int\int m(|D_yu_1^\epsilon|^2+|D_yu_2^\epsilon|^2)dxdyds$$

$$+\int_t^T e^{\alpha s}\int\int((u_1^\epsilon)^2+(u_2^\epsilon)^2)(E+m\alpha)dxdyds$$

$$+\int_t^T e^{\alpha s}\int\int D_xm(u_1^\epsilon D_xu_1^\epsilon+u_2^\epsilon D_xu_2^\epsilon)dxdy$$

$$=e^{\alpha T}\int\frac{1}{2}\phi^2(x)e^{-2G(x,T)}dx.$$

$$(5.4.38)$$

On the other hand, from (5.4.34) we also deduce †

$$e^{\alpha t}\int\frac{1}{2}((u_1)^2+(u_2)^2)(x,t)dx+\int_t^T e^{\alpha s}\int(|D_xu_1|^2+|D_xu_2|^2)dxds$$

$$+\int_t^T e^{\alpha s}\int((u_1)^2+(u_2)^2)(\hat{E}+\alpha)dxds=e^{\alpha T}\int\frac{1}{2}\phi^2e^{-G(x,T)}dx.$$

$$(5.4.39)$$

Now multiply (5.4.28) by mu_1^ϵ, mu_2^ϵ respectively. We also multiply (5.4.34) by mu_1^ϵ, mu_2^ϵ. After integrating and adding up we obtain

$$e^{\alpha t}\int\int m(u_1^\epsilon u_1+u_2^\epsilon u_2)(x,y,t)dxdy$$

$$+2\int_t^T e^{\alpha s}\int\int m(D_xu_1^\epsilon D_xu_1+D_xu_2^\epsilon D_xu_2)dxdyds$$

$$+2\int_t^T e^{\alpha s}\int\int(u_1u_1^\epsilon+u_2u_2^\epsilon)(E+m\alpha)dxdyds$$

$$+\int_t^T e^{\alpha s}\int\int D_xmD_x(u_1^\epsilon u_1+u_2^\epsilon u_2)dxdyds$$

$$(5.4.40)$$

$$+\int_t^T e^{\alpha s}\int\int(f-\bar{f})m(D_xu_1u_1^\epsilon+D_xu_2u_2^\epsilon)dxdyds$$

$$-\int_t^T e^{\alpha s}\int\int(\gamma-\bar{\gamma})m(u_1u_1^\epsilon+u_1u_2^\epsilon)dxdyds$$

$$+\int_t^T e^{\alpha s}\int\int(u_2^\epsilon u_1-u_1^\epsilon u_2)(h-\bar{h})\beta mdxdyds=\int\phi^2(x)e^{-2G(x,T)}dx.$$

---

† Here $\hat{E}=\int_Y E\,dY$.

Combining (5.4.38), (5.4.39), (5.4.40) yields

$$e^{\alpha t} \int \int \frac{1}{2} m \left[(u_1^\epsilon - u_1)^2 + (u_2^\epsilon - u_2)^2 \right] (x, y, t) dt$$

$$+ \int_t^T e^{\alpha s} \int \int m \left[|D_x(u_1^\epsilon - u_1)|^2 + |D_x(u_2^\epsilon - u_2)|^2 \right] dx dy ds$$

$$+ \frac{1}{\epsilon} \int_t^T e^{\alpha s} \int \int m(|D_y u_1^\epsilon|^2 + |D_y u_2^\epsilon|^2) dx dy ds$$

$$+ \int_t^T e^{\alpha s} \int \int \left[(u_1^\epsilon - u_1)^2 + (u_2^\epsilon - u_2)^2 \right] (E + m\alpha) dx dy ds$$

$$+ \frac{1}{2} \int_t^T e^{\alpha s} \int \int D_x m D_x \left[(u_1^\epsilon - u_1)^2 + (u_2^\epsilon - u_2)^2 \right] dx dy ds \tag{5.4.41}$$

$$= \int e^{\alpha s} \int \int (f - \bar{f}) m (D_x u_1 u_1^\epsilon + D_x u_2 u_2^\epsilon) dx dy ds$$

$$+ \int_t^T e^{\alpha s} \int \int (\gamma - \bar{\gamma}) m (u_1 u_1^\epsilon + u_2 u_2^\epsilon) dx dy ds$$

$$- \int_t^T e^{\alpha s} \int \int (u_2^\epsilon u_1 - u_1^\epsilon u_2)(h - \bar{h}) \beta dm dx dy ds.$$

The right hand side of (5.4.38) tends to 0 as ϵ tends to 0. On the other hand, in view of the discussion in the proof of Lemma 5.4.2 we can minorize the left hand side of (5.4.38) by the quantity

$$\sigma_0 \left\{ \int \int \left[(u_1^\epsilon - u_1)^2 + (u_2^\epsilon - u_2)^2 \right] (x, y, t) dx dy \right.$$

$$+ \int_t^T \int \int \left[|D_x(u_1^\epsilon - u_1)|^2 + |D_x(u_2^\epsilon - u_2)|^2 + \frac{1}{\epsilon}(|D_y u_1^\epsilon|^2 + |D_y u_2^\epsilon|^2) \right.$$

$$\left. + (u_1^\epsilon - u_1)^2 + (u_2^\epsilon - u_2)^2 \right] dx dy ds \right\}$$

where σ_0 is a positive constant. This quantity thus tends to 0 as ϵ tends to 0 and therefore (5.4.37) obtains. ∎

5.4.4   Convergence result

We begin with

Lemma 5.4.5 *Assume that $h(x, y)$ satisfies*

$$|h(x, y)| \leq h_0 + H|x|, \text{ with } H \text{ not too large (see (5.4.55))} \tag{5.4.42}$$

and

$$\Pi_0(\exp \lambda |x|^2) < \infty, \text{ for } \lambda > 0 \text{ sufficiently large} \tag{5.4.43}$$

then one has

$$E(\eta_T^\epsilon)^2 \leq C_T. \tag{5.4.44}$$

Proof

We have

$$(\eta_T^\epsilon)^2 = \exp\left[2\int_0^T h(x_s^\epsilon, y_s^\epsilon)dz - 2s\int_0^T |h(x_s^\epsilon, y_s^\epsilon)|^2 ds\right]$$

$$\times \exp\left[(2s-1)\int_0^T |h(x_s^\epsilon, y_s^\epsilon)|^2 ds\right]$$

hence, for $s > 1$,

$$E(\eta_T^\epsilon)^2 \leq \left\{E\exp\left[\frac{s(2s-1)}{s-1}\int_0^T |h(x_s^\epsilon, y_s^\epsilon)|^2 ds\right]\right\}^{(s-1)/s}$$

In the interval $(1, \infty)$ the function $2s(2s-1)/(s-1)$ admits a minimum at $s_0 = 1 + 1/\sqrt{2}$, equal to $(2 + \sqrt{2})^2$.

It is therefore convenient to take this value of s. To prove (5.4.44) it is sufficient to prove that

$$E\exp[(2+\sqrt{2})^2 TH^2|x_t^\epsilon|^2] \leq C_T, \quad \forall t \in (0, T) \tag{5.4.45}$$

where we have made use of Jensen's inequality.

Consider the test function

$$\phi_t(\tau; x) = \exp[P_t(\tau)|x|^2 + \rho_t(\tau)], \quad \tau \in [0, t]$$

where $P_0(\tau) > 0$ and

$$P_t(t) = \delta, \quad \rho_t(t) = 0. \tag{5.4.46}$$

We are going to prove that if δ is not too large then

$$\frac{\partial \phi}{\partial \tau} + f(x, y) \cdot D_x \phi + \Delta_x \phi \leq 0, \quad \forall x, y, \tau. \tag{5.4.47}$$

After a calculation it is easy to check that (5.4.47) will hold provided one has (omitting to write the parameter t)

$$\dot{P} + Pf_0(2 + \mu) + 4P^2 = 0$$

$$\dot{\rho} + \frac{Pf_0}{\mu} + 2nP = 0 \tag{5.4.48}$$

where μ is > 0 arbitrary and f_0 is such that

$$|f(x, y)| \leq f_0(|x| + 1).$$

We can solve (5.4.48) with the initial conditions (5.4.46) (at $\tau = t$) and obtain

$$\frac{1}{P_t(\tau)} = \frac{1}{\delta}e^{-(2+\mu)f_0(t-\tau)} - \frac{4}{(2+\mu)f_0}\left(1 - e^{-(2+\mu)f_0(t-\tau)}\right) \tag{5.4.49}$$

$$\rho_t(\tau) = \left(\frac{f_0}{\mu} + 2n\right)\int_\tau^t P(\sigma)d\sigma. \tag{5.4.50}$$

To get $P_t(\tau) > 0$ for $\tau \in [0, t]$, which is necessary to verify (5.4.47), we must have

$$\frac{1}{\delta} > \frac{4}{(2+\mu)f_0}(e^{(2+\mu)f_0 t} - 1)$$

and thus if we impose

$$\frac{1}{\delta} > \frac{4}{(2+\mu)f_0}(e^{(2+\mu)f_0 T} - 1) \tag{5.4.51}$$

we can assert that (5.4.47) holds for $0 \le \tau \le t \le T$. By Ito's formula we then deduce

$$\phi_t(\tau; x_\epsilon(\tau)) - \phi_t(0; \xi) \le \sqrt{2} \int_0^\tau D_x \phi_t(s, x_\epsilon(s)) dw(s)$$

hence

$$E\left[\exp(\delta|x_\epsilon(t)|^2)|\xi\right] \le \phi_t(0; \xi) \tag{5.4.52}$$

and thus if

$$\Pi_0 \exp(P_T(0)|x|^2) < \infty \tag{5.4.53}$$

where

$$P_T(0) = \exp[e^{(2+\mu)f_0 T}] \cdot \left(\frac{1}{\delta} - \frac{4}{(2+\mu)f_0}\left\{\exp\left[e^{(2+\mu)f_0 T}\right] - 1\right\}\right)^{-1} \tag{5.4.54}$$

we can assert from (5.4.52) that

$$E \exp(\delta|x_\epsilon(t)|^2) \le C_T.$$

Comparing with (5.4.45) we have to take $\delta = (2+\sqrt{2})^2 T H^2$ and thus must assume that

$$(2+\sqrt{2})^2 T H^2 < \frac{(2+\mu)f_0}{4(e^{(2+\mu)f_0 T} - 1)}, \quad \mu > 0 \tag{5.4.55}$$

and (5.4.53) holds for this choice of δ.

The result (5.4.44) has thus been proved. ∎

Remark 5.4.1 We can take H and $\lambda > 0$, and in (5.4.42), (5.4.43) arbitrary, provided T is sufficiently small and (5.4.44) will still hold. ∎

We can now turn to the main result of this section.

We shall make an additional assumption on the initial values of x^ϵ, y^ϵ:

q_0 has a density $q_0(x, y)$ with respect to Lebesgue measure on $R^n \times Y$
and $q_0 e^{G(x,0)} \in L^2(R^n \times Y)$. (5.4.56)

We then state

Theorem 5.4.1 *We assume (5.4.1), (5.4.2), (5.4.3), (5.4.42), (5.4.43), (5.4.56); then one has*

$$p^\epsilon(T)(\phi) \to p(T)(\psi) \tag{5.4.57}$$

$\forall \psi(x)$ *continuous bounded, in* $L^2(\Omega, \mathcal{Z}^T, P)$ *weakly, where*

$$p(t)(\psi) = E\left[\psi(x_t)\eta_t | \mathcal{Z}^t\right] \tag{5.4.58}$$

with the definition

$$dx_t = \bar{f}(x_t)dt + \sqrt{2}dw_t, \quad x_0 = \xi$$

$$\eta = \exp\left[\int_0^t \bar{h}(x_s)dz_s - \frac{1}{2}\int_0^t |\bar{h}(x_s)|^2 ds\right]. \tag{5.4.59}$$

Proof

We first check that

$$E|p^\epsilon(T)(\psi)|^2 \le C. \tag{5.4.60}$$

Indeed since ψ is bounded

$$|p^\epsilon(T)(\psi)| \le \|\psi\| E[\eta_T^\epsilon | \mathcal{Z}^T]$$

hence

$$E|p^\epsilon(T)(\psi)|^2 \le \|\psi^2\| E|E[\eta_T^\epsilon | \mathcal{Z}^T]|^2$$
$$\le \|\psi\|^2 E(\eta_T^\epsilon)^2 \le C_T$$

by virtue of Lemma 5.4.5.

The set of linear combinations of variables $\sum c_i \theta_i(T)$, where $\theta_i(T)$ corresponds to (5.4.15) with $\beta = \beta_i$, forms a dense subspace of $L^2(\Omega, \mathcal{Z}^T, P)$. This follows from the density result of Lemma 4.1.4. Therefore to prove (5.4.57), it is sufficient to prove that

$$Ep^\epsilon(T)(\psi)\theta(T) \to Ep(T)(\psi)\theta(T) \quad \forall \beta(\cdot).$$

It is sufficient to prove this result for $\psi = \phi \in C_0^\infty(R^n)$. According to the duality argument of § 5.4.2 (see (5.4.16)) this means

$$q_0(v_1^\epsilon(0)) \to q_0(v_1(0)), \tag{5.4.61}$$

where the pair $v_1(x,t)$, $v_2(x,t)$ is the solution of

$$-\frac{\partial v_1}{\partial t} - \bar{f} \cdot D_x v_1 - \Delta_x v_1 + v_2 \bar{h} \cdot \beta = 0$$

$$-\frac{\partial v_2}{\partial t} - \bar{f} \cdot D_x v_2 - \Delta_x v_2 - v_1 \bar{h} \cdot \beta = 0 \tag{5.4.62}$$

$$v_1(x,T) = \phi(x), \quad v_2(x,T) = 0.$$

Comparing with (5.4.34) it is clear that

$$v_1(x,t) = u_1(x,t)e^{G(x,t)}$$
$$v_2(x,t) = u_2(x,t)e^{G(x,t)}.$$

Therefore thanks to (5.4.56), (5.4.61) amounts to,

$$\int\int u_1^\epsilon(x,y,0)e^{G(x,0)}q_0(x,y)dxdy \to \int u_1(x,0)e^{G(x,0)}p_0(x)dx \tag{5.4.63}$$

where p_0 is the density of Π_0; $p_0(x) = \int_Y q_0(x,y)dy$. Since by (5.4.56) $q_0(x,y)e^{G(x,0)} \in L^2(R^n \times Y)$, the result (5.4.63) follows immediately from the second part of (5.4.37) (Lemma 5.4.4). The proof has been completed. ∎

Corollary 5.4.1 *Let $\bar{P}$ be the probability on $(\Omega, \mathcal{A})$ defined by the Radon–Nikodym derivative*

$$\left.\frac{d\bar{P}}{dP}\right|_{\mathcal{F}^t} = \eta_t$$

then one has

$$E^\epsilon \Pi^\epsilon(T)(\psi)\xi_T \rightarrow \bar{E}\Pi(T)(\psi)\xi_T, \quad \forall \xi_T \in L^2(\Omega, \mathcal{Z}^T, P)$$

where

$$\Pi(T)(\psi) = \frac{p(T)(\psi)}{p(T)(1)}.$$

Proof

Indeed

$$
\begin{aligned}
E^\epsilon \Pi^\epsilon(T)(\psi)\xi_T &= E^\epsilon \psi(x^\epsilon(T))\xi_T \\
&= E\psi(x^\epsilon(T))\eta_T^\epsilon \xi_T \\
&= Ep^\epsilon(T)(\psi)\xi_T \\
&\rightarrow Ep(T)(\psi)\xi_T \\
&= \bar{E}\Pi(T)(\psi)\xi_T
\end{aligned}
$$

and the desired result has been proven. ■

6   Some explicit solutions of the Zakai equation

Introduction

We have seen in Chapter 4 that one can recover the classical Kalman filter from
the Zakai equation (see Theorem 4.4.1). In fact one obtains more, namely an explicit
solution of the Zakai equation, and the conditional probability is gaussian. It is clearly
important to look for other examples when one can solve explicitly the Zakai equation.
The few cases that are known in the literature belong essentially to three classes.
The first one, considered by Benes and Karatzas (1983) and Makowski (1986), is the
following:

$$
\begin{aligned}
dx &= (F_t x_t + f_t)dt + dw_t \\
dz_t &= (H_t x_t + h_t)dt + db_t \\
x(0) &= \xi, \quad z(0) = 0
\end{aligned}
\tag{6.1}
$$

where the initial distribution Π_0 of ξ is arbitrary. The second class, introduced by
Benes (1981) with a convenient generalization by Shukhman (1985) and Zeitouni and
Bobrovsky (1984), is of the type

$$
\begin{aligned}
dx &= (F_t x_t + f_t + g(x))dt + dw_t \\
dz_t &= (H_t x_t + h_t)dt + db_t \\
x(0) &= \xi, \quad z(0) = 0
\end{aligned}
\tag{6.2}
$$

where the initial distribution is a gaussian, and g satisfies a special condition of gradient type.

The third class concerns the 'conditional gaussian' case introduced by Lipster and
Shiryaev (1977). Using the notation of § 4.5.5, it will amount to the following model

$$
\begin{aligned}
dx_t &= (F_t(z_t)x_t + f_t(z_t))dt + \sigma(z_t)dw_t \\
x_0 &= \xi \\
dz_t &= (H_t(z_t)x_t + h_t(z_t))dt + \alpha(z_t)dw_t + db_t \\
z_0 &= 0
\end{aligned}
\tag{6.3}
$$

where ξ is gaussian.

There is a possibility of mixing which yields new models for which an explicit solution is obtained.

Haussmann and Pardoux (1986) have given a unified treatment of all the known cases. They have introduced a further possibility of correlation: namely, in (6.3) the state equation is replaced by

$$
\begin{aligned}
\mathrm{d}x_t =& (F_t(z_t) + f_t(z_t))\mathrm{d}t + \sigma(z_t)\mathrm{d}w_t \\
& + \sum_{j=1}^{m}(G_t^j(z_t)x_t + g_t^j(z_t))\mathrm{d}z_t^j.
\end{aligned}
\tag{6.4}
$$

We shall study the cases (6.1), (6.2), (6.3), relying on the theory of Chapter 4, namely the existence and uniqueness result for the weak formulation of the Zakai equation (cf. Theorems 4.1.1 and 4.2.1, and § 4.5.5). The case (6.4) is more complex and will be dealt with independently.

Attempts to derive general conditions to guarantee the existence of finite dimensional sufficient statistics have been made in the literature, using Lie algebraic methods (see Hazewinkel and Markus (1982) and Sussman (1982), among other works in this direction).

6.1   Non gaussian initial condition

6.1.1   Setting of the problem

The model is that of § 4.4.1, namely

$$\begin{aligned}
\mathrm{d}x_t &= (F_t x_t + f_t)\mathrm{d}t + \mathrm{d}w_t \\
\mathrm{d}z_t &= (H_t x_t + h_t)\mathrm{d}t + \mathrm{d}b_t \\
x(0) &= \xi, \quad z(0) = 0
\end{aligned} \tag{6.1.1}$$

except that the initial distribution Π_0 of ξ is arbitrary (but satisfies the assumptions of Theorem 4.1.1, namely the moments up to the third exist).

It is possible to derive the conditional probability directly as has been shown by Makowski (1986), but we would like to derive it from the Zakai equation. Benes and Karatzas (1983) have used the robust form, which introduces restrictive assumptions such as the existence of a density for Π_0 and regularity in t of the coefficients. We shall see that if we use the weak formulation no additional assumption is needed.

6.1.2   Solution of the Zakai equation

We shall need some notation. We shall consider the Riccati equation

$$\begin{aligned}
\dot{\Sigma} + \Sigma H^* R^{-1} H \Sigma - Q - F\Sigma - \Sigma F^* &= 0 \\
\Sigma(0) &= 0
\end{aligned} \tag{6.1.2}$$

and let $m_t(x)$ be the solution of

$$\begin{aligned}
\mathrm{d}m_t &= (F(m_t) + f)\mathrm{d}t + \Sigma H^* R^{-1}[\mathrm{d}z - (Hm_t + h)\mathrm{d}t] \\
m_0 &= x.
\end{aligned} \tag{6.1.3}$$

The equations (6.1.2), (6.1.3) correspond to the Kalman filter relative to (6.1.1) when the initial condition is known and equal to x. We shall also use

$$\begin{aligned}
s_t(x) = \exp\Big\{ &\int_0^t (m_t^*(x)H^* + h^*)R^{-1}\mathrm{d}z \\
&- \frac{1}{2}\int_0^t (m_t^*(x)H^* + h^*)R^{-1}(Hm_t(x) + h)\mathrm{d}x \Big\}.
\end{aligned} \tag{6.1.4}$$

We shall prove

Theorem 6.1.1 *Assume (6.1.1) then the solution of the Zakai equation (4.1.37) (unique in the sense of Theorem 4.2.1) is given by*

$$p(t)(\psi) = \int \left[s_t(x) \int \psi(m_t(x) + \Sigma_t^{1/2}\xi) \frac{\exp(-\frac{1}{2}|\xi|^2)}{(2n)^{n/2}} d\xi \right] \Pi_0(dx). \qquad (6.1.5)$$

Proof

Because of uniqueness, it is sufficient to verify (4.1.37). To begin with, let us verify that the process

$$p(t)(1) = \int s_t(x)\Pi_0(dx) \qquad (6.1.6)$$

satisfies the relation

$$p(t)(1) = 1 + \int_0^t p(\tau)(x^*H_\tau^* + h_\tau^*)R_\tau^{-1}dz \qquad (6.1.7)$$

where from (6.1.5)

$$p(\tau)(x^*H^* + h^*) = \int s_\tau(x)(m_\tau^*(x)H_\tau^* + h_\tau^*)\Pi_0(dx).$$

Hence we must verify

$$\int s_t(x)\Pi_0(dx) = 1 + \int_0^t \left\{ \int [s_\tau(x)(m_\tau^*(x)H_\tau^* + h_\tau^*)]\,\Pi_0(dx) \right\} R_\tau^{-1}dz$$

which follows from

$$s_t(x) = 1 + \int_0^t s_\tau(x)(m_\tau^*(x)H_\tau^* + h_\tau^*)R_\tau^{-1}dz.$$

More generally, if we compare with the calculations made in Theorem 4.4.1, we can see that since the pair Σ_t, m_t satisfies the same equations as P_t, $\hat{x}_t$ (except for the initial conditions), then these calculations remain applicable here. This permits us to verify that the functional (6.1.5) satisfies

$$dp(t)(\psi(t)) = p(t) \left[\frac{\partial\psi}{\partial t} + D\psi^*(Fx + f) + \frac{1}{2}\mathrm{tr}\,D^2\psi Q \right] dt$$
$$+ p(t)\left[\psi(x^*H^* + h^*)\right] R^{-1}dz.$$

On the other hand evaluating (6.1.5) at time $t = 0$, yields

$$p(0)(\psi) = \int \psi(x)\Pi_0(dx)$$

which is also the desired result. ∎

6.1.3   Some additional results

We can write (6.1.5) in a more convenient form, making the dependence on x of $m_t(x)$ and $s_t(x)$ more apparent. Let us consider the matrix differential equation

$$\dot{\Phi} = (F - \Sigma H^* R^{-1} H)\Phi$$
$$\Phi(0) = I \tag{6.1.8}$$

and let β_t be the solution of

$$d\beta_t = (F\beta_t + f)dt + \Sigma H^* R^{-1}[dz - (H\beta_t + h)dt]$$
$$\beta_0 = 0 \tag{6.1.9}$$

then one has

$$m_t(x) = \Phi_t x + \beta_t. \tag{6.1.10}$$

Next, from (6.1.4), one has

$$s_t(x) = \gamma_t \exp\left[-\frac{1}{2}(x^* S_t x - 2x^* \rho_t)\right] \tag{6.1.11}$$

where

$$\gamma_t = \exp\left[\int_0^t (\beta_s^* H_s^* + h_s^*) R_s^{-1} dz - \frac{1}{2}\int_0^t (\beta_s^* H_s^* + h_s^*) R_s^{-1}(H_s\beta_s + h_s)ds\right]$$

$$S_t = \int_0^t \Phi_s^* H_s^* R_s^{-1} H_s \Phi_s ds \tag{6.1.12}$$

$$\rho_t = \int_0^t \Phi_s^* H_s^* R_s^{-1}[dz_s - (H_s\beta_s + h_s)ds]$$

hence (6.1.5) yields

$$p(t)(\psi) = \gamma_t \int\int \psi(\Phi_t x + \beta_t + \Sigma_t^{1/2}\xi)$$
$$\times \exp\left[-\frac{1}{2}(|\xi|^2 + x^* S_t x - 2x^* \rho_t)\right]\frac{1}{(2n)^{n/2}}\Pi_0(dx)d\xi. \tag{6.1.13}$$

Let us compute the conditional mean and the conditional variance. We can either use (6.1.13), or make use of the Kushner equation (see § 6.3.2). The general equations are recalled. First, for the conditional mean $\hat{x}_t$ one has

$$d\hat{x}_t = \hat{g}dt + \widehat{[(x - \hat{x})h^*]}R^{-1}(dz - \hat{h}dt)$$
$$\hat{x}_0 = x_0 \tag{6.1.14}$$

where $\hat{\phi}$ is short for $\Pi(t)(\phi)$, whenever ϕ is any function of x.

Next let Γ_t be the conditional variance

$$\Gamma_t = \pi(t)\left[(x - \hat{x}_t)(x - \hat{x}_t)^*\right] = \widehat{[(x - \hat{x}_t)x^*]}$$

one has

$$\mathrm{d}\Gamma_t = \left[\widehat{[g(x-\hat{x})^*]} + \widehat{[(x-\hat{x})g^*]} + [\widehat{\sigma Q \sigma^*}] - \widehat{[(x-\hat{x})h^*]} R^{-1} \widehat{[h(x-\hat{x})^*]} \right] \mathrm{d}t$$
$$+ \left[\widehat{(x-\hat{x})(x-\hat{x})^*(h-\hat{h})^*} \right] R^{-1}(\mathrm{d}z - \hat{h}\mathrm{d}t) \tag{6.1.15}$$
$$\Gamma_0 = P_0$$

where x_0, P_0 are respectively the mean and the variance of the initial distribution Π_0.

In the present case (6.1.14) yields

$$\mathrm{d}\hat{x}_t = (F_t\hat{x}_t + f_t)\mathrm{d}t + \Gamma_t H_t^* R_t^{-1}[\mathrm{d}z - (H_t\hat{x}_t + h_t)\mathrm{d}t] \tag{6.1.16}$$
$$\hat{x}_0 = x_0$$

$$\mathrm{d}\Gamma_t = (F_t\Gamma_t + \Gamma_t F_t^* + Q - \Gamma_t H^* R^{-1} H \Gamma_t)\mathrm{d}t$$
$$+ [(x-\hat{x})\widehat{(x-\hat{x})}^*(x-\hat{x})^*]H^* R^{-1}[\mathrm{d}z - (H\hat{x}_t + h_t)\mathrm{d}t] \tag{6.1.17}$$
$$\Gamma_0 = P_0.$$

From formula (6.1.13) we see that only two *sufficient* statistics play a role, namely β_t and ρ_t. Note that (6.1.13) yields also

$$\hat{x}_t = \frac{p(t)(x)}{p(t)(1)}$$
$$\hat{x}_t = \beta_t + \Phi_t b_t(\rho_t) \tag{6.1.18}$$

where

$$b_t(\eta) = \frac{\int x \exp[-\frac{1}{2}(x^* S_t x - 2x^*\eta)]\Pi_0(\mathrm{d}x)}{\int \exp[-\frac{1}{2}(x^* S_t x - 2x^*\eta)]\Pi_0(\mathrm{d}x)}. \tag{6.1.19}$$

From (6.1.18), we see that it is convenient to take as sufficient statistics the pair $\hat{x}_t$, ρ_t. It is then possible to derive a system of stochastic differential equations with solutions $\hat{x}_t$, ρ_t. For that purpose it is convenient to compute Γ_t from (6.1.13). After some calculations we deduce the following formula

$$\Gamma_t = \Sigma_t + \Phi_t(G_t(\rho_t) - b_t(\rho_t)b_t^*(\rho_t))\Phi_t^* \tag{6.1.20}$$

where we have set

$$G_t(\eta) = \frac{\int\int xx^* \exp[-\frac{1}{2}(x^* S_t x - 2x^*\eta)]\Pi_0(\mathrm{d}x)}{\int \exp[-\frac{1}{2}(x^* S_t x - 2x^*\eta)]\Pi_0(\mathrm{d}x)}. \tag{6.1.21}$$

Since from the third relation (6.1.12) we deduce

$$\mathrm{d}\rho_t = \Phi_t^* H_t^* R_t^{-1} H_t \Phi_t b_t(\rho_t)\mathrm{d}t + \Phi_t^* H_t^* R_t^{-1}[\mathrm{d}z - (H_t\hat{x}_t + h_t)\mathrm{d}t] \tag{6.1.22}$$
$$\rho_0 = 0,$$

the system (6.1.16), (6.1.22) taking account of (6.1.20) is indeed a pair of stochastic differential equations (recall that $\mathrm{d}z_t - (H_t\hat{x}_t + h_t)\mathrm{d}t = \mathrm{d}\nu_t$ where ν_t is a Wiener process with covariance R_t; ν_t is the innovation process, cf. § 4.3.3).

We can express the quantity

$$[\widehat{*(\underline{x}-x)_*(\underline{x}-x)(\underline{x}-x)}]$$
$$= \frac{\int [\Phi_t(x-b_t(\rho_t))(x-b_t(\rho_t))^*\Phi_t^*(x-b_t(\rho_t))^*\Phi_t^*] \exp[-\frac{1}{2}(x^* S_t x - 2x^*\rho_t)]\Pi_0(\mathrm{d}x) \cdot}{\int \exp[-\frac{1}{2}(x^* S_t x - 2x^*\rho_t)]\Pi_0(\mathrm{d}x)}$$

$$\tag{6.1.23}$$

Remark 6.1.1 We have implicitly assumed in the above calculations that the functions $b_t(\eta)$, $G_t(\eta)$ are well defined. This can be taken as a restriction on Π_0. ∎

6.1.4   The gaussian case

It is of course worthwhile to particularize the formulas obtained in the preceding paragraph to the case when Π_0 is gaussian, to recover the formulas of the Kalman filter (see § 4.4.2). Let us assume now that Π_0 is a gaussian with mean x_0 and covariance Π_0. We then have

$$b_t(\eta) = \frac{\int (x_0 + P_0^{1/2}\xi) \exp\left\{ -\frac{1}{2}\left[(x_0 + P_0^{1/2}\xi)^* S_t (x_0 + P_0^{1/2}\xi) + |\xi|^2 - 2(x_0 + P_0^{1/2}\xi)^*\eta \right] \right\} d\xi}{\int \exp\left\{ -\frac{1}{2}\left[(x_0 + P_0^{1/2}\xi)^* S_t (x_0 + P_0^{1/2}\xi) + |\xi|^2 - 2(x_0 + P_0^{1/2}\xi)^*\eta \right] \right\} d\xi}$$

which yields

$$b_t(\eta) = x_0 - P_0^{1/2}(I + P_0^{1/2} S_t P_0^{1/2})^{-1} P_0^{1/2}(S_t x_0 - \eta). \qquad (6.1.24)$$

It is thus an affine function of η.

Furthermore

$$G_t(\eta) = b_t(\eta)b_t^*(\eta) + P_0^{1/2}(I + P_0^{1/2} S_t P_t^{1/2})^{-1} P_0^{1/2} \qquad (6.1.25)$$

hence Γ_t is independent of η and is given by

$$\Gamma_t = \Sigma_t + \Phi_t P_0^{1/2}(I + P_0^{1/2} S_t P_t^{1/2})^{-1} P_0^{1/2}. \qquad (6.1.26)$$

One can also check that the quantity (6.1.23) vanishes and, from (6.1.17), it follows that Γ_t coincides with the solution of the Riccati equation of the Kalman filter. Since Γ_t is deterministic, the statistic ρ_t is now unnecessary.

6.2 Explicit solution in the case of a non linear drift

6.2.1 Setting of the problem

Consider now the system

$$dx_t = [(F_t x_t + f_t + g_t(x_t)] \, dt + dw_t$$
$$x(0) = \xi,$$
$$dz_t = (H_t x_t + h_t) dt + db_t \tag{6.2.1}$$
$$z_0 = 0$$

where the assumptions on $g_t(x)$ will be made precise later. We assume that the initial distribution Π_0 (probability law of ξ) is general (not necessarily gaussian).

6.2.2 A formal calculation

To understand the assumptions to be made on g, we begin by a formal calculation, trying to find a solution of the Zakai equation written in the strong form, i.e.,

$$dp + \left\{ \text{div}[p(g + Fx + f)] - \frac{1}{2} \text{tr } QD^2 p \right\} dt = p(x^* H^* + h^*) R^{-1} dz. \tag{6.2.2}$$

We look for a solution of the form

$$p(x,t) = \exp\left[\Phi(x,t) - \frac{1}{2} x^* M_t x + r_t^* x + \rho_t \right] \tag{6.2.3}$$

where $\Phi(x,t)$, M_t are deterministic (M_t is symmetric), but r_t, ρ_t may be random. In fact, when $g = 0$, we know that p is of the form (6.2.3) with $\Phi = 0$. We write

$$dr_t = k_t dt + H_t^* R_t^{-1} dz_t$$
$$d\rho_t = \mu_t dt + h_t^* R_t^{-1} dz_t \tag{6.2.4}$$

the choice of the coefficients of dz_t being made to cancel out the right hand side of (6.2.2) (it is the only possible choice). By Ito's calculus we have indeed for the function defined by (6.2.3),

$$dp = p\left\{ \left[\frac{\partial \Phi}{\partial t} - \frac{1}{2} x^* \dot{M}_t x + k_t^* x + \mu_t + \frac{1}{2}(x^* H^* + h^*) \times R^{-1}(Hx + h) \right] dt \right.$$
$$\left. + (x^* H^* + h^*) R^{-1} \right\} dz$$

and

$$\mathrm{D}p = p(\mathrm{D}\Phi - M_t x + r_t)$$

$$\mathrm{D}^2 p = p(\mathrm{D}\Phi - M_t x + r_t)(\mathrm{D}\Phi - M_t x + r_t)^* + p(\mathrm{D}^2\Phi - M_t)$$

hence if we want to satisfy (6.2.2) we must verify

$$\frac{\partial \Phi}{\partial t} - \frac{1}{2} x^* \dot{M}_t x + k_t^* x + \mu_t + \frac{1}{2}(x^* H^* + h^*) R^{-1}(Hx + h)$$
$$+ (g^* + x^* F^* + f^*)(\mathrm{D}\Phi - M_t x + r_t) + \operatorname{div} g + \operatorname{tr} F \qquad (6.2.5)$$
$$- \frac{1}{2}(\mathrm{D}\Phi - M_t x + r_t)^* Q(\mathrm{D}\Phi - M_t x + r_t) - \frac{1}{2}\operatorname{tr} Q(\mathrm{D}^2\Phi - M_t) = 0.$$

We collect the random terms, namely

$$k_t^* x + \mu_t + (g^* + x^* F^* + f^* - \mathrm{D}\Phi^* Q + x^* M_t Q) r_t - \frac{1}{2} r_t^* Q r_t$$

which must be deterministic according to our choices. The dependence in x also implies

$$g_t(x) = Q \mathrm{D}\Phi(x, t) \qquad (6.2.6)$$

and

$$k_t + (F^* + M_t Q) r_t = \alpha_t$$
$$\mu_t + f^* r_t - \frac{1}{2} r_t^* Q r_t = \beta_t \qquad (6.2.7)$$

where α_t, β_t are deterministic. Now from (6.2.6) we get

$$\operatorname{div} g = \operatorname{tr} Q \mathrm{D}^2 \Phi, \qquad g^* \mathrm{D}\Phi = \mathrm{D}\Phi^* Q \mathrm{D}\Phi$$

hence (6.2.5) yields

$$\frac{\partial \Phi}{\partial t} + \frac{1}{2}\operatorname{tr} Q \mathrm{D}^2\Phi + \frac{1}{2}\mathrm{D}\Phi^* Q \mathrm{D}\Phi + (x^* F^* + f^*)\mathrm{D}\Phi$$
$$= \frac{1}{2} x^* (\dot{M} - H^* R^{-1} H + F^* M + MF + MQM) x$$
$$+ x^* (-\alpha + Mf - H^* R^{-1} h) - \beta - \frac{1}{2} h^* R^{-1} h - \operatorname{tr} F - \frac{1}{2}\operatorname{tr} QM$$

hence the function Φ must satisfy

$$\frac{\partial \Phi}{\partial t} + \frac{1}{2}\operatorname{tr} Q \mathrm{D}^2\Phi + \frac{1}{2}\mathrm{D}\Phi^* Q \mathrm{D}\Phi + (x^* F^* + f^*)\mathrm{D}\Phi$$
$$= \frac{1}{2} x^* \Lambda_t x + x^* \sigma_t + \delta_t \qquad (6.2.8)$$

where Λ_t is symmetric. If this is the case we can complete the choices of r_t and ρ_t in (6.2.3) as follows

$$\dot{M} + MF + F^* M + MQM = \Lambda_t + H^* R^{-1} H \qquad (6.2.9)$$

$$dr_t + \left[(F^* + MQ)r + \sigma - Mf + H^* R^{-1} h\right] dt = H^* R^{-1} dz \qquad (6.2.10)$$

$$d\rho + \left[f^* r_t - \frac{1}{2} r_t^* Q r_t + \delta + \frac{1}{2} h^* R^{-1} h + \frac{1}{2}\operatorname{tr} QM + \operatorname{tr} F\right] dt = h^* R^{-1} dz. \qquad (6.2.11)$$

We are now in a position to figure out the assumption on g. It reads

> There exists $\Phi(x,t)$ which is $C^{2,1}$ and Λ_t symmetric, σ_t, δ_t
> such that (6.2.6), (6.2.8) are satisfied and $\qquad\qquad$ (6.2.12)
> $\Lambda_t + H_t^* R_t^{-1} H_t \geq 0.$

We then define M_t symmetric, $M_t \geq 0$, and r_t, ρ_t by (6.2.9), (6.2.10), (6.2.11) with the initial conditions

$$M_0 = 0, \quad r_0 = 0, \quad \rho_0 = 0. \qquad\qquad (6.2.13)$$

6.2.3 Examples where (6.2.12) is satisfied

If we consider functions Φ of the form

$$\Phi(x,t) = \log \chi(x,t)$$

then χ must satisfy

$$\frac{\partial \chi}{\partial t} + \frac{1}{2}\operatorname{tr} QD^2\chi + (x^*F^* + f^*)D\chi = \chi\left(\frac{1}{2}x^*\Lambda x + x^*\sigma + \delta\right). \qquad (6.2.14)$$

If for example

$$\chi(x,t) = \frac{1}{2}x^* Rx + x^* c + d$$

with

$$\dot{R} + F^* R + RF - \delta R = 0$$
$$\dot{c} + F^* c - \delta c + Rf = 0$$
$$\dot{d} - \delta d + \frac{1}{2}\operatorname{tr} QR + f^* c = 0$$

then (6.2.14) is satisfied with $\Lambda = 0$, $\sigma = 0$.

In that case one has

$$g_t(x) = \frac{Q_t(R_t x + c_t)}{\frac{1}{2}x^* R_t x + x^* c_t + d_t}. \qquad\qquad (6.2.15)$$

We can also satisfy (6.2.14) with exponentials, as follows

$$\chi(x,t) = \alpha_1(t)\exp e(x,t) + \alpha_2(t)\exp(-E(x,t)) \qquad (6.2.16)$$

where

$$E(x,t) = \frac{1}{2}x^* R_t x + x^* c_t + d_t \qquad\qquad (6.2.17)$$

and

$$\dot{R} + F^* R + RF = 0$$
$$\dot{c} + F^* c + Rf = 0$$
$$\dot{d} + f^* c + \frac{1}{2}\operatorname{tr} RQ = \frac{1}{2}\frac{d}{dt}\left(\log\frac{\alpha_2}{\alpha_1}\right)$$

with the choices

$$\Lambda = RQR$$

$$\sigma = RQc$$

$$\delta = \frac{1}{2}c^*Qc + \frac{1}{2}\frac{d}{dt}(\log \alpha_1\alpha_2).$$

In that case one has

$$g_t(x) = \frac{\alpha_1(t)\exp E(x,t) - \alpha_2(t)\exp(-E(x,t))}{\alpha_1(t)\exp E(x,t) + \alpha_2(t)\exp(-E(x,t))}(R_t x + c_t). \qquad (6.2.18)$$

6.2.4   Exact solution of the Zakai equation

We introduce some notation as in § 6.1.2. Let Σ_t be the solution of the Riccati equation

$$\dot{\Sigma} = \Sigma(\Lambda + H^*R^{-1}H)\Sigma - Q - F\Sigma - \Sigma F^* = 0$$
$$\Sigma(0) = 0 \qquad (6.2.19)$$

and let $m_t(x)$ be the solution of

$$dm_t = [(F - \Sigma\Lambda)m_t + f - \Sigma\sigma]\,dt + \Sigma H^*R^{-1}[dz - (Hm_t + h)dt]$$
$$m(0) = x. \qquad (6.2.20)$$

We shall also need the quantity

$$s_t(x) = \exp\left\{\int_0^t (m_r^*(x)H^* + h^*)R^{-1}dz\right.$$
$$-\frac{1}{2}\int_0^t \left[m_r^*(x)(\Lambda + H^*R^{-1}H)m_r(x) + 2m_r^*(x)(H^*R^{-1}h + \sigma)\right. \qquad (6.2.21)$$
$$\left.+ h^*R^{-1}h + \operatorname{tr}\Sigma\Lambda + 2\delta_r\right]\bigg\}.$$

We can then prove the following.

Theorem 6.2.1 *Assume (6.2.12), then the solution of the Zakai equation for the model (6.2.1) is given by*

$$p(t)(\psi) = \int\left\{\exp(-\Phi(x,0))s_t(x)\right.$$
$$\int\left[\frac{\psi(m_t(x) + \Sigma_t^{1/2}\xi)\exp(\Phi(m_t(x) + \Sigma_t^{1/2}\xi, t) - \frac{1}{2}|\xi|^2)}{(2n)^{n/2}}\right]d\xi\bigg\}\Pi_0(dx).$$
$$\qquad (6.2.22)$$

Proof

At time $t = 0$, we have from (6.2.22)

$$p(0)(\psi) = \int \psi(x)\Pi_0(dx).$$

hence it is sufficient to verify the relation

$$
\begin{aligned}
dp(t)(\psi(t)) =& p(t)\left[\frac{\partial \psi}{\partial t} + D\psi^*(QD\Phi + Fx + f) + \frac{1}{2}\text{tr } D^2\psi Q\right] dt \\
& + p(t)(\psi(x^*H^* + h^*))R^{-1}dz.
\end{aligned}
\tag{6.2.23}
$$

We compute for x frozen, setting $\chi = \psi \exp \Phi$, $\Gamma_t = \Sigma_t^{1/2}$ †

$$
\begin{aligned}
d(s_t\chi(m_t + \Sigma_t^{1/2}\xi,t)) =& s_t\Big\{\chi_t + D\chi^*[(F - \Sigma\Lambda)m + f - \Sigma\sigma] \\
& + D\chi^*\dot{\Gamma}_t\xi + \frac{1}{2}\text{tr } D^2\chi\Sigma H^*R^{-1}H\Sigma \\
& - \chi(\frac{1}{2}m^*\Lambda m + m^*\sigma + \frac{1}{2}\text{tr } \Sigma\lambda + \delta)\Big\} \\
& + s_t\left[\chi(m^*H^* + h^*) + D\chi^*\Sigma H^*\right]R^{-1}dz
\end{aligned}
\tag{6.2.24}
$$

where the space argument is evaluated at $m_t + \Sigma_t^{1/2}\xi$.

We next use

$$
\begin{aligned}
&\int D\chi^*(m_t + \Sigma_t^{1/2}\xi,t)\dot{\Gamma}_t\xi \exp\left(-\frac{1}{2}|\xi|^2\right) d\xi \\
&= \frac{1}{2}\int \text{tr } D^2\chi(m_t + \Sigma_t^{1/2}\xi,t)\dot{\Sigma}_t \exp\left(-\frac{1}{2}|\xi|^2\right) d\xi
\end{aligned}
$$

hence

$$
\begin{aligned}
& ds_t\int \chi(m_t + \Sigma_t^{1/2}\xi,t) \exp\left(-\frac{1}{2}|\xi|^2\right) d\xi \\
=& s_t\int \Big\{\chi_t + D\chi^*[(F - \Sigma\Lambda)m + f - \Sigma\sigma] + \frac{1}{2}\text{tr } D^2\chi(\dot{\Sigma}_t + \Sigma H^*R^{-1}H\Sigma) \\
& - \chi\left(\frac{1}{2}m^*\Lambda m + m^*\sigma + \frac{1}{2}\text{tr } \Sigma\lambda + \delta\right)\Big\} \exp\left(-\frac{1}{2}|\xi|^2\right) d\xi \\
& + s_t\left\{\int [\chi(m^*H^* + h^*) + D\chi^*\Sigma H^*]\exp\left(-\frac{1}{2}|\xi|^2\right)\right\} R^{-1}dz.
\end{aligned}
\tag{6.2.25}
$$

We also use

$$
\begin{aligned}
&\int D\chi^*(m_t + \Sigma_t^{1/2}\xi,t)\Sigma_t H^* \exp\left(-\frac{1}{2}|\xi|^2\right) d\xi \\
&= \int \chi\xi^*\Sigma_t^{1/2}H^* \exp\left(-\frac{1}{2}|\xi|^2\right) d\xi
\end{aligned}
\tag{6.2.26}
$$

which permits us to check easily that the terms in dz on the left and right hand sides of (6.2.23) coincide.

---

† We shall mimic the derivation of Theorem 4.4.1.

Therefore comparing (6.2.23) and (6.2.25), everything amounts to proving the following:

$$
\int \left\{ \chi_t + D\chi^*[(F - \Sigma\lambda)m + f - \Sigma\sigma] + \frac{1}{2}\mathrm{tr}\, D^2\chi(\dot{\Sigma}_t + \Sigma H^* R^{-1} H \Sigma) \right.
$$
$$
\left. - \chi\left(\frac{1}{2}m^*\Lambda m + m^*\sigma + \frac{1}{2}\mathrm{tr}\,\Sigma\Lambda + \delta\right) \right\} \exp\left(-\frac{1}{2}|\xi|^2\right) d\xi
$$
$$
= \int \exp\left\{ \phi\left[\psi_t + D\psi^*(QD\Phi + Fm + f + F\Sigma^{1/2}\xi) + \frac{1}{2}\mathrm{tr}\, D^2\psi Q\right] \right\}
$$
$$
\exp\left(-\frac{1}{2}|\xi|^2\right) d\xi.
$$
(6.2.27)

We shall use

$$
\chi_t = \exp[\Phi(\psi_t + \psi\Phi_t)]
$$
$$
D\chi = \exp[\Phi(D\psi + \psi D\Phi)]
$$
$$
D^2\chi = \exp[\Phi(D^2\psi + \psi D^2\Phi + \psi D\Phi D\Phi^* + 2D\Phi D\psi^*)].
$$

Taking account of (6.2.8), we have

$$
\int \chi(\Phi_t + \frac{1}{2}\mathrm{tr}\, QD^2\Phi + \frac{1}{2}D\Phi^* QD\Phi - f^* D\Phi - \delta_t)(m_t + \Sigma_t^{1/2}\xi, t) \exp\left(-\frac{1}{2}|\xi|^2\right) d\xi
$$
$$
= \int \chi\left[\frac{1}{2}(m + \Sigma^{1/2}\xi)^*\Lambda(m + \Sigma^{1/2}\xi) + \sigma^*(\Sigma^{1/2}\xi + m) - D\Phi^* F(\Sigma^{1/2}\xi + m)\right]
$$
$$
\exp\left(-\frac{1}{2}|\xi|^2\right) d\xi
$$
(6.2.28)

and

$$
\frac{1}{2}\int \chi\xi^*\Sigma^{1/2}\Lambda\Sigma^{1/2}\xi \exp\left(-\frac{1}{2}|\xi|^2\right) d\xi = \frac{1}{2}\int (\chi\mathrm{tr}\,\Sigma\Lambda + \mathrm{tr}\, D^2\chi\Sigma\Lambda\Sigma)
$$
$$
\exp\left(-\frac{1}{2}|\xi|^2\right) d\xi
$$
(6.2.29)

and

$$
\int \chi(\sigma^* + m^*\Lambda)\Sigma^{1/2}\xi \exp\left(-\frac{1}{2}|\xi|^2\right) d\xi
$$
$$
= \int D\chi^*\Sigma(\Lambda m + \sigma) \exp\left(-\frac{1}{2}|\xi|^2\right) d\xi - \int \chi D\Phi^* F\Sigma^{1/2}\xi \exp\left(-\frac{1}{2}|\xi|^2\right) d\xi
$$
$$
= \int [-D\xi^* + \exp(\Phi D\psi^*)]F\Sigma^{1/2}\xi \exp\left(-\frac{1}{2}|\xi|^2\right) d\xi
$$
$$
= \int [-\mathrm{tr}\, D^2\chi F\Sigma + \exp(\Phi D\psi^*)F\Sigma^{1/2}\xi] \exp\left(-\frac{1}{2}|\xi|^2\right) d\xi.
$$
(6.2.30)

Collecting results the left hand side of (6.2.27) becomes

$$
\int \left(\exp\left\{ \Phi[\psi_t + D\psi^* F\Sigma^{1/2}\xi + D\psi^*(Fm + f)] \right\} \right.
$$
$$
+ \frac{1}{2}\mathrm{tr}\, D^2\chi(\dot{\Sigma} + \Sigma(H^* R^{-1} H + \Lambda)\Sigma - F\Sigma - \Sigma F^*)
$$
$$
\left. - \chi\left(\frac{1}{2}\mathrm{tr}\, QD^2\Phi + \frac{1}{2}D\Phi^* QD\Phi\right) \right) \exp\left(-\frac{1}{2}|\xi|^2\right) d\xi
$$
(6.2.31)

and taking account of (6.2.19), this quantity is equal to the right hand side of (6.2.27). The proof has been completed. ∎

6.3 The conditionally gaussian case

6.3.1 Setting of the problem in the diffusion case

We consider the following model

$$dx_t = (F_t(z_t)x_t + f_t(z_t))dt + \sigma_t(z_t)dw_t$$

$$x_0 = \xi,$$

$$dz_t = (H_t(z_t)x_t + h_t(z_t))dt + \alpha_t(z_t)dw_t + db_t$$

$$z_0 = 0.$$

$$(6.3.1)$$

We assume

$F_t(z)$, $H_t(z)$, $\sigma_t(z)$, $\alpha_t(z)$, $f_t(z)$, $h_t(z)$ are Borel functions of t, z, which are C^2 in z, bounded with bounded derivatives.   (6.3.2)

The process x_t, z_t is a diffusion process, but not gaussian, of course. The smoothness assumptions are made in order to be able to apply the 'uniqueness' argument for the filtering equation of § 4.5.5, (4.5.48).

We define

$$D_t(z) = R_t + \alpha_t(z)Q_t\alpha_t^*(z)$$

where R_t, Q_t are the covariance matrices of b_t, w_t respectively.

We use the framework of § 4.5.5, which we recall briefly. Let

$$B_t(z) = Q_t - Q_t\alpha_t^*(z)D_t^{-1}(z)\alpha_t(z)Q_t.$$

By virtue of the smoothness assumptions, the process x_t, z_t is well defined on a convenient probability system $(\Omega, \mathcal{A}, P, \mathcal{F}^t)$ on which w_t, b_t are given as independent Wiener processes.

We define next three processes ρ_t, $\tilde{z}_t$, $\tilde{w}_t$ as follows:

$$d\rho = -\rho(x_t^* H_t^*(z_t) + h_t^*(z_t))D^{-1}(z_t)(\alpha_t(z_t)dw_t + db_t)$$

$$\rho(0) = 1$$

$$d\tilde{z}_t = D^{-1/2}(z_t)dz_t, \quad \tilde{z}_0 = 0$$

$$d\tilde{w}_t = B(z_t)^{-1/2}\left\{dw_t - Q_t\alpha_t^*(z_t)D^{-1}(z_t)[dz_t - (H_t(z_t)x_t + h(z_t))dt]\right\}$$

$$\tilde{w}_0 = 0.$$

We perform the change of probability

$$\left.\frac{d\tilde{P}}{dP}\right|_{\mathcal{F}^t} = \rho_t$$

and for the system $(\Omega, \mathcal{A}, \tilde{P}, \mathcal{F}^t)$ the processes $\tilde{w}_t$, $\tilde{z}_t$ appear as independent Wiener processes.

If we consider $\eta_t = 1/\rho_t$, the unnormalized conditional probability is defined by the operator

$$p(t)(\psi(t)) = \tilde{E}\left[\psi(x_t, z_t, t)\eta_t | \mathcal{Z}^t\right]$$

for a test function depending on x, z.

We assume also that

$$\xi \text{ is } \mathcal{F}^0 \text{ measurable and gaussian, with probability} \qquad (6.3.3)$$
$$\text{denoted by } \Pi_0 \text{ (mean } x_0, \text{ covariance } P_0).$$

The operator $p(t)$ is the solution of (cf. (4.5.49))

$$
\begin{aligned}
&p(t)(\psi(t)) \\
&= \Pi_0(\psi(\cdot, 0, 0)) \\
&\quad + \int_0^t p(s)\Big[\psi_s + D_x\psi^*(F(z)x + f(z)) + D_z\psi^*(H(z)x + h(z)) \\
&\quad + \frac{1}{2}\mathrm{tr}\, D_x^2\psi\sigma(z)Q\sigma^*(z) + \mathrm{tr}\, \sigma(z)Q\alpha^*(z)D_{zx}^2\psi + \frac{1}{2}\mathrm{tr}\, D(z)D^2\psi\Big]\,ds \\
&\quad + \int_0^t p(s)\left[D_x\psi^*\sigma(z)Q\alpha^*(z) + \psi(x^*H^*(z) + h^*(z)) + D_z\psi^*D(z)\right] D^{-1}(z_s)\,dz_s.
\end{aligned}
$$
$$(6.3.4)$$

There is, moreover, uniqueness of the solution of (6.3.4) in a convenient space of stochastic processes, which can be made precise along the lines of § 4.2.2.

6.3.2 Solution of (6.3.4)

We shall exhibit a solution of (6.3.4), which will necessarily coincide with the unnormalized conditional probability (by the uniqueness property).

We define

$$p(t)(\psi(t)) = s_t\left[\int \psi(\hat{x}_t + P_t^{1/2}\xi, z_t, t)\frac{\exp(-\frac{1}{2}|\xi|^2)}{(2n)^{n/2}}\,d\xi\right] \qquad (6.3.5)$$

where

$$
\begin{aligned}
d\hat{x}_t &= [F_t(z_t)\hat{x}_t + f_t(z_t)]dt + [P_tH_t^*(z_t) + \sigma(z_t)Q_t\alpha^*(z_t)]D(z_t) \\
&\quad \times [dz_t - (H_t(z_t)\hat{x}_t + h_t(z_t))dt] \\
\hat{x}_0 &= x_0
\end{aligned}
$$
$$(6.3.6)$$

and

$$
\begin{aligned}
&\dot{P}_t + [P_tH_t^*(z_t) + \sigma_t(z_t)Q_t\alpha_t^*(z_t)]D^{-1}(z_t)[P_tH(z_t) + \alpha(z_t)Q\sigma^*(z_t)] \\
&\quad - \sigma(z_t)Q\sigma^*(z_t) - F(z_t)P_t - P_tF^*(z_t) = 0, \qquad P(0) = P_0
\end{aligned}
$$
$$(6.3.7)$$

and

$$s_t = \exp\left\{ \int_0^t [\hat{x}^* H^*(z) + h^*(z)] D^{-1}(z)[H(z)\hat{x} + h(z)] \right.$$
$$\left. - \frac{1}{2}\int_0^t [\hat{x}H^*(z) + h^*(z)]D^{-1}(z)[H(z)\hat{x} + h(z)]\mathrm{d}s \right\}.$$

We can state

Theorem 6.3.1 *Under the assumption (6.3.2), the unnormalized conditional probability is given explicitly by the formula (6.3.5), where $\hat{x}_t$, P_t are respectively the conditional mean and covariance and are the solutions of (6.3.6), (6.3.7).*

Proof

By direct checking of formula (6.3.5) in the relation (6.3.4), the desired result obtains.

∎

6.3.3   Setting of the problem in the non diffusion case

The previous method is limited to the case where the coefficients F_t, f_t, ... depend on the observation z_t at time t, but do not depend on past observations. In fact, one can consider a much more general class of conditionally gaussian processes, as introduced by Lipster and Shiryayev (1977).

Consider the space $C([0,T]; R^m)$ equipped with its Borel σ algebra $\mathcal{B}$ (m is the dimension of the observation). Let $C_T = C([0,T]; R^m)$ to simplify the notation, and if $z \in C_T$, let $\mathcal{B}^t = \sigma(z_s, s \leq t)$. We assume to begin with that

$F_t(z)$, $f_t(z)$, $\sigma_t(z)$, $H_t(z)$, $h_t(z)$, $\alpha_t(z)$ are Borel functions on $[0,T] \times C_T$,

with values respectively in $\mathcal{L}(R^n; R^m)$, R^n, $\mathcal{L}^2(R^n; R^n)$, $\mathcal{L}(R^n; R^m)$, R^m,   (6.3.8)

$\mathcal{L}(R^n, R^m)$, and for fixed t, $\mathcal{B}^t$ measurable.

$$F_t, \sigma_t, H_t \text{ are uniformly bounded.} \tag{6.3.9}$$

Q_t, R_t are symmetric matrices on R^n, R^m respectively;

$Q_t \geq 0$, $R_t \geq \bar{r}I$; $\bar{r} > 0$, and Q_t, R_t are L^∞ functions of t.   (6.3.10)

Define

$$D_t(z) = R_t + \alpha_t(z)Q_t\alpha_t^*(z)$$
$$B_t(z) = Q_t - Q_t\alpha_t^*(z)D_t^{-1}(z)\alpha_t(z)Q_t$$

and assume that

$$|D_t^{1/2}(z) - D_t^{1/2}(\tilde{z})|^2 \leq L_1 \int_0^t |z_s - \tilde{z}_s|^2 \mathrm{d}K(s) + L_2|z_t - \tilde{z}_t|^2 \tag{6.3.11}$$

$$|D_t(z)| \leq L_1 \int_0^t (1 + |z_s|^2)\mathrm{d}K(s) + L_2(1 + |z_t|^2). \tag{6.3.12}$$

We consider a probability space $(\Omega, \mathcal{A}, \tilde{P}, \mathcal{F}^t)$, on which are defined two independent standard Wiener processes, $\tilde{w}_t$, $\tilde{z}_t$, with values in R^n and R^m respectively. Let also

$$\xi \text{ gaussian variable, with values in } R^n, \mathcal{F}^0 \text{ measurable,}$$
$$\text{with mean } x_0, \text{ and covariance matrix } P_0. \tag{6.3.13}$$

We begin the construction of the model. We first solve the Ito equation

$$\mathrm{d}z = D_t^{1/2}(z)\mathrm{d}\tilde{z}, \quad z(0) = 0. \tag{6.3.14}$$

The solution of (6.3.14) is uniquely defined in the space $L^2(\Omega, \mathcal{A}, \tilde{P}; C_T)$, by virtue of the assumptions (6.3.11), (6.3.12), (cf. Lipster and Shiryaev 1977, p. 129, vol 1).

Let $\tilde{\mathcal{Z}}^t$, $\mathcal{Z}^t$ be the σ-algebras generated by the processes z and $\tilde{z}$ respectively. We have

$$\tilde{\mathcal{Z}}^t = \mathcal{Z}^t \tag{6.3.15}$$

by the invertibility of $D_t(z)$.

We must make precise a growth condition on the functions introduced in (6.3.8), except of course for (6.3.9). This is done as follows; we assume that

$$\tilde{E} \int_0^T |f_t(z)|^2 \mathrm{d}t, \qquad \tilde{E} \int_0^T |h_t(z)|^2 \mathrm{d}t\dagger \tag{6.3.16}$$

are finite.

Note that, by virtue of (6.3.12) one has also

$$\tilde{E} \int_0^T \mathrm{tr}\, D_t(z)\mathrm{d}t, \quad \tilde{E} \int_0^T \mathrm{tr}\, \alpha_t(z)Q_t\alpha_t^*(z)\mathrm{d}t < \infty. \tag{6.3.17}$$

One then defines x_t by solving the equation

$$\begin{aligned}
\mathrm{d}x_t = &\left\{ \left[F_t(z) - \sigma_t(z)Q_t\alpha_t^*(z)D_t^{-1}(z)H_t(z) \right] x_t \right.\\
&+ f_t(z) - \sigma_t(z)Q_t\alpha_t^*(z)D_t^{-1}(z)h_t(z) \right\} \mathrm{d}t\\
&+ \sigma_t(z)Q_t\alpha_t^*(z)D_t^{-1/2}(z)\mathrm{d}\tilde{z}_t + \sigma_t(z)B_t^{1/2}(z)\mathrm{d}\tilde{w}
\end{aligned} \tag{6.3.18}$$
$$x(0) = \xi.$$

Note that the matrices $F_t - \sigma_t Q_t\alpha_t^* D_t^{-1} H_t$ and $\sigma_t Q_t\alpha_t^* D_t^{-1}$ are uniformly bounded. It is then easy to check that x_t is well defined and that

$$\tilde{E} \sup_{0 \le t \le T} |x_t|^2 < C \tag{6.3.19}$$

where we have made use of the assumptions, in particular (6.3.16) and (6.3.9).

We next define the processes w_t, b_t, which are not Wiener processes, by the formulas

$$\begin{aligned}
\mathrm{d}w_t =&\, Q_t\alpha_t^*(z)D_t^{-1}(z)[\mathrm{d}z_t - (H_t(z)x_t + h_t(z))\mathrm{d}t]\\
&+ B_t^{1/2}(z)\mathrm{d}\tilde{w}_t
\end{aligned} \tag{6.3.20}$$
$$w(0) = 0$$

---

† This assumption will be strengthened later on.

$$db_t = dz_t - [H_t(z)x_t + h_t(z)]dt - \alpha_t(z)dw_t$$
$$b(0) = 0. \tag{6.3.21}$$

With these definitions we can write x_t, z_t as follows

$$dx_t = (F_t(z)x_t + f_t(z))dt + \sigma_t(z)dw_t$$
$$x_0 = \xi,$$
$$dz_t = (H_t(z)x_t + h_t(z))dt + \alpha_t(z)dw_t + db_t \tag{6.3.22}$$
$$b(0) = 0$$

but w, b are not Wiener processes.

We now define the process η_t by

$$d\eta_t = \eta_t(x_t^* H_t^*(z) + h_t^*(z))D_t^{-1/2}(z)d\tilde{z}_t$$
$$\eta(0) = 1 \tag{6.3.23}$$

and assume that

$$\tilde{E}\eta_t = 1. \tag{6.3.24}$$

The assumption (6.3.24) can be interpreted as a condition of finite expectation, for a functional of the processes f_t and h_t.

We can then define a probability P by setting

$$\left.\frac{dP}{d\tilde{P}}\right|_{\mathcal{F}^t} = \eta_t \tag{6.3.25}$$

and we have

Lemma 6.3.1 *For the system* $(\Omega, \mathcal{A}, P, \mathcal{F}^t)$, *the processes* w_t, b_t *are independent* $\mathcal{F}^t$ *Wiener processes, with covariance matrices* Q_t, R_t *respectively.*

Proof

Let $\phi \in L^\infty(0, T; R^m)$, $\psi \in L^\infty(0, T; R^n)$ and consider

$$I_t(\phi, \psi) = \int_t^T \phi^* db + \int_t^T \psi^* dw. \tag{6.3.26}$$

We have to prove that

$$E\left[\exp\left(iI_t(\phi, \psi)\right)|\mathcal{F}^t\right] = \exp\left(-\frac{1}{2}\int_t^T (\phi^* R\phi + \psi^* Q\psi)ds\right). \tag{6.3.27}$$

Let θ_t be $\mathcal{F}^t$ measurable and bounded, since $I_t(\phi, \psi)$ is $\mathcal{F}^t$ measurable and, by definition of the probability P, the equality (6.3.27) amounts to

$$\tilde{E}\theta_t\eta_T \exp\left(iI_t(\phi, \psi)\right) = \tilde{E}\theta_t\eta_t \exp\left(-\frac{1}{2}\int_t^T (\phi^* R\phi + \psi^* Q\psi)ds\right). \tag{6.3.28}$$

But we can write, after some calculations,

$$\eta_T \exp\left(\mathrm{i}I_t(\phi,\psi)\right)$$

$$= \exp\left(\int_0^T \left\{\mathrm{i}\mathcal{J}_t\left[\phi^*(I-\alpha Q\alpha^* D^{-1}) + \psi^* Q\alpha^* D^{-1}\right]\right.\right.$$

$$\left. + (x^* H^* + h^*)D^{-1}\right\}D^{1/2}\mathrm{d}\tilde{z} + \mathrm{i}\mathcal{J}_t(-\phi^*\alpha + \psi^*)B^{1/2}\mathrm{d}\tilde{w}.$$

$$- \frac{1}{2}\int_0^T \left\{\mathrm{i}\mathcal{J}_t\left[\phi^*(I-\alpha Q\alpha^* D^{-1}) + \psi^* Q\alpha^* D^{-1}\right] + (x^* H^* + h^*)D^{-1}\right\}$$

$$\times D\left\{\mathrm{i}\mathcal{J}_t\left[(I-D^{-1}\alpha Q\alpha^*)\phi + D^{-1}\alpha Q\psi\right] + D^{-1}(Hx+h)\right\}\mathrm{d}t$$

$$\left. + \frac{1}{2}\int_0^T \left[\mathcal{J}_t(-\phi^*\alpha + \psi^*)B(-\alpha^*\phi + \psi)\right]\mathrm{d}t\right)\exp\left(-\frac{1}{2}\int_t^T (\phi^* R\phi + \psi^* Q\psi)\mathrm{d}s\right)$$

where $\mathcal{J}_t$ denotes the indicator function of the interval (t, T). Multiplying by θ_t and taking the mathematical expectation, one deduces easily that (6.3.28) holds. ∎

6.3.4   Unnormalized probability

The unnormalized probability is by definition the operator

$$p(t)(\psi) = \tilde{E}\left[\psi(x_t)\eta_t|\mathcal{Z}^t\right] \tag{6.3.29}$$

for any test function ψ. Although we can no longer rely on an equation for finding $p(t)$, we shall obtain an exact formula, through a direct checking.

By analogy with (6.3.6), (6.3.7) consider the pair of equations

$$\dot{P}_t + [P_t H_t^*(z) + \sigma_t(z)Q_t\alpha_t^*(z)]D_t^{-1}(z)[H_t(z)P_t + \alpha_t(z)Q_t\sigma_t^*(z)]$$
$$- \sigma_t(z)Q_t\sigma_t^*(z) - F_t(z)P_t - P_t F_t^*(z) = 0 \tag{6.3.30}$$
$$P(0) = P_0$$

and

$$\mathrm{d}\hat{x}_t = (F_t(z)\hat{x}_t + f_t(z))\mathrm{d}t + [P_t H_t^*(z) + \sigma_t(z)Q_t\alpha_t^*(z)]$$
$$\times D_t^{-1}(z)\left\{\mathrm{d}z_t - [H_t(z)\hat{x}_t + h_t(z)]\mathrm{d}t\right\} \tag{6.3.31}$$
$$\hat{x}(0) = x_0.$$

Clearly for any fixed z one can solve (6.3.30) as a deterministic equation. To derive bounds, it is useful to interpret P_t in relation to a control problem. This can be done as follows. Consider the backward state dynamics

$$-\frac{\mathrm{d}q}{\mathrm{d}t} = F_t^*(z)q + H_t^*(z)\xi \tag{6.3.32}$$
$$q(T) = \bar{q}$$

where ξ is a control, and $\bar{q}$ is a given datum (note that z enters parametrically).

We want to minimize the cost functional

$$J_{\bar{q}}(\xi) = q(0)^* P_0 q(0) + \int_t^T \xi^* D_t(\xi) dt$$

$$+ \int_0^T q^* \sigma_t(z) Q_t \sigma_t^*(z) q dt + 2 \int_0^T q^* \sigma_t(z) Q_t \alpha_t^*(z) \xi dt. \tag{6.3.33}$$

We can write a necessary condition of optimality as follows, using standard variational arguments. Let us consider the two point boundary value problem.

$$\dot{\phi} = (F - \sigma Q \alpha^* D^{-1} H)\phi - \sigma B \sigma^* \psi$$

$$\phi(0) = -P_0 \psi(0)$$

$$-\dot{\psi} = (F^* - H^* D^{-1} \alpha Q \sigma^*)\psi + H^* D^{-1} H \phi \tag{6.3.34}$$

$$\psi(T) = \bar{q}$$

then ψ is the optimal state, and ϕ the adjoint variable, whereas the optimal control is given by the formula

$$\hat{\xi} = D^{-1}(H\phi - \alpha Q \sigma^* \psi). \tag{6.3.35}$$

Moreover, we have by a decoupling argument

$$\phi(t) = -P(t)\psi(t) \tag{6.3.36}$$

where P is the solution of (6.3.30). Also

$$\inf J_{\bar{q}}(\xi) = \bar{q}^* P(T)\bar{q}. \tag{6.3.37}$$

Note that we can also write

$$J_{\bar{q}}(\xi) = q(0)^* P_0 q(0) + \int_0^T \xi^* R \xi dt$$

$$+ \int_0^T (q^* \sigma + \xi^* \alpha) Q(\sigma^* q + \alpha^* \xi) dt$$

which proves that $P(T) \geq 0$.

In addition

$$\bar{q}^* P(T)\bar{q} \leq J_{\bar{q}}(0) \leq C_T |\bar{q}|^2$$

where C_T depends only on the bounds of F_t and σ_t. Hence the solution of (6.3.30) is bounded uniformly with respect to z.

Of course the previous argument is formal. It is made rigorous by considering small intervals of time (i.e. T small), then obtaining the a priori estimates which allow the extension to any horizon.

Since P_t is bounded, the matrix $F_t - (P_t H_t^* + \sigma_t Q_t \alpha_t^*) D_t^{-1} H_t$ is bounded uniformly, and thus we can solve (6.3.31) without difficulty.

We can then define the operator

$$p(t)(\psi) = s_t \left[\int \psi(\hat{x}_t + P_t^{1/2}\xi) \frac{\exp(-\frac{1}{2}|\xi|^2)}{(2n)^{n/2}} d\xi \right] \tag{6.3.38}$$

where

$$
\begin{aligned}
s_t = \exp \bigg[& \int_0^t (\hat{x}^* H^* + h^*) D^{-1/2} \mathrm{d}\tilde{z} \\
& - \frac{1}{2} \int_0^t (\hat{x}^* H^* + h^*) D^{-1}(H\hat{x} + h) \mathrm{d}s \bigg]
\end{aligned}
\tag{6.3.39}
$$

and we want to prove

Theorem 6.3.2 *We assume (6.3.8), (6.3.9), (6.3.10), (6.3.11), (6.3.12), (6.3.13), (6.3.16), (6.3.24); then the equality (6.3.38) holds, where the left hand side must be interpreted as the unnormalized conditional probability defined by (6.3.29).*

Proof

We follow a method due to Lipster and Shiryaev (1977), also used by Haussmann and Pardoux (1980). The main point is to prove that

$$
\tilde{E}\left[\eta_T \exp(\mathrm{i}\rho^* x_T)|\mathcal{Z}^T\right] = k_T \exp\left(\mathrm{i}\rho^* \mu_T - \frac{1}{2}\rho^* \Gamma_t \rho\right)
\tag{6.3.40}
$$

where μ_T, Γ_T and k_T are $\mathcal{Z}^T$ measurable, μ_T and Γ_T take values in R^n and $\mathcal{L}(R^n; R^m)$ respectively. Suppose (6.3.40) is proven; then

$$
E\left[\exp(\mathrm{i}\rho^* x_T)|\mathcal{Z}^T\right] = \exp\left(\mathrm{i}\rho^* \mu_T - \frac{1}{2}\rho^* \Gamma_T \rho\right)
\tag{6.3.41}
$$

since necessarily $k_T = \tilde{E}\left[\eta_T|Z^T\right]$ and thus the conditional probability of x_T is a gaussian. It would remain to identify μ_T and Γ_T as $\hat{x}_T$ and P_T respectively. This will be done afterwards. Now turning to (6.3.18) we can write x_t as

$$
x_t = \bar{x}_t + y_t
$$

where $\bar{x}_t$ is $\mathcal{Z}^t$ measurable and y_t is the solution of

$$
\begin{aligned}
\mathrm{d}y_t &= \left[F_t(z) - \sigma_t(z)Q_t\alpha_t^*(z)D_t^{-1}(z)H_t(z)\right] y_t + \sigma_t(z)B_t^{1/2}(z)\mathrm{d}\tilde{w} \\
y(0) &= \xi.
\end{aligned}
$$

Next we can write

$$
\begin{aligned}
\eta_T \exp(\mathrm{i}\rho^* x_T) = \chi_T \exp \bigg\{ & \mathrm{i}\rho^* y_T + \int_0^T y_t^* H^* D^{-1/2}[\mathrm{d}\tilde{z} - \mathrm{D}^{1/2}(H\bar{x}_t + h_t)\mathrm{d}t] \\
& - \frac{1}{2} \int_0^T y_t^* H^* D^{-1} H y_t \mathrm{d}t \bigg\} \exp(\mathrm{i}\rho^* \bar{x}_T)
\end{aligned}
\tag{6.3.42}
$$

where χ_T is $\mathcal{Z}^T$ measurable and does not depend on ρ. Now define p_t by

$$
\begin{aligned}
-\mathrm{d}p &= (F^* - H^* D^{-1}\alpha Q\sigma^*)p + H^* D^{-1/2}[\mathrm{d}\tilde{z} - D^{-1/2}(H\bar{x}_t + h_t)\mathrm{d}t] \\
p(0) &= 0
\end{aligned}
\tag{6.3.42'}
$$

then the first integral inside the bracket on the right hand side of (6.3.42) reads

$$
\int_0^T p_t^* \sigma_t B_t^{1/2} \mathrm{d}\tilde{w} - p_T^* y_T.
$$

Now writing y_t as

$$y_t = \Phi(t,0)\left[\xi + \int_0^t \Phi(0,s)\sigma_s B_s^{1/2}\mathrm{d}\tilde{w}\right] = \Phi_t \zeta_t$$

where $\Phi(t,s)$ is the fundamental matrix corresponding to $F - \sigma Q\sigma^* D^{-1}H$, we see finally that we can write the quantity within brackets in (6.3.42), denoted X_T, as

$$X_T = (\mathrm{i}\rho^* - p_T^*)\Phi_T \zeta_T + \int_0^T p_t^* \sigma_t B_t^{1/2}\mathrm{d}\tilde{w}$$

$$- \frac{1}{2}\int_0^T \zeta_t^* \Phi_t^* H_t^* H_t^{-1} H_t \Phi_t \zeta_t \mathrm{d}t \qquad (6.3.43)$$

which is of the form

$$X_T = b_T^* \zeta_T + \int_0^T g^*(t)\mathrm{d}\tilde{w} - \frac{1}{2}\int_0^T \zeta_t^* S_t \zeta_t \mathrm{d}t \qquad (6.3.44)$$

where b_t is $\mathcal{Z}^T$ measurable, $g(t)$, ζ_t are $\mathcal{Z}^t$ measurable, and

$$\zeta_t = \xi + \int_0^t \Lambda(s)\mathrm{d}\tilde{w}$$

with Λ, $\mathcal{Z}^t$ adapted. Now, since $\tilde{z}$ and $\tilde{w}$, ξ are independent, we compute $\tilde{E}\left[\exp X_T | \mathcal{Z}^T\right]$ by freezing the value of z and taking the expectation with respect to ξ, and $\tilde{w}$. This can be done using the result of Lemma 6.3.3 below. Since b_T depends linearly on ρ, the result (6.3.41) obtains.

It remains to identify μ_T and Γ_T as $\hat{x}_t$ and P_t. Now from the result (6.3.40) we can assert that

$$p(t)(\psi(t)) = k_t\left[\int \psi(\mu_t + \Gamma_t^{1/2}\xi, z, t)\frac{\exp(-\frac{1}{2}|\xi|^2)\mathrm{d}\xi}{(2n)^{n/2}}\right] \qquad (6.3.45)$$

where $\psi(x,z,t)$ is defined on $R^n \times C([0,T];R^m) \times [0,T]$, is Borel in all the arguments and

$$\forall t, x, \quad z \to \psi(x,\cdot,t) \text{ is } \mathcal{B}^t \text{ measurable.}$$

Of course the left hand side of (6.3.45) is by definition

$$p(t)(\psi(t)) = \tilde{E}\left[\psi(x,z,t)\eta_t | \mathcal{Z}^t\right].$$

We shall now derive an equation which looks like the equation of non linear filtering, although cannnot be used as such, because of lack of uniqueness. Nevertheless, it will be sufficient to find the equations for the moments μ_t, Γ_t as well as k_t.

We consider a smooth function of x only, ϕ bounded as well as its derivatives. Let also

$$\theta_t = \exp\left(\int_0^t \mathrm{i}\beta_s \mathrm{d}\tilde{z}_s + \frac{1}{2}\int_0^t |\beta_s|^2 \mathrm{d}s\right)$$

where $\beta \in L^\infty(0,T;R^m)$. We compute the Ito differential of $\theta_t \eta_t \phi(x_t)$, which yields

$$\mathrm{d}\theta_t \eta_t \phi(x_t) = \theta\eta\left(\left[\mathrm{D}_x\phi^*(Fx_t + f + \mathrm{i}\sigma Q\alpha^* D^{-1/2}\beta)\right.\right.$$

$$\left. + \frac{1}{2}\mathrm{tr}\,\mathrm{D}_x^2\phi Q + \phi\mathrm{i}\beta^* D^{-1/2}(Hx_t + h)\right]\mathrm{d}t + \mathrm{D}_x\phi^*\sigma B^{1/2}\mathrm{d}\tilde{w}$$

$$\left. + \left\{\mathrm{D}_x\phi^*\sigma Q\alpha^* D^{-1/2} + \phi[\mathrm{i}\beta^* + (x^* H^* + h^*)D^{-1/2}]\right\}\mathrm{d}\tilde{z}\right).$$

An approximation is necessary to justify that we can take the expectation and assert that the expectation of the stochastic integrals vanish (one has to replace η by $\eta/1 + \epsilon\eta$; details are left to the reader). We deduce that

$$
\begin{aligned}
\tilde{E}\theta_t\eta_t\phi(x_t) = {}&\Pi_0(\phi) + \tilde{E}\int_0^t \theta_s\eta_s\Big[D_x\phi^*(Fx_s + f + i\sigma Q\alpha^* D^{-1/2}\beta) \\
&+ \frac{1}{2}\mathrm{tr}\, D_x^2\phi Q + \phi i\beta^* D^{-1/2}(Hx_s + h)\Big]\mathrm{d}s.
\end{aligned}
\tag{6.3.46}
$$

Note that by virtue of (6.3.25) the right hand side integral is well defined.

Using the notation of the operator $p(t)$, we can write (6.3.46) as

$$
\begin{aligned}
\tilde{E}\theta_t p(t)(\phi) = {}&\Pi_0(\phi) \\
&+ \tilde{E}\int_0^t \theta_s p(s)\Big[D_x\phi^*(F_s x + f_s) + \frac{1}{2}\mathrm{tr}\, D_x^2\phi Q_s\Big]\mathrm{d}s \\
&+ \tilde{E}\int_0^t i\theta_s\beta^* D_s^{-1/2}\left[p(s)(\phi(H_s x + h_s)) + \alpha_s Q_s\sigma_s^* p(s)(D_x\phi)\right]\mathrm{d}s \\
= {}&\Pi_0(\phi) \\
&+ \tilde{E}\theta_t\bigg\{\int_0^t p(s)\Big[D_x\phi^*(F_s x + f_s) + \frac{1}{2}\mathrm{tr}\, D_x^2\phi Q_s\Big]\mathrm{d}s \\
&+ \int_0^t p(s)\left[(x^* H_s^* + h_s^*)\phi + D_x\phi^*\sigma_s Q_s\alpha_s^*\right] D_s^{-1/2}\mathrm{d}\tilde{z}\bigg\}
\end{aligned}
$$

and thus we have proven that the operator $p(t)$ given by (6.3.15) satisfies the relation

$$
\begin{aligned}
p(t)(\phi) = {}&\Pi_0(\phi) + \int_0^t p(s)\Big[D_x\phi^*(F_s x + f_s) + \frac{1}{2}\mathrm{tr}\, D_x^2\phi\sigma_s Q_s\sigma_s^*\Big]\mathrm{d}s \\
&+ \int_0^t p(s)\left[(x^* H_s^* + h_s^*)\phi + D_x\phi^*\sigma_s Q_s\alpha_s^*\right] D_s^{-1/2}\mathrm{d}\tilde{z}.
\end{aligned}
\tag{6.3.47}
$$

We emphasize again that the equation (6.3.47) does not characterize $p(t)$. Nevertheless, since we know the explicit form of $p(t)$, it will be sufficient to identify k_t, μ_t, Γ_t. Note also the technical difficulty that the integrands in the stochastic integral on the right hand side of (6.3.47) are not L^2 functions. We will proceed slightly formally, avoiding the details or approximations which should be performed in order to justify the derivatives fully. Take $\phi = 1$ in (6.3.47) which yields

$$
\begin{aligned}
k_t &= 1 + \int_0^t p(s)\,(x^* H_s^* + h_s^*)\, D_s^{-1/2}\mathrm{d}\tilde{z} \\
&= 1 + \int_0^t k(s)(\mu_s^* H_s^* + h_s^*)D_s^{-1/2}\mathrm{d}\tilde{z}
\end{aligned}
$$

hence

$$
k_t = \exp\left[\int_0^t (\mu_s^* H_s^* + h_s^*)D_s^{-1/2}\mathrm{d}\tilde{z} - \frac{1}{2}\int_0^t (\mu_s^* H_s^* + h_s^*)D_s^{-1}(H_s\mu_s + h)\mathrm{d}s\right].
\tag{6.3.48}
$$

Next take $\phi = x$ (although not bounded), which yields

$$
\begin{aligned}
k_t \mu_t =& x_0 + \int_0^t p(s)(F_s x + f_s)\mathrm{d}s + \int_0^t p(s)(xx^* H_s^* + xh_s^* + \sigma_s Q_s \alpha_s^*)\mathrm{D}_s^{-1/2}(z)\mathrm{d}\tilde{z} \\
=& x_0 + \int_0^t k_s(F_s \mu_s + f_s)\mathrm{d}s \\
& + \int_0^t k_s(\Gamma_s H_s^* + \mu_s \mu_s^* H_s^* + \mu_s h_s^* + \sigma_s Q_s \alpha_s^*)\mathrm{D}_s^{-1/2}(z)\mathrm{d}\tilde{z}.
\end{aligned}
$$

Using (6.3.48), after some calculation, yields

$$
\begin{aligned}
\mathrm{d}\mu_t =& (F_t \mu_s + f_t)\mathrm{d}t + (\Gamma_t H_t^* + \sigma_t Q_t \alpha_t^*)\mathrm{D}_t^{-1/2}\mathrm{d}\tilde{z} \\
& - (\Gamma_t H_t^* + \sigma_t Q_t \alpha_t^*)\mathrm{D}_t^{-1}(H_t \mu_t + h_t)\mathrm{d}t
\end{aligned} \tag{6.3.49}
$$

$$
\mu(0) = x_0.
$$

Next, using $\phi = xx^*$, we obtain

$$
\begin{aligned}
k_t(\Gamma_t + \mu_t \mu_t^*) =& P_0 + x_0 x_0^* + \int_0^t k_s \left[F_s(\Gamma_s + \mu_s \mu_s^*) + (\Gamma_s + \mu_s \mu_s^*)F_s^* \right. \\
& + f_s \mu_s^* + \mu_s f_s^* + \sigma_s Q_s \sigma_s^* \right] \mathrm{d}s \\
& + \int_0^t k_s \left[(\Gamma_s + \mu_s \mu_s^*)(\mu_s^* H_s^* + h_s^*)\mathrm{D}_s^{-1/2}\mathrm{d}\tilde{z} \right. \\
& + \mu_s \mathrm{d}\tilde{z}^* \mathrm{D}_s^{-1/2}(H_s \Gamma_s + \alpha_s Q_s \sigma_s^*) \\
& + (\Gamma_s H_s^* + \sigma_s Q_s \alpha_s^*)\mathrm{D}_s^{-1/2}\mathrm{d}\tilde{z}\mu_s^* \left. \right]
\end{aligned} \tag{6.3.50}
$$

in which we have used the fact that the odd moments of the gaussian law vanish.

After some lengthy but easy calculations, using (6.3.49), (6.3.48) in (6.3.50), one obtains that Γ_t satisfies exactly equation (6.3.30) and thus

$$
\Gamma_t = P_t
$$

which implies, comparing (6.3.49) and (6.3.31), that $\mu_t = \hat{x}_t$. Finally comparing (6.3.48) and (6.3.49), one obtains $k_t = s_t$.

The proof has been completed.   ∎

The justification of the computation of $\tilde{E}[\exp X_T | \mathcal{Z}^T]$ follows from the following result, due to Lipster and Shiryaev (1977); (cf. also Haussmann and Pardoux 1986).

Lemma 6.3.2 *Let w be a standard Wiener process, ξ a gaussian random variable independent from w, and let*

$$
X_T = b_T^* \zeta_T + \int_0^T g^*(t)\mathrm{d}w - \frac{1}{2}\int_0^T \zeta_s^* S_t \zeta_t \mathrm{d}t
$$

where

$$
\zeta_t = \xi + \int_0^t \Lambda(s)\mathrm{d}w
$$

and g, S, λ, b *are deterministic functions.*

Define $\Gamma(t)$ and $q(t)$ by the formulas

$$\dot{\Gamma} = -S + \Gamma\Lambda\Lambda^*\Gamma, \quad \Gamma(T) = 0$$
$$\dot{q} = \Gamma\lambda\Lambda^*q + \Gamma\Lambda g, \quad q(T) = b_T$$

then one has the formula

$$
E \exp X_T = \left(\exp\left\{ q_0^* x_0 - \frac{1}{2} x_0^* \Gamma_0 x_0 \right.\right.
$$
$$
+ \frac{1}{2}(x_0^*\Gamma_0 - q_0^*)P_0^{1/2}(I + P_0^{1/2}\Gamma_0 P_0^{1/2})^{-1}P_0^{1/2}(\Gamma_0 x_0 - q_0)
$$
$$
\left.\left. + \frac{1}{2}\int_0^T [(g^* + q^*\Lambda)(g + \Lambda^*q) - tr\,\Gamma\Lambda\Lambda^*]\,dt \right\} \right) \tag{6.3.51}
$$
$$
\times \frac{1}{\det(I + P_0^{1/2}\Gamma_0 P_0^{1/2})}
$$

where x_0 and P_0 are the mean and covariance of ξ.

Proof

By the definition of Γ, one has

$$
X_T = b_T^*\zeta_T + \int_0^T g^*dw - \frac{1}{2}\xi^*\Gamma_0\xi - \int_0^T \zeta^*\Gamma_t\Lambda dw
$$
$$
- \frac{1}{2}\int_0^T \zeta_t^*\Gamma\Lambda\Lambda^*\Gamma\zeta_t dt - \frac{1}{2}\int_0^T tr\,\Gamma\Lambda\Lambda^* dt.
$$

Define a new probability $\bar{P}$ by the Radon–Nikodym derivative

$$
\frac{d\bar{P}}{dP} = \exp\left(-\int_0^T \zeta_t^*\Gamma_t\Lambda dw - \frac{1}{2}\int_0^T \zeta_t^*\Gamma\Lambda\Lambda^*\Gamma\zeta_t dt \right)
$$

and set

$$
\bar{w} = w + \int_0^t \lambda^*\Gamma\zeta ds.
$$

By Girsanov's theorem, $\bar{w}$ is a standard Wiener process for $\bar{P}$. Moreover

$$
b_T^*\zeta_T + \int_0^T g^*dw = \int_0^T (g^* + q^*\Lambda)d\bar{w} + q_0^*\xi
$$

hence

$$
E \exp X_T = \bar{E} \exp\left\{ \int_0^T (g^* + q^*\Lambda)d\bar{w} + q_0^*\xi - \frac{1}{2}\xi^*\Gamma_0\xi \right\}
$$
$$
\times \exp\left(-\frac{1}{2}\int_0^T tr\,\Gamma\Lambda\Lambda^* dt \right).
$$

Since $\bar{w}$ and ξ are independent, we can compute separately $E \exp \int_0^T (g^* + q^*\Lambda)d\bar{w}$ and $\bar{E} \exp(q_0^*\xi - \frac{1}{2}\xi^*\Gamma_0\xi)$.

The desired formula (6.3.51) is easily deduced. ∎

We can turn to the assumption (6.3.24), which can be made more precise. We write, following (6.3.42),

$$\eta_T = \chi_T \exp\left(-p_T^* y_T + \int_0^T p_t^* \sigma_t B_t^{1/2} d\tilde{w} - \frac{1}{2} \int_0^T y_t^* H^* D^{-1} H y_t dt \right)$$

with

$$\chi_T = \exp\left[\int_0^T (\bar{x}_t^* H_t^* + h_t^*) D_t^{-1/2} d\tilde{z} - \frac{1}{2} \int_0^T (\bar{x}_t^* H_t^* + h_t^*) D_t^{-1} (H_t \bar{x}_t + h_t) dt \right]$$

and p_t has been defined in (6.3.42'). In fact, also

$$\begin{aligned} p_t &= -\Phi^*(0,t) \int_0^t \Phi^*(s,0) H^* D^{-1/2} [d\tilde{z} - D^{-1/2} (H\bar{x}_s + h_s) ds] \\ &= -\Phi^*(0,t) b_t \end{aligned}$$

then

$$-p_T^* y_T = b_T^* \zeta_T$$

(see (6.3.44) for the definition of ζ_t).

We can then perform the calculation of $\tilde{E}\eta_T$ by first taking the expectation with respect to the pair $\xi, w(\cdot)$. This leads to formula (6.3.51). It remains an expression depending on z, with bounded quantities, except for h_t and f_t. In this way (6.3.24) can be made explicit, through a very complicated (although algebraic) formula. We shall not elaborate it completely.

6.3.5   A more general set up

Haussmann and Pardoux (1986) have considered a more general set up for the conditionally gaussian case. We present it in this paragraph. Compared with § 6.3.3 there are some simplifications to avoid too lengthy development. Let us consider as in (6.3.8)

$F_t(z)$, $f_t(z)$, $\sigma_t(z)$, $H_t(z)$, $h_t(z)$, Borel functions on $[0,T] \times C_T$

with values respectively in $\mathcal{L}(R^n; R^m)$, R^n, $\mathcal{L}(R^n; R^m)$, $\mathcal{L}(R^n; R^m)$, R^m (6.3.52)

and, for each t, $\mathcal{B}^t$ measurable.

For $j = 1, \ldots, m$, $G_t^j(z)$, g_t^j are Borel functions on $[0,T] \times C_T$ with values in $\mathcal{L}(R^n; R^n)$ and R^n respectively and, for each t, $\mathcal{B}^t$ measurable.   (6.3.53)

$$F_t, H_t, G_j^j \text{ are bounded.} \qquad (6.3.54)$$

Consider a probability space $(\Omega, \mathcal{A}, \tilde{P}, \mathcal{F}^t)$ on which are defined two independent standard Wiener processes w_t, z_t, with values in R^n and R^m respectively. Let also

ξ be a gaussian random variable, with values in R^n, $\mathcal{F}^0$ measurable, with mean x_0 and covariance matrix P_0.   (6.3.55)

We assume† that

$$\tilde{E}\int_0^T |f_t(z)|^2 dt, \quad \tilde{E}\int_0^T |h_t(z)|^2 dt, \quad \tilde{E}\int_0^T |g_t^j(z)|^2 dt,$$
$$\tilde{E}\int_0^T \operatorname{tr} \sigma_t \sigma_t^*(z) dt < \infty.$$

(6.3.56)

One then defines the process x_t (the state) by

$$dx_t = (F_t(z)x_t + f_t(z))dt + \sigma_t(z)dw + \sum_{j=1}^m (G_t^j(z)x_t + g_t^j)dz_t^j$$

(6.3.57)

$$x(0) = \xi.$$

One next defines b_t by

$$db_t = dz_t - (H_t(z)x_t + h_t(z))dt$$
$$b(0) = 0$$

(6.3.58)

and η_t by

$$d\eta_t = \eta_t(x_t^* H_t^*(z) + h_t^*(z))dz_t$$
$$\eta(0) = 1.$$

(6.3.59)

Note that we can assert that

$$\tilde{E}\sup_{0 \le t \le T} |x_t|^2 \le C.$$

(6.3.60)

We also assume, as in (6.3.24), that

$$\tilde{E}\eta_t = 1$$

(6.3.61)

and thus, defining the probability P by

$$\left.\frac{dP}{d\tilde{P}}\right|_{\mathcal{F}^t} = \eta_t,$$

we can make use of the Girsanov theorem to assert that, for the system $\Omega, \mathcal{A}, P, \mathcal{F}^t$, w_t and b_t are independent standard $\mathcal{F}^t$ Wiener processes (the situation is simpler than that of § 6.3.3). We begin by proving that the conditional probability of x_t, given $\mathcal{Z}^t = \sigma(z(s), s \le t)$, is for this system a gaussian, which, in a similar way to that in Theorem 6.3.2, amounts to proving the following result.

Lemma 6.3.3 *One has the property*

$$\tilde{E}\left[\eta_T \exp(i\rho^* x_T)|\mathcal{Z}^T\right] = k_T \exp\left(i\rho^* \mu_T - \frac{1}{2}\rho^* \Gamma_T \rho\right)$$

(6.3.62)

where k_T, μ_T, Γ_T are $\mathcal{Z}^T$ measurable and do not depend on ρ.

---

† We do not need σ_t bounded in this set up. It was needed in (6.3.18) because of the second term in front of x_t, which does not exist here.

Proof

We split x_t in two parts

$$x_t = \bar{x}_t + y_t$$

where $\bar{x}_t$ is $\mathcal{Z}^T$ measurable, and y_t is the solution of

$$dy_t = F_t y_t dt + \sum_{j=1}^{m} G_t^j y_t dz_t^j + \sigma_t dw$$

$$y(0) = \xi$$

and we can write
$$\eta_T \exp(i\rho^* x_T)$$

$$= \chi_T \exp(i\rho^* \bar{x}_T) \exp \left\{ i\rho^* y_T + \int_0^T y_t^* H_t^* [dz - (H_t \bar{x}_t + h_t)dt] - \frac{1}{2} \int_0^T |H_t y_t|^2 dt \right\}.$$

$$(6.3.63)$$

Define p_t by

$$-dp_t = \left(F_t^* - \sum_{j=1}^{m} (G^j)^*(G_j)^* \right) p dt + H_t^* [dz - (H_t \bar{x}_t + h_t)dt]$$

$$+ \sum_j (G^j)^*(p dz^j - (H^*)^j dt)$$

$$(6.3.64)$$

$$p(0) = 0$$

(where $(H^*)^j$ is the jth column of the matrix H^*) and we have again

$$\int_0^T y_t^* H_t^* [dz - (H_t \bar{x}_t + h_t)dt] = -p_T^* y_T + \int_0^T p^* \sigma dw.$$

Define the fundamental matrix $\Phi(t)$ by

$$d\Phi = F_t \Phi dt + \sum_j G_t^j \Phi dz^j$$

$$\Phi(0) = I$$

which possesses an inverse given by the solution of

$$d\Psi = \Psi \left(-F_t + \sum_j G_t^j G_t^j \right) dt - \sum_j \Psi G_t^j dz^j$$

$$\Psi(0) = I.$$

Therefore we can write

$$y_t = \Phi(t) \left[\xi + \int_0^t \Psi(s)\sigma(s)dw \right] = \Phi_t \zeta_t$$

and the quantity within brackets in the right hand side of (6.3.63), still denoted by X_T, reads

$$X_T = (i\rho^* - p_T^*)\Phi_T \zeta_T + \int_0^T p_t^* \sigma_t dw - \frac{1}{2} \int_0^T \zeta_t^* \Phi_t^* H_t^* H_t \Phi_t \zeta_t dt$$

and is again of the form (6.3.44).

This is sufficient to infer the desired result, by the same argument as in Theorem 6.3.2. ∎

It remains to identify the conditional mean μ_t and the conditional covariance Γ_t.

To guess the equations, we proceed with a formal calculation, along the lines of Theorem 6.3.2. Define the unnormalized conditional probability

$$p(t)(\psi(t)) = \tilde{E}\left[\psi(x_t, z, t)\eta_t | \mathcal{Z}^t\right]. \tag{6.3.65}$$

We know that

$$p(t)(\psi(t)) = k_t \left[\int \psi(\mu_t + \Gamma_t^{1/2}\xi, z, t)\frac{\exp(-\frac{1}{2}|\xi|^2)}{(2n)^{n/2}}d\xi\right] \tag{6.3.66}$$

We now derive an equation for $p(t)$, sufficient to obtain the moment equations.

Let

$$\theta_t = \exp\left(\int_0^t i\beta_s dz_s + \frac{1}{2}\int_0^t |\beta_s|^2 ds\right)$$

and for ϕ smooth we compute the Ito differential of $\theta_t\eta_t\phi(x_t)$, which yields

$$d\theta\eta\phi(x_t) = \theta\eta\left\{\left[D_x\phi^*(Fx + f + \sum_j(i\beta_j + x^*(H^*)^j + h_j)(G^j x + g^j)\right]\right.$$

$$+ \frac{1}{2}\mathrm{tr}\,D_x^2\phi\left[\sigma\sigma^* + \sum_j(G^j x + g^j)(x^*(G^j)^* + (g^j)^*)\right]$$

$$+ \phi i\beta^*(Hx + h) + D_x\phi^*\left(\sigma dw + \sum_j(G^j x + g^j)dz^j\right)$$

$$\left. + \phi(i\beta^* + x^*H^* + h^*)dz\right\}.$$

Taking the mathematical expectation yields

$$\tilde{E}\theta_t\eta_t\phi(x_t)$$

$$= \Pi_0(\phi) + \tilde{E}\int_0^t \theta\eta\left[D_x\phi^*\left(Fx + f + \sum_j(i\beta_j + x^*(H^*)^j + h_j)(G^j x + g^j)\right)\right.$$

$$\left. + \frac{1}{2}\mathrm{tr}\,D_x^2\phi\left(\sigma\sigma^* + \sum_j(G^j x + g^j)(x^*(G^j)^* + (g^j)^*)\right) + \phi i\beta^*(Hx + h)\right]ds$$

hence as for (6.3.47) we get

$$p(t)(\phi) = \Pi_0(\phi) + \int_0^t p(s)\left[D_x\phi^*\left(Fx + f + \sum_j(x^*(H^*)^j + h_j)(G^j x + g^j)\right)\right.$$

$$\left. + \frac{1}{2}\mathrm{tr}\,D_x^2\phi\left(\sigma\sigma^* + \sum_j(G^j x + g^j)(x^*(G^j)^* + (g^j)^*)\right)\right]$$

$$+ \int_0^t \sum_j p(s)\left[\phi(x^*(H^*)^j + h_j) + D_x\phi^*(G^j x + g^j)\right]dz^j.$$

$$\tag{6.3.67}$$

Taking first $\phi = 1$, we get

$$k_t = 1 + \int_0^t k_s(\mu_s^* H_s^* + h_s^*)\mathrm{d}s$$

hence

$$k_t = \exp\left[\int_0^t (\mu_s^* H_s^* + h_s^*)\mathrm{d}z - \frac{1}{2}\int_0^t |H_s\mu_s + h_s|^2\mathrm{d}s\right]. \qquad (6.3.68)$$

It is convenient at this stage to use (6.3.68) and (6.3.67) to derive an equation for the conditional probability itself:

$$\Pi(t)(\phi) = E\left[\phi(x_t)|\mathcal{Z}^t\right] = \frac{p(t)(\phi)}{k_t}.$$

After some calculations, one obtains the relation

$$\begin{aligned}
\mathrm{d}\Pi(t)(\psi) =\Pi(t)&\left[D_x\phi^*\left(Fx + f + \sum_j (x^*(H^*)^j + h_j)(G^j x + g^j)\right)\right.\\
&\left.+\frac{1}{2}\mathrm{tr}\,D_x^2\phi\left(\sigma\sigma^* + \sum_j (G^j x + g^j)(x^*(G^j)^* + (g^j)^*)\right)\right]\mathrm{d}t \qquad (6.3.69)\\
&+\sum_j \left\{\Pi(t)(\phi(x^*(H^*)^j + h_j)) - \Pi(t)(\phi)(\mu_t^*(H^*)^j + h_j)\right.\\
&\left.+\Pi(t)(D_x\phi^*(G^j x + g^j))\right\}(\mathrm{d}z^j - (\mu_t^*(H^*)^j + h_j)).
\end{aligned}$$

Applying (6.3.69) with $\phi(x) = x$, we obtain

$$\begin{aligned}
\mathrm{d}\mu_t =&\left[F_t\mu_t + f_t + G_t^j \Gamma_t(H_t^*)^j - \Gamma_t H_t^*(H_t\mu_t + h_t)\right]\mathrm{d}t\\
&+ \sum_j [\Gamma_t(H^*)^j + G_t^j \mu_t + g_t^j]\mathrm{d}z^j \qquad (6.3.70)\\
\mu(0) =&x_0
\end{aligned}$$

and applying (6.3.69) with $\phi = xx^*$, we obtain

$$\begin{aligned}
\mathrm{d}\Gamma_t + \mathrm{d}\mu_t\mu_t^* =\Pi(t)&\left[x(x^*F^* + f^*) + (Fx + f)x^*\right.\\
&+ \sum_j (x^*(H^*)^j + h_j)(x(x^*(G^j)^* + (g^j)^*) + (G^j x + g^j)x^*)\\
&\left.+ \sigma\sigma^* + \sum_j (G^j x + g^j)(x^*(G^j)^* + (g^j)^*)\right]\mathrm{d}t\\
&+ \sum_j \left\{\Pi(t)(xx^*(x^*(H^*)^j + h_j)) - \Pi(t)(xx^*)(\mu_t^*(H^*)^j + h_j)\right.\\
&\left.+ \Pi(t)(x(x^*(G^j)^* + (g^j)^*) + (G^j x + g^j)x^*)\right\}\\
&\times (\mathrm{d}z^j - (\mu_t^*(H^*)^j + h_j)\mathrm{d}t)
\end{aligned}$$

and thus performing some easy (although tedious) calculations yields

$$d\Gamma_t = [\Gamma_t F_t^* + F_t \Gamma_t + \sum_j G_t^j \Gamma_t (G_t^j)^* - \Gamma_t H_t^* H_t \Gamma_t + \sigma_t \sigma_t^*] dt$$
$$+ \sum_j [\Gamma_t (G_t^j)^* + G_t^j \Gamma_t] dz_t^j \tag{6.3.71}$$
$$\Gamma(0) = P_0.$$

Unlike the situation of (6.3.30), it does not seem possible to interpret (6.3.71) in the context of a control problem. However, we may claim that there exists an a *priori* bound which arises from the interpretation of Γ_T as the conditional covariance. We can deduce an exact formula for Γ_t, from Lemma 6.3.4. Indeed recall the matrices Φ_t and Ψ_t introduced in Lemma 6.3.4; set

$$\Lambda_t = \Psi_t \sigma_t, \quad S_t = \Phi_t^* H_t^* H_t \Phi_t.$$

We solve the Riccati equation

$$\dot{K} = -S + K \Lambda \Lambda^* K, \quad K(T) = 0$$

and the linear equation

$$\dot{q} = K \Lambda \Lambda^* q, \quad q(T) = \rho$$

then we can assert that

$$\rho^* \Gamma_T \rho = \int_0^T q^* \Lambda q \, dt + q_0^* P_0^{1/2} (I + P_0^{1/2} K_0 P_0^{1/2})^{-1} P_0^{1/2} q_0. \tag{6.3.72}$$

Since Λ_t, S_t are bounded on any finite interval of time (not uniformly in z though, because Φ_t, Ψ_t have diffusion terms), we can assert that this is the case for Γ_t.

For more details on this case, refer to the articles of Haussmann and Pardoux (1986).

7 Some explicit controls for systems with partial observation

Introduction

In Chapters 2 and 3 the LQG (linear quadratic gaussian) and LEG (linear exponential gaussian) control problems, in the fully as well as partially observable cases, have been considered and solved. As already pointed out, the solution was reached directly, without use of the more elaborate theories of optimal stochastic control and non linear filtering theory.

In this chapter, we make use of these theories to present a few cases, known in the literature, for which explicit solutions can be provided.

As a general starting point the observation will be defined a *priori* as a Wiener process, and we shall make use of the Girsanov transformation to recover the model we would like. It should be emphasized that this rules out the most general treatment of the LQG problem (see § 3.1.2), but it contains the development of § 2.4.1 as a particular case.

The general separation principle is studied in Sections 7.1–7.3. Particular cases are considered in Sections 7.4-7.6. The concept of the approximated separation principle is developed in Section 7.7.

7.1 The separation principle

7.1.1 Setting of the problem

Let us consider

F_t, f_t, H_t, h_t, bounded functions with values in $\mathcal{L}(R^n; R^m)$, R^n, $\mathcal{L}(R^n; R^m)$, R^m. $\qquad(7.1.1)$

$g_t(v)$, a Borel map on $[0, T] \times U_{ad}$, with values in R^n, where U_{ad} is a subset of R^k (the set of controls), $\qquad(7.1.2)$

$$|g_t(v)| \leq c(1 + |v|).$$

Also let $l_t(x, v)$ be a Borel map on $[0, T] \times R^n \times U_{ad}$, and $k(x)$ Borel on R^n, such that

$$|l_t(x, v)| \leq c(1 + |x|^2 + |v|^2)$$
$$|k(x)| \leq c(1 + |x|^2). \qquad(7.1.3)$$

(These growth conditions can be relaxed somewhat.) Now let $(\Omega, \mathcal{A}, P, \mathcal{F}^t)$ be the type of system, on which are defined two independent Wiener processes w_t, z_t, with values in R^n and R^m respectively, whose covariance matrices are Q_t, R_t; we assume

$$R_t \geq \bar{r}I, \quad \bar{r} > 0. \qquad(7.1.4)$$

Let $\mathcal{Z}^t = \sigma(z(s), \ s \leq t)$, and $L^2_Z(0, T; R^k)$ be the set of square integrable stochastic processes, adapted to $\mathcal{Z}^t$, with values in R^k.

The set of *pre-admissible controls* is defined by

$$\tilde{\mathcal{U}} = \left\{ v \in L^2_Z(0, T; R^k), \quad v_t \in U_{ad}, \text{ a.e. a.s.} \right\}. \qquad(7.1.5)$$

Given $v \in \mathcal{U}$, it is clear from (7.1.2) that

$$g_t(v_t) \in L^2_Z(0, T; R^n).$$

We can define the process x_t by the equation

$$dx_t = (F_t x_t + f_t + g_t(v_t))dt + dw_t$$
$$x(0) = \xi \qquad(7.1.6)$$

where

ξ is $\mathcal{F}^0$ measurable, gaussian, with mean x_0 and covariance matrix P_0. $\qquad(7.1.7)$

Clearly $x(\cdot) \in L^2_{\mathcal{F}^t}(0, T; R^n)$ and

$$E \sup_{0 \le t \le T} |x_t|^2 < \infty.$$

Next define η^v_t by

$$d\eta^v_t = \eta^v_t (x^*_t H^*_t + h^*_t) R^{-1}_t dz_t, \quad \eta^v_0 = 1 \qquad (7.1.8)$$

i.e.

$$\eta^v_t = \exp \left(\int_0^t x^* H^* R^{-1} dz - \frac{1}{2} \int_0^t x^* H^* R^{-1} Hx ds \right).$$

We proceed along the lines of § 2.4.1. The process η^v_t satisfies

$$E\eta^v_t \le 1$$

and not necessarily

$$E\eta^v_t = 1.$$

As in (2.4.24) it is somewhat natural to restrict the admissible controls to satisfy the property

$$\mathcal{U} = \left\{ v(\cdot) \in \tilde{\mathcal{U}} \,\middle|\, E \int_0^T \eta^v_t |v(t)|^2 dt < \infty \right\}. \qquad (7.1.9)$$

For admissible controls one can check that

$$\eta^v_t \text{ is a } \mathcal{F}^t, P \text{ martingale.} \qquad (7.1.10)$$

The proof is quite similar to that of Lemma 2.4.2, and will be omitted.

Thanks to (7.1.10) we can define a new probability by setting

$$\left. \frac{dP^v}{dP} \right|_{\mathcal{F}^t} = \eta^v_t. \qquad (7.1.11)$$

We next define the cost function

$$J(v(\cdot)) = E^v \left[\int_0^T l_t(x_t, v_t) dt + m(x_T) \right]. \qquad (7.1.12)$$

Remark 7.1.1 In Christopeit and Helmes (1983), one assumes first that η^v_t is a martingale and then the condition

$$E^v \int_0^T |v(t)|^2 dt < \infty.$$

Although natural, this is slightly redundant.

7.1.2   The separated problem

We know from the Liptser and Shiryaev theory (1977) — see § 6.3.3 — that the conditional probability law of under $\mathcal{Z}^t$ is a gaussian. The conditional mean is the solution of the Kalman filter

$$d\hat{x}_t = (F_t \hat{x}_t + f_t + g_t(v_t))dt + P_t H^*_t R^{-1}_t [dz_t - (H_t \hat{x}_t + h_t)dt]$$

$$\hat{x}(0) = x_0, \qquad (7.1.13)$$

whereas P_t is the solution of the Riccati equation

$$\dot{P}_t = F_t P_t + P_t F_t^* - P_t H_t^* R_t^{-1} H_t P_t + Q_t$$
$$P(0) = P_0.$$
$$(7.1.14)$$

We may also proceed as in Lemma 2.4.3, writing

$$x_t = \alpha_t + x_{1t}$$

where

$$d\alpha_t = (F_t \alpha_t + f_t)dt + dw_t$$
$$\alpha(0) = \xi$$

and proving that

$$\hat{x}_t = x_{1t} + \hat{\alpha}_t$$

where $\hat{\alpha}_t$ is the Kalman filter corresponding to α_t. Of course, under P^v, z appears as the solution of

$$dz_t = (H_t x_t + h_t)dt + db_{v,t}$$
$$z(0) = 0$$
$$(7.1.15)$$

where $b_{v,t}$ is a $\mathcal{F}^t$ Wiener process with covariance R. Moreover the innovation

$$I_{v,t} = z_t - \int_0^t (H_s \hat{x}_s + h_s)ds$$

is a $\mathcal{Z}^t$ Wiener process with covariance R (cf. (2.4.37)), and thus $\hat{x}_t$ appears as the solution of

$$d\hat{x}_t = (F_t \hat{x}_t + f_t + g_t(v_t))dt + P_t H_t^* R_t^{-1} dI_v$$
$$\hat{x}(0) = x_0.$$
$$(7.1.16)$$

On the other hand, we can write the functional $J(v(\cdot))$ as follows:

$$J(v(\cdot)) = E^v \left[\int_0^T \tilde{l}_t(\hat{x}_t, v_t)dt + \tilde{m}(\hat{x}_T) \right]$$
$$(7.1.17)$$

where

$$\tilde{l}_t(x, v) = \int l_t(x + P_t^{1/2}\xi, v) \frac{\exp(-\frac{1}{2}|\xi|^2)}{(2n)^{n/2}} d\xi$$
$$(7.1.18)$$

$$\tilde{m}_t(x) = \int m(x + P_t^{1/2}\xi) \frac{\exp(-\frac{1}{2}|\xi|^2)}{(2n)^{n/2}} d\xi.$$
$$(7.1.19)$$

It is clear from (7.1.3) that $\tilde{l}_t(x, v)$ and $\tilde{m}(x)$ have the same properties as l_t, m.

Therefore the problem is reduced to a problem of stochastic control with full observation, with the minor restriction that the source of noise I_v is not given *a priori*, although it is indeed a $\mathcal{Z}^t$, P^v Wiener process with covariance $R(\cdot)$.

In Chapter 2, we have previously introduced the concept of the *separation* principle (see the Comments section at the end of the chapter). Briefly speaking, the idea consists of solving (7.1.16), (7.1.17) viewing I_v as a given Wiener process, and showing that the solution thus obtained, as a feedback on the Kalman filter, is indeed the optimal one for the original problem.

Christopeit and Helmes (1982, 1983) have formalized this question and have given a framework in which conditions on the solution of (7.1.16), (7.1.17) to be fulfilled ensure that one obtains in this way a solution of the original problem. They have applied this framework in three examples. One is of course the LQG problem, considered in Chapter 2; the other two cases correspond to situations where U_{ad} is bounded. When U_{ad} is bounded the Bellman equation can be solved in a suitable way. This is why we shall deal with this case directly.

7.2   The Bellman equation for the separated problem when U_{ad} is bounded

7.2.1   Notation

We make the assumptions

$$U_{ad} \text{ is bounded} \tag{7.2.1}$$

$$P_t H_t^* R_t^{-1} H_t P_t \geq \alpha I, \quad \alpha > 0. \tag{7.2.2}$$

Naturally (7.2.2) is very restrictive, since it implies the invertibility of H_t. Define next the 'Hamiltonian'

$$\tilde{H}(x, t, p) = \inf_{v \in U_{ad}} \left\{ \tilde{l}_t(x, v) + p \cdot g_t(v) \right\} \tag{7.2.3}$$

and set

$$\tilde{a}(t) = \frac{1}{2} P_t H^* R_t^{-1} H_t P_t.$$

We consider the second order differential operator

$$\tilde{A}(t) = -\sum_{ij} \tilde{a}_{ij}(t) \frac{\partial^2}{\partial x_i \partial x_j} - \sum_j (F_{ij}(t) x_j + f_i(t)) \frac{\partial}{\partial x_i.} \tag{7.2.4}$$

Then the Bellman equation related to the problem (7.1.16), (7.1.17) reads

$$-\frac{\partial u}{\partial t} + \tilde{A}(t) u - \tilde{H}(x, t, Du) = 0$$
$$u(x, T) = \tilde{m}(x). \tag{7.2.5}$$

Naturally, to study (7.2.5), one has to make precise the functional space in which the solution of (7.2.5) is sought. This will require the use of Sobolev spaces with weights to deal with the fact that the space variable runs in the whole of R^n and the data grow to $+\infty$ at infinity. A weight is defined by

$$\Pi_s(x) = (1 + |x|^2)^{-s/2}. \tag{7.2.6}$$

We define

$$L_{\Pi_s}^r = \{ z | \Pi_s z \in L^r(R^n) \}$$
$$W_{\Pi_s}^{1,r} = \left\{ z \,\middle|\, z \in L_{\Pi_s}^r, \; \frac{\partial z}{\partial x_i} \in L_{\Pi_s}^r \right\}$$

and $W_{\Pi_s}^{2,r}$ in a similar way.

Considering functions $z(x,t)$ we shall use the spaces

$$W_{\Pi_s}^{2,1,r} = \left\{ z \,\bigg|\, z, \frac{\partial z}{\partial x_i}, \frac{\partial z}{\partial t}, \frac{\partial^2 z}{\partial x_i \partial x_j} \in L^r(0,T; L_{\Pi_s}^r) \right\}$$

and in a similar way $W_{\Pi_s}^{1,0,r}$.

For $r = 2$, we write $H_{\Pi_s}^{2,1}$, $H_{\Pi_s}^{1,0}$.

7.2.2   Study of (7.2.5)

Clearly since U_{ad} is bounded

$$|l_t(x,v)| \le K(1 + |x|^2)$$

and for $s > n/2 + 2$ one has $1 + |x|^2 \in L_{\Pi_s}^2$. We first prove the following result.

Theorem 7.2.1 *Assume (7.1.1), (7.1.2), (7.1.3), (7.1.4), (7.2.1), (7.2.2); then (7.2.5) has a unique solution in $H_{\Pi_s}^{1,0} \cap L^\infty(0,T; L_{\Pi_s}^2)$, $\forall s > n/2 + 2$.*

Proof

Let us first consider the linear equation

$$-\frac{\partial z}{\partial t} + \tilde{A}(t)z = f \tag{7.2.7}$$
$$z(x,T) = \tilde{m}$$

where we assume that

$$f \in L^2(0,T; L_{\Pi_s}^2). \tag{7.2.8}$$

Let us show that there exists a unique solution of (7.2.7) such that

$$z \in H_{\Pi_s}^{1,0} \cap L^\infty(0,T; L_{\Pi_s}^2). \tag{7.2.9}$$

We can first check that there exists an a priori estimate. Indeed multiplying (7.2.7) by $\Pi_s^2 z$ and integrating over R^n yields

$$-\frac{1}{2}\frac{d}{dt}|z(t)|_{\Pi_s}^2 + \sum \int_{R^n} \tilde{a}_{ij} \frac{\partial z}{\partial x_j}\left(\frac{\partial z}{\partial x_i}\Pi_s^2 - 2z\frac{\Pi_s^2 x_i}{1+|x|^2}\right) dx$$
$$-\sum \int_{R^n} F_{ij}x_j \cdot z\frac{\partial z}{\partial x_i}\Pi_s^2 dx = \int_{R^n} fz\Pi_s^2 dx. \tag{7.2.10}$$

But

$$-\sum \int_{R^n} F_{ij}x_j \cdot z\frac{\partial z}{\partial x_i}\Pi_s^2 dx = \frac{1}{2}\int_{R^n} \mathrm{tr}\, Fz^2\Pi_s^2 dx$$
$$- s\int_{R^n} z^2\frac{Fx\cdot x}{1+|x|^2}\Pi_s^2 dx. \tag{7.2.11}$$

We deduce from (7.2.11) used in (7.2.10) that

$$-\frac{1}{2}\frac{d}{dt}|z(t)|_{\Pi_s}^2 + \frac{\alpha}{2}\|z(t)\|_{\Pi_s}^2 \le c_s(|z(t)|_{\Pi_s}^2 + |f(t)|_{\Pi_s}^2).$$

Therefore it follows easily that

$$\|z\|_{L^\infty(0,T;L^2_{\Pi_s})} \le c_s[|\tilde{k}|_{L^2_{\Pi_s}} + |f|_{L^2(0,T;L^2_{\Pi_s})}]$$
$$\|z\|_{H^{1,0}_{\Pi_s}} \le c_s[|\tilde{k}|_{L^2_{\Pi_s}} + |f|_{L^2(0,T;L^2_{\Pi_s})}]$$

(7.2.12)

where the constant c_s depends only on (besides s) the norms of F, $\tilde{a}$ and α.

From such an *a priori* estimate it is easy to deduce the result (7.2.9). One first solves the problem by assuming $\tilde{k}$, f are bounded, in which case the maximum principle can be used to prove that the solution is bounded (note that the Green function is a gaussian) and thus the preceding calculation is justified. By linearity it extends to the case (7.2.12).

Now, to deal with the non linear problem, we shall rely on a contraction fixed point argument. Indeed, from (7.2.3) one has

$$|\tilde{H}(x,t,p_1) - \tilde{H}(x,t,p_2)| \le K|p_1 - p_2|.$$

(7.2.13)

Therefore consider $\zeta \in H^{1,0}_{\Pi_s}$ and z to be the solution of

$$-\frac{\partial z}{\partial t} + \tilde{A}(t)z = \tilde{H}(x,t,\zeta)$$
$$z(x,t) = \tilde{m}.$$

(7.2.14)

Clearly $\tilde{H}(x,t,\zeta) \in L^2_{\Pi_s}$ and thus from (7.2.7), (7.2.9) there exists a unique solution of (7.2.14) in $H^{1,0}_{\Pi_s} \cap L^\infty(0,T;L^2_{\Pi_s})$. Let ζ_1, ζ_2 be two elements of $H^{1,0}_{\Pi_s}$ and z_1, z_2 be the corresponding solutions of (7.2.14). One can deduce as in the calculation (7.2.10) that

$$-\frac{1}{2}\frac{d}{dt}|z_1 - z_2|^2_{\Pi_s} + \frac{\alpha}{2}\|z_1 - z_2\|^2_{\Pi_s}$$
$$\le c_0|z_1 - z_2|^2_{\Pi_s} + c_1\|\zeta_1 - \zeta_2\|_{\Pi_s}|z_1 - z_2|_{\Pi_s}$$
$$\le \delta\|\zeta_1 - \zeta_2\|^2_{\Pi_s} + c\left(\frac{1}{\delta} + 1\right)\|z_1 - z_2\|^2_{\Pi_s}$$

(7.2.15)

where δ is arbitrary. It follows that

$$-\frac{1}{2}\frac{d}{dt}e^{kt}|z_1 - z_2|^2_{\Pi_s} + \frac{k}{2}e^{kt}|z_1 - z_2|^2 + \frac{\alpha}{2}e^{kt}\|z_1 - z_2\|^2$$
$$\le \delta e^{kt}\|\zeta_1 - \zeta_2\|^2 + ce^{kt}\left(\frac{1}{\delta} + 1\right)|z_1 - z_2|^2.$$

Taking $\delta = \alpha/4$ and choosing $k = 2c(1 + 4/\delta)$, we obtain from this inequality

$$\int_0^t e^{kt}\|z_1 - z_2\|^2_{\Pi_s}\,ds \le \frac{1}{2}\int_0^T e^{kt}\|\zeta_1 - \zeta_2\|^2\,dt.$$

Since $\int_0^T e^{kt}\|z\|^2_{\Pi_s}\,ds$ is a norm on $L^2(0,T;H^1_{\Pi_s})$ equivalent to $\int_0^T \|z\|^2\,ds$, the map $\zeta \mapsto z$ is a contraction.

This suffices to imply the existence and uniqueness of the solution of (7.2.5) in $H^{1,0}_{\Pi_s} \cap L^\infty(0,T;L^2_{\Pi_s})$. ∎

Remark 7.2.1 It is clear from the proof that Theorem 7.2.1 extends to drifts of the form $g_0(x) + g_1(x, v)$ where g_0 is Lipschitz and g_1 is bounded.

We can then assert a regularity result.

Theorem 7.2.2 *We make the assumptions of Theorem 7.2.1 and*

$$H_t, R_t \text{ are continuous on } [0, T] \tag{7.2.16}$$

$$m \text{ is twice differentiable and} \tag{7.2.17}$$

$$|Dm(x)|, |D^2 m(x)| \le C(1 + |x|^2).$$

Then the solution of (7.2.5) belongs to $W_{\Pi_s}^{2,1,q}$, where $\infty > q \ge 2$ and s is to be chosen depending on q.

Proof

We shall make use of the results of Ladyzenskaya, Solonnikov and Ural'tseva (1968) concerning solutions of linear parabolic equations in unbounded domains (see Theorem 9.1, pp. 341, 342 of that work).

We write (7.2.5) as

$$-\frac{\partial u}{\partial t} - a_{ij} \frac{\partial^2 u}{\partial x_i \partial x_j} = (Fx + f) \cdot Du + H(x, t, Du) \tag{7.2.18}$$

where we have omitted the symbol $\sim$ to simplify the notation.

Let us set for $\sigma > n/2 + 2$

$$\phi_\sigma = u \Pi_\sigma;$$

after multiplication by Π_σ and performing some easy calculations we obtain for ϕ_σ the equation

$$-\frac{\partial \phi_\sigma}{\partial t} - a_{ij} \frac{\partial^2 \phi_\sigma}{\partial x_i \partial x_j} - a_i \frac{\partial \phi_\sigma}{\partial x_i} - a_0 \phi_\sigma$$
$$= \Pi_\sigma((Fx + f) \cdot Du + H(x, t, Du)) \tag{7.2.19}$$
$$\phi_\sigma(x, T) = \tilde{m} \Pi_\sigma$$

where

$$a_i = \frac{2\sigma a_{ij} x_j}{1 + |x|^2}$$

$$a_0 = \frac{\sigma}{1 + |x|^2} \left[a_{ii} + (\sigma - 2) a_{ij} \frac{x_i x_j}{1 + |x|^2} \right].$$

Note that

$$a_i \in C^0([0, T]; L^r(R^n)) \quad \forall r > n \tag{7.2.20}$$
$$a_0 \in C^0([0, T]; L^s(R^n)) \quad \forall s > \frac{n}{2}.$$

Suppose that we know that at some stage

$$\Pi_{\sigma_0}[(Fx + f) \cdot Du + H(x, t, Du)] \in L^{q_0}(R^n \times [0, T]), \text{ for some } q_0 \ge 2 \tag{7.2.21}$$

then we can apply the theorem of LSU quoted above to claim that $\phi_{\sigma_0} \in W^{2,1,q_0}$, hence $u \in W^{2,1,q_0}_{\Pi_{\sigma_0}}$. Indeed from (7.2.17) $\tilde{k}\Pi_{\sigma_0} \in W^{2,q_0}$, and we can choose in (7.2.20)

$$r = \begin{cases} \max(q_0, n+2) & \text{if } q_0 \neq n+2 \\ n+2+\epsilon & \text{if } q_0 = n+2 \end{cases}$$

$$r = \begin{cases} \max(q_0, n/2+1) & \text{if } q_0 \neq n/2+1 \\ n/2+1+\epsilon & \text{if } q_0 = n/2+1. \end{cases}$$

The assumptions of the theorem are satisfied.

Once we know $u \in W^{2,1,q}_{\Pi_{\sigma_0}}$ we can use the embedding theorems (cf. Ladyzenskaya, Solonnikov and Ural'tseva 1968, p. 80). Two cases must be considered: $q_0 > n+2$ and $q_0 \leq n+2$. If $q_0 > n+2$, then

$$\Pi_{\sigma_0} Du \in C^{\lambda, \lambda/2}\dagger, \quad \text{with } \lambda = 1 - \frac{n+2}{q_0}.$$

In particular

$$Du \leq c(1 + |x|^2)^{\delta_0/2}.$$

Next consider (7.2.19) with $\sigma_1 > \sigma_0 + 1 + n/2$, then the function

$$\Pi_{\sigma_1}((Fx+f) \cdot Du + H(x, t, Du)) \text{ belongs to } L^q, \forall q, \ 2 \leq q < \infty.$$

Since also $\tilde{k}\Pi_{\sigma_1} \in L^q$, $\forall q$, $2 \leq q < \infty$, the above theorem of LSU yields $\phi_{\sigma_1} \in W^{2,1,q}$, $\forall q, 2 \leq q < \infty$, hence $u \in W^{2,1,q}_{\Pi_{\sigma_1}}$, $\forall q, 2 \leq q < \infty$, which is what we wanted to prove.

Suppose $q_0 \leq n+2$. The embedding theorem of LSU yields

$$\Pi_{\sigma_0} Du \in L^{q_1}, \quad \text{with } q_1 = \frac{q_0(n+2)}{n+2-q_0},$$

the value $+\infty$ being possible. If $q_1 = +\infty$ we are in fact in the previous case, so suppose $q_0 < n+2$, hence $q_1 < \infty$.

It follows that

$$\Pi_{\sigma_0+1}((Fx+f) \cdot Du + H(x, t, Du)) \in L^{q_1}$$

and thus we are back in the case (7.2.21) with $\sigma_1 = \sigma_0 + 1$ and q_1 instead of q_0.

Clearly we can start with $q_0 = 2$ and $\sigma_0 > n/2+3$. This follows from Theorem 7.2.1. Then $q_1 = 2(n+2)/n$. If $n = 1, 2$, it is finished with any $\sigma_1 > n+4$. If $n > 2$, then we proceed with q_1, $\sigma_1 = \sigma_0 + 1$ and define $q_2 = q_1(n+2)/(n+2-q_1)$.

After a finite number of steps we reach a situation with $q_k \geq n+2$. This follows from the fact that

$$0 < \frac{1}{q_{k+1}} = \frac{1}{q_k} - \frac{1}{n+2}$$

which forbids an infinite number of steps. ∎

---

† The space of Hölder continuous functions. $\phi \in C^{\lambda, \frac{\lambda}{2}}$ means

$$\sup_{\substack{x, x' \in R^n \\ t, t' \in [0, T]}} \left[\frac{|\phi(x, t) - \phi(x', t)|}{|x - x'|^\lambda} + \frac{|\phi(x, t) - \phi(x, t')|}{|t - t'|^{\lambda/2}} \right] < \infty$$

and ϕ bounded, with $\lambda = 1 - (n+2)/q_0$.

From the regularity theorem, it follows in particular that

$$\Pi_\sigma u, \ \Pi_\sigma Du \in C^{\lambda,\lambda/2} \quad 1 > \lambda > 0,$$
$$\text{for some convenient } \sigma > n/2 + 2. \tag{7.2.22}$$

Remark 7.2.2 We can replace $Fx + f$ by $g_0(x)$ Lipschitz, and $g(v)$ by $g_1(x,v)$ bounded. Theorem 7.2.2 will remain valid.   ∎

7.3 Solution of the stochastic control problem with partial information when U_{ad} is bounded

7.3.1 Solution of the separated problem

Consider now in addition to the assumptions of Theorem 7.2.2 that

$$g_t(v), l_t(x, v) \text{ are continuous functions of } x, v. \tag{7.3.1}$$

$$\text{The minimum of } \tilde{l}_t(x, v) + p \cdot g_t(v) \text{ in } v \in U_{ad} \text{ is unique.} \tag{7.3.2}$$

Then, by virtue of (7.3.1), (7.3.2), we can define a function $\hat{V}(x, t, p)$ such that

$$\hat{V} \text{ is Borel with respect to all arguments, continuous with respect to } x, p \text{ and minimizes } \tilde{l}_t(x, v) + p \cdot g_t(v), \text{ for } v \in U_{ad}. \tag{7.3.3}$$

Setting

$$\hat{v}(x, t) = \hat{V}(x, t, \mathrm{D}u) \tag{7.3.4}$$

we obtain a Borel map, continuous with respect to x, with values in U_{ad}. This follows, of course, from the regularity result on u (see (7.2.22)).

Moreover $\hat{v}(x, t)$ satisfies

$$\tilde{H}(x, t, \mathrm{D}u) = \tilde{l}_t(x, \hat{v}(x, t)) + \mathrm{D}u(x, t) \cdot g_t(\hat{v}(x, t)) \quad \forall x, t. \tag{7.3.5}$$

We turn now to solving (7.1.13) with the feedback thus obtained. This amounts to the equation

$$\begin{aligned}
\mathrm{d}\hat{y} =& (F_t\hat{y}_t + f_t + g_t(\hat{v}(\hat{y}_t, t))\mathrm{d}t \\
& + P_t H_t^* R_t^{-1}[\mathrm{d}z_t - (H_t\hat{y}_t + h_t)\mathrm{d}t] \tag{7.3.6} \\
\hat{y}(0) =& x_0.
\end{aligned}$$

The difficulty is that we want to show that $\hat{y}$ has a strong solution, which for us means a process $\hat{y}_t$ adapted to $\mathcal{Z}^t$. Only if this is possible can we claim that $\hat{v}(\hat{y}_t, t)$ is an admissible control (at least a preadmissible control (see (7.1.5)), but in fact an admissible control, since U_{ad} is bounded and thus $\mathcal{U} = \tilde{\mathcal{U}}$, see (7.1.9)). Let us notice that if we could consider the problem with *full* information, that is

$$\begin{aligned}
\mathrm{d}\hat{x}_t =& (F_t\hat{x}_t + f_t + g_t(v_t))\mathrm{d}t + P_t H_t^* R_t^{-1}\mathrm{d}I_v \\
\hat{x}(0) =& x_0 \\
J(v(\cdot)) =& E^v\left[\int_0^T \tilde{l}_t(\hat{x}_t, v_t)\mathrm{d}t + \tilde{m}(\hat{x}_T)\right]
\end{aligned}$$

with v_t adapted to $\mathcal{Z}^t$, $I_{v,t}$ a process adapted to $\mathcal{Z}^t$, which is a Wiener process for a probability P^v which we may choose conveniently (not the one defined by (7.1.11)). The feedback $\hat{v}(x,t)$ would then be optimal without the strongest assumption of uniqueness (7.3.2). This is a consequence of the classical theory of stochastic control, which allows weak sense trajectories (see for instance Bensoussan 1982). Unfortunately, this is not sufficient in our framework since, in the way we stated the problem, we need strong sense trajectories. We shall return to this point in Section 7.6. But, thanks to our assumptions, which imply that $\hat{v}$ is continuous with respect to x, we can define $\hat{y}_t$ in a strong sense, as we shall see in the next paragraph.

7.3.2 Strong solutions of (7.3.6)

We have the following

Proposition 7.3.1 *The equation (7.3.6) has a strong solution, i.e. a process satisfying*

$$\hat{y}_t \text{ is adapted to } \mathcal{Z}^t$$

$$E \sup_{0 \le t \le I} |\hat{y}_t|^2 \le C, \tag{7.3.7}$$

and

$$
\hat{y}_t = x_0 + \int_0^t [(F_s - P_s H_s^* R_s^{-1} H_s)\hat{y}_s + g_s(\hat{v}(\hat{y}_s, s)) + f_s - P_s H_s^* R_s^{-1} h_s] ds
$$
$$
+ \int_0^t P_s H_s^* R_s^{-1} dz_s, \quad \forall t, \text{ a.s.} \tag{7.3.8}
$$

Proof

Define μ_t by

$$d\mu_t = [(F_t - P_t H_t^* R_t^{-1} H_t)\mu_t + f_t - P_t H_t^* R_s^{-1} h_t] dt + P_t H_t^* R_t^{-1} dz_t$$

$$\mu(0) = x_0$$

then clearly solving (7.3.8) amounts to solving

$$\frac{d\xi_t}{dt} = (F_t - P_t H_t^* R_t^{-1} H_t)\xi_t + g_t(\hat{v}(\mu_t + \xi_t, t)) \tag{7.3.9}$$

$$\xi(0) = 0$$

which we consider as an ordinary differential equation depending parametrically on μ_t. To simplify notation write

$$\phi_t(\mu_t + \xi_t) = g_t(\hat{v}(\mu_t + \xi_t, t))$$

and $\phi_t(x)$ satisfies

$$\phi_t(x) \text{ is Borel in } x, t, \text{ continuous in } x, \text{ and bounded.} \tag{7.3.10}$$

Therefore we have

$$\frac{d\xi_t}{dt} = A_t \xi_t + \phi_t(\xi_t + \mu_t) \tag{7.3.11}$$

$$\xi(0) = 0$$

where, in this context

$$A_t = F_t - P_t H_t^* R_t^{-1} H_t. \dagger$$

---

† Not to be confused with the operator $\tilde{A}_t$.

The solution of (7.3.9) can be obtained through the classical Carathéodory techniques (see Coddington and Levinson 1955). It amounts to defining a sequence ξ_t^k such that

$$
\begin{aligned}
\frac{d\xi_t^k}{dt} &= A_t \xi_t^k + \phi_t(\xi_{t-k}^k + \mu_t), \quad k < t < T \\
\frac{d\xi_t^k}{dt} &= A_t \xi_t^k + \phi_t(\mu_t), \quad 0 < t < k \\
\xi_0^k &= 0.
\end{aligned}
\tag{7.3.12}
$$

Clearly ξ_t^k remains in a bounded subset of $H^1(0,T)$. Hence we can extract a subsequence $\xi_t^{k'}$ such that

$$
\begin{aligned}
\xi_t^{k'} \to \xi_t \quad &\text{in } H^1(0,T) \text{ weakly and} \\
&\text{in } C^0([0,T]) \text{ strongly.}
\end{aligned}
$$

Let

$$
\tilde{\xi}_t^k = \begin{cases} \xi_{t-k}^k & \text{if } t \geq k \\ 0 & \text{if } 0 < t \leq k. \end{cases}
$$

Clearly also

$$
\begin{aligned}
\xi_t^{k'} \to \xi_t \quad &\text{in } H^1(0,T) \text{ weakly and} \\
&\text{in } C^0([0,T]) \text{ strongly.}
\end{aligned}
$$

Note that (7.3.12) can be written as

$$
\frac{d\xi_t^k}{dt} = A_t \xi_t^k + \phi_t(\tilde{\xi}_t^k + \mu_t)
$$

and it is easy to pass to the limit, obtaining (7.3.11).

The process ξ_t^k is clearly $\mathcal{Z}^t$ adapted, and thus the solution ξ_t obtained in the limit of $\xi_t^{k'}$ is also $\mathcal{Z}^t$ adapted.

The process $\hat{y}_t = \xi_t + \mu_t$ satisfies the conditions (7.3.7) and (7.3.8). ∎

7.3.3 Optimality result

Given a solution of (7.3.7), (7.3.8), called a strong solution of (7.3.6), we define

$$
\hat{v}_t = \hat{v}(\hat{y}_t, t)
\tag{7.3.13}
$$

which is an admissible control (see (7.1.5), (7.1.9) and recall that $\mathcal{U} = \tilde{\mathcal{U}}$).

Our main result is the following.

Theorem 7.3.1 *We assume (7.1.1), (7.1.2), (7.1.3), (7.1.4), (7.2.1), (7.2.2), (7.2.16), (7.2.17), (7.3.1), (7.3.2); a process $\hat{v}_t$ defined by (7.3.13), where $\hat{y}_t$ is a strong solution of (7.3.6), is optimal for the problem (7.1.6), (7.1.12).*

Proof

Let us first check that

$$
u(x_0, 0) \leq J(v(\cdot))
\tag{7.3.14}
$$

for any $v(\cdot)$ admissible, where $u(x,t)$ is the solution of the Bellman equation (7.2.5) and x_0 the initial value of $\hat{x}_t$ (7.1.16). Take the ball of radius R centered on x_0, denoted $B_R(x_0)$, and let T_R be the first exit time of $\hat{x}_t$ from the boundary of B_R.

The function $u \in L^q(0,T;W^{2,q}(B_R))$ and $\partial u/\partial t \in L^q(0,T;L^q(B_R))$, $\forall q, 2 \leq q < \infty$. Therefore we can use the integrated form of Ito's formula (see for instance Bensoussan and Lions 1978, Ch. 2) to deduce that

$$E^v u(\hat{x}(T \wedge T_R), T \wedge T_R) = u(x_0,0) + E^v \int_0^{T \wedge T_R} \left(\frac{\partial u}{\partial t} - \tilde{A}(t)u + g_t(v_t) \cdot Du(\hat{x},t) \right) dt$$

where the right hand side is well defined, even though $\partial u/\partial t - \tilde{A}(t)u$ is only $L^q(B_R \times (0,T))$.

Now, from (7.2.5), it follows,

$$E^v u(\hat{x}(T \wedge T_R), T \wedge T_R) = u(x_0,0) + E^v \int_0^{T \wedge T_R} (-\tilde{H} + g_t(v_t) \cdot Du)(\hat{x}_t,t)dt.$$

The function $\tilde{H}(x,t,Du) - g_t(v_t) \cdot Du$ satisfies

$$\tilde{H}(x,t,Du) - g_t(v_t) \cdot Du \leq \tilde{l}_t(x,v_t)$$

hence we obtain

$$E^v u(\hat{x}(T \wedge T_R), T \wedge T_R) \geq u(x_0,0) - E^v \int_0^{T \wedge T_R} l_t(\hat{x}_t,v_t)dt.$$

As $R \to +\infty$, $T_R \to +\infty$ and

$$u(\hat{x}(T \wedge T_R), T \wedge T_R) \to u(\hat{x}(T),T) = \tilde{m}(\hat{x}_T).$$

Moreover we know that

$$|u(x,t)| \leq C(1 + |x|^2)^{\delta/2} \quad \text{for a convenient } \sigma$$

and

$$|u(\hat{x}(T \wedge T_R), T \wedge T_R)| \leq C \sup_{0 \leq t \leq T} (1 + |\hat{x}_t|^2)^{\delta/2}.$$

The term on the right hand side has a finite P^v expectation (recalling in particular that $g_t(v_t)$ is bounded). Therefore, from Lebesgue's theorem, we can infer that

$$E^v u(\hat{x}(T \wedge T_R), T \wedge T_R) \to E^v \tilde{m}(\hat{x}_T),$$

and thus (7.3.14) is proven.

Now let us check that

$$u(x_0,0) = J(\hat{v}(\cdot)) \tag{7.3.15}$$

where $\hat{v}_t$ is defined by (7.3.13).

Consider $\hat{v}_t$, define x_t by (7.1.6), $\eta_t^{\hat{v}}$ by (7.1.8) and $P^{\hat{v}}$ by (7.1.11). Since the Kalman filter is given by (7.1.13), we can assert, comparing with (7.3.6), that the Kalman filter coincides with $\hat{y}_t$ and thus, from (7.1.17),

$$J(\hat{v}(\cdot)) = E^{\hat{v}} \left[\int_0^T \tilde{l}_t(\hat{y}_t, \hat{v}_t)dt + \tilde{m}(\hat{y}_T) \right].$$

Noting that

$$\tilde{H}(\hat{y}_t, t, \mathrm{D}u(\hat{y}_t, t)) = \tilde{l}_t(\hat{y}_t, \hat{v}_t) + \mathrm{D}u(\hat{y}_t, t) \cdot g_t(\hat{v}_t)$$

and proceeding as above, we easily check that

$$u(x_0, 0) = J(\hat{v}(\cdot))$$

and the desired result has been proven. ∎

7.4 Solution of the Bellman equation in some particular cases, with bounded controls

We discuss here two well-known examples where explicit solutions can be given.

7.4.1 Predicted-miss problem

We assume here

$U_{ad} = [-1, +1]^k$

$m(x) = \phi(\theta \cdot x)$

where θ is a given vector and $\phi(s)$ is even, $\phi \in C^2$, $\phi'(s) \geq 0$

for $s > 0$, $\phi'(0) = 0$; $|\phi(s)|, |\phi'(s)|, |\phi''(s)| \leq C(1 + s^2)$ (7.4.1)

$l_t(x, v) = 0$

$g_t(v) = G_t v$, G_t is a Borel bounded $m \times k$ matrix.

This problem in the fully observable case has been studied by several authors, in particular Benes (1975, 1976), Benes and Karatzas (1982), Clark and Davis (1979), Haussmann (1981), Ruzicka (1977) and for the partially observable case by Christopeit and Helmes (1982). Writing the Bellman equation (7.2.5) we obtain

$$-\frac{\partial u}{\partial t} - \tilde{a}_{ij}\frac{\partial^2 u}{\partial x_i \partial x_j} - (F_t x + f_t) \cdot Du + \sum_i \left| \sum_j G_{t,ji} \frac{\partial u}{\partial x_j} \right| = 0 \tag{7.4.2}$$

$$u(x, T) = \tilde{\phi}(\theta \cdot x)$$

where we have defined

$$\tilde{\phi}(s) = \int \phi(s + \theta^* P_T^{1/2} \xi) \frac{\exp(-\frac{1}{2}|\xi|^2)}{(2n)^{n/2}} d\xi$$

$$= \int_{-\infty}^{+\infty} \phi(s + (\theta^* P_T \theta)^{1/2} \lambda) \frac{(\exp -\frac{1}{2}\lambda^2)}{\sqrt{2\pi}} d\lambda. \tag{7.4.3}$$

Lemma 7.4.1 $\tilde{\phi}(s)$ is even, C^2, $\tilde{\phi}'(s) \geq 0$ for $s \geq 0$, $\tilde{\phi}'(0) = 0$, $|\tilde{\phi}(s)|, |\tilde{\phi}'(s)|, |\tilde{\phi}''(s)| \leq C(1 + s^2)$.

Proof

We may assume $(\theta^* P_T \theta)^{1/2} > 0$. Otherwise $\tilde{\phi} = \phi$. Let $a = \theta^* P_T \theta$, then one has

$$\tilde{\phi}(s) = a^{-1/2} \int_{-\infty}^{+\infty} \phi(\lambda) \exp\left(-\frac{1}{2}\frac{(\lambda - s)^2}{a}\right) d\lambda.$$

Clearly $\tilde{\phi}(s) = \tilde{\phi}(-s)$.

Let us check that $\tilde{\phi}(s_2) \geq \tilde{\phi}(s_1)$ if $s_2 \geq s_1 \geq 0$. We write

$$\tilde{\phi}(s_2) - \tilde{\phi}(s_1) = \int_{-\infty}^{s_1} + \int_{s_2}^{+\infty} + \int_{s_1}^{s_2}$$
$$= I_1 + I_2 + I_3$$

and

$$I_1 + I_2 = a^{-1/2} \int_0^{+\infty} (\phi(s_2 + \mu) - \phi(\mu_1 - s_1))$$
$$\left(\exp -\frac{1}{2a}\mu^2 - \exp -\frac{1}{2a}(\mu + s_2 - s_1)^2 \right) d\mu$$
$$\geq 0$$

and

$$I_3 = a^{-1/2} \int_{s_1}^{s_2} \phi(\lambda) \left[\exp - \left(\frac{1}{2}\frac{(\lambda - s_2)^2}{a} \right) - \exp \left(-\frac{1}{2}\frac{(\lambda - s_1)^2}{a} \right) \right] d\lambda$$
$$= a^{-1/2} \int_0^{\Delta} \left[\exp \left(-\frac{1}{2}(\lambda - \Delta)^2 \right) - \exp \left(-\frac{1}{2}(\lambda + \Delta)^2 \right) \right]$$
$$\times \left[\phi \left(\lambda + \frac{s_1 + s_2}{2} \right) - \phi \left(\frac{s_1 + s_2}{2} - \lambda \right) \right] d\lambda$$

where $\Delta = (s_2 - s_1)/2$ and $I_3 \geq 0$. Therefore $\tilde{\phi}'(s) \geq 0$. The rest is obvious. ∎

We look for a solution of (7.4.2) of the form

$$u(x, t) = z(r_t^* x + \nu_t, t) \tag{7.4.4}$$

where $z(\lambda, t)$ is a function of a one dimensional variable λ, and t; in addition r_t and ν_t can be chosen. Let us check that such a choice is possible, with convenient functions z and θ_t.

Substituting in (7.4.2) yields

$$-\frac{\partial z}{\partial t} - \frac{\partial^2 z}{\partial \lambda^2} r_t^* \tilde{a}_t r_t - \frac{\partial z}{\partial \lambda} [\dot{r}_t^* x + \dot{\nu}_t + r_t^*(F_t x + f_t)]$$
$$+ \sum_i |(G_t^* r_t)_i| \left| \frac{\partial z}{\partial \lambda} \right| = 0$$
$$z(\lambda, T) = \tilde{\phi}(\lambda), \quad r_T = \theta, \quad \nu_T = 0$$

and thus if we define r_t, ν_t as

$$\dot{r}_t + F_t^* r_t = 0, \quad r_T = \theta$$
$$\dot{\nu}_t + f_t^* r_t = 0, \quad \nu_T = 0 \tag{7.4.5}$$

and z as the solution of the problem

$$-\frac{\partial z}{\partial t} - \frac{\partial^2 z}{\partial \lambda^2} \sigma_t^2 + \left| \frac{\partial z}{\partial \lambda} \right| \gamma_t = 0$$
$$z(\lambda, T) = \tilde{\phi}(\lambda) \tag{7.4.6}$$

where $\sigma_t^2 = r_t^* \tilde{a}_t r_t$, $\gamma_t = \sum_i |(G_t^* r_t)_i|$, we obtain that (7.4.4) is a solution of (7.4.2).

It remains to solve (7.4.6).

Consider the linear PDE

$$-\frac{\partial \chi}{\partial t} - \frac{\partial^2 \chi}{\partial \lambda^2} \sigma_t^2 + \frac{\partial \chi}{\partial \lambda} \gamma_t \frac{\lambda}{|\lambda|} = 0 \tag{7.4.7}$$

$$\chi(\lambda, T) = \tilde{\phi}(\lambda).$$

It has a unique solution in $W_{\Pi_s}^{2,1,q}(R \times (0,T))$, for some convenient s. This follows from Theorem 7.2.2 applied in a very particular case. The function χ, and its space derivative $\partial \chi / \partial \lambda$, are such that

$$\Pi_s \chi, \ \Pi_s \frac{\partial \chi}{\partial \lambda} \ \text{are Hölder continuous in } R^n \times [0, T].$$

Then one has

Lemma 7.4.2 *The solution z of (7.4.6) coincides with χ, the solution of (7.4.7).*

Proof

It is enough to show that

$$\frac{\partial \chi}{\partial \lambda} \geq 0 \text{ for } \lambda \geq 0$$

$$\frac{\partial \chi}{\partial \lambda} \leq 0 \text{ for } \lambda \leq 0. \tag{7.4.8}$$

Note first that $\chi(-\lambda, t)$ is also a solution of (7.4.7). Hence by uniqueness

$$\chi(\lambda, t) = \chi(-\lambda, t)$$

which implies that χ is even. Therefore only the first part of (7.4.8) has to be proven.

Call $\psi = \partial \chi / \partial \lambda$, then $\psi \in W_{\Pi_s}^{1,0,q}(R \times (0,T))$, in particular $\Pi_s \psi$ is Hölder continuous. From (7.4.7) one has

$$-\frac{\partial \psi}{\partial t} - \frac{\partial^2 \psi}{\partial \lambda^2} \sigma_t^2 + \frac{\partial \psi}{\partial \lambda} \gamma_t = 0, \quad \lambda > 0 \tag{7.4.9}$$

where $\partial \psi / \partial t$, $\partial^2 \psi / \partial \lambda^2$ are of course to be taken in the distributional sense.

One has also

$$\psi(0, t) = 0 \tag{7.4.10}$$

and

$$\psi(\lambda, T) = \tilde{\phi}'(\lambda), \quad \lambda > 0. \tag{7.4.11}$$

Note that $\tilde{\phi}'(\lambda) \geq 0$, for $\lambda > 0$. The conditions (7.4.9), (7.4.10), (7.4.11) imply, by maximum principle considerations, that $\psi \geq 0$. The desired result has been proven. ∎

We can then state the following.

Theorem 7.4.1 *We make the assumptions (7.1.1), (7.1.4), (7.2.2), (7.2.16), and (7.4.1); then the solution of the Bellman equation (7.4.2) is given by the formula*

$$u(x, t) = \chi(r_t^* x + \nu_t, t) \tag{7.4.12}$$

where r_t, ν_t are defined by (7.4.5) and $\chi(\lambda, t)$ is the solution of the linear equation (7.4.7). ∎

We can deduce from the preceding result an 'optimal' feedback

$$\hat{v}_i(x,t) = -\frac{(G_t^* r_t)_i}{|(G_t^* r_t)_i|} \frac{r_t^* x + \nu_t}{|r_t^* x + \nu_t|}. \tag{7.4.13}$$

It is optimal in the sense that

$$\tilde{H}(x,t,Du) = \inf_{v \in U_{ad}} Du \cdot G_t v$$

$$= \sum_i \frac{\partial u}{\partial x_i} \hat{v}_i(x,t), \quad \forall x,t. \tag{7.4.14}$$

However, the function $\hat{v}(x,t)$ defined by (7.4.13) is not *continuous* with respect to x. This follows from the fact that the minimum in (7.4.14) cannot be unique for all values of x,t (in particular if $Du = 0$, any value of v is possible).

Therefore Theorem 7.3.1 cannot be applied as such to assert that $\hat{v}(x,t)$ defined by (7.4.13) yields an optimal control for the original problem.

What is missing is the possibility of defining a strong solution of (7.2.28) with that feedback $\hat{v}$. We shall return to this question in the next section, but before that we shall consider another case where we can solve the Bellman equation.

7.4.2   Minimum-distance problem

The problem has been studied by Benes (1975), Ikeda and Watanabe (1981), Ruzicka (1977), Christopeit and Helmes (1982). It concerns the following case:

$$U_{ad} = \{v \in R^k | \, |v| \leq 1\}$$
$$l_t(x,v) = 0, \quad m(x) = |x|^2 \tag{7.4.15}$$
$$g_t(v) = G_t v$$

and moreover if $\Phi(t,s)$ is the fundamental matrix corresponding to F_t then one assumes the properties of 'rotational invariance'

$$\Phi(T,t)\tilde{a}_t \Phi^*(T,t) = \gamma_t^2 I, \quad \gamma_t^2 \geq c_0 > 0 \tag{7.4.16}$$

$$\Phi(T,t)G_t G_t^* \Phi^*(T,t) = \beta_t^2 I, \tag{7.4.17}$$

γ_t, β_t continuous, and we choose β_t, $\gamma_t > 0$.

If we note that

$$\tilde{m}(x) = |x|^2 + \text{tr } P_T$$

the Bellman equation for this problem is given as follows

$$-\frac{\partial u}{\partial t} - \tilde{a}_{ij} \frac{\partial^2 u}{\partial x_i \partial x_j} - (F_t x + f_t) \cdot Du + |G_t^* Du| = 0$$

$$u(x,T) = |x|^2 + \text{tr } P_T. \tag{7.4.18}$$

We look for a solution of (7.4.18) of the form

$$u(x,t) = z(|M_t x + r_t|, t) \tag{7.4.19}$$

where M_t and r_t are conveniently chosen. In fact we shall see that

$$M_t = \Phi(T, t)$$
$$r_t = \int_t^T \Phi(T, s) f_s ds \qquad (7.4.20)$$

are good choices.

After some tedious but easy calculations, taking into account the relations (7.4.16), (7.4.17), one can check that $z(\lambda, t)$ should be the solution of

$$-\frac{\partial z}{\partial t} - \gamma_t^2 \frac{\partial^2 z}{\partial \lambda^2} - \frac{\partial z}{\partial \lambda} \gamma_t^2 \frac{(n-1)}{\lambda} + \beta_t \left| \frac{\partial z}{\partial \lambda} \right| = 0 \qquad (7.4.21)$$
$$z(\lambda, T) = \lambda^2 + \operatorname{tr} P_T.$$

The equation presents a singularity at 0, and thus differs from the form of the general framework (7.2.5).

We shall use a different weight, which removes the singularity. Define

$$\omega_s(\lambda) = |\lambda|^{s/2} (1 + \lambda^2)^{-s/2}. \qquad (7.4.22)$$

We notice the following relations:

$$\omega_s'(\lambda) = \frac{s\lambda}{1 + \lambda^2} \left(\frac{\omega_{s-4}}{2(1 + \lambda^2)} - \omega_s \right) \qquad (7.4.23)$$

$$\omega_s''(\lambda) = \frac{s(s-2)}{4(1 + \lambda^2)^2} \omega_{s-4} + s(s+2) \omega_{s+4} - \frac{(s+1)s}{1 + \lambda^2} \omega_s. \qquad (7.4.24)$$

Now multiply (7.4.21) by $z\omega_s^2$. After integrating by parts, one obtains (performing some calculations)

$$-\frac{1}{2} \frac{d}{dt} |z(t)|_{\omega_s}^2 + \gamma_t^2 \left| \frac{\partial z}{\partial \lambda} \right|_{\omega_s}^2 + \beta_t \int z \left| \frac{\partial z}{\partial \lambda} \right| \omega_s^2$$
$$+ \frac{1}{2} \gamma_t^2 (n-1) \int z^2 \left[-\frac{1}{\lambda^2} \omega_s^2 + \frac{1}{2} 2\omega_s \frac{s\lambda}{\lambda 1 + \lambda^2} \left(\frac{\omega_{s-4}}{2(1 + \lambda^2)} - \omega_s \right) \right] = 0. \qquad (7.4.25)$$

But

$$-\frac{1}{\lambda^2} \omega_s^2 + \frac{2s\omega_s}{1 + \lambda^2} \left(\frac{\omega_{s-4}}{2(1 + \lambda^2)} - \omega_s \right) = -\frac{\omega_{s-2}^2}{(1 + \lambda^2)^2} + \frac{s\omega_s \omega_{s-4}}{(1 + \lambda^2)^2} - \frac{2s\omega_s^2}{1 + \lambda^2}$$
$$= \frac{(s-1)\omega_{s-2}^2}{(1 + \lambda^2)^2} - \frac{2s\omega_s^2}{1 + \lambda^2}.$$

and thus (7.4.25) implies for $s \geq 2$

$$-\frac{1}{2} \frac{d}{dt} |z(t)|_{\omega_s}^2 + \gamma_t^2 \left| \frac{\partial z}{\partial \lambda} \right|_{\omega_s}^2 \leq \gamma_t s(n-1) |z|_{\omega_s}^2 - \beta_t \int z \left| \frac{\partial z}{\partial \lambda} \right| \omega_s^2.$$

Therefore also

$$-\frac{1}{2} \frac{d}{dt} |z(t)|_{\omega_s}^2 + \alpha \|z(t)\|_{\omega_s}^2 \leq c |z(t)|_{\omega_s}^2.$$

and thus we deduce that for $s > 5$

$$z \in L^2(0, T; H^1_{\omega_s}) \cap L^\infty(0, T; L^2_{\omega_s}). \tag{7.4.26}$$

Now take any ω_σ, $\sigma > 5$ and set

$$\phi_\sigma = z\omega_\sigma.$$

One deduces from (7.4.21) the relation

$$-\frac{\partial \phi_\sigma}{\partial t} - \gamma_t^2 \frac{\partial^2 \phi_\sigma}{\partial \lambda^2} + \frac{\lambda \gamma_t^2}{1 + \lambda^2} \left(\frac{\sigma - n + 1}{1 + \lambda^2} \phi'_{\sigma - 4} - 2\sigma \phi'_\sigma \right)$$

$$+ \gamma_t^2 \left[\frac{\phi_{\sigma - 4}}{2(1 + \lambda^2)^2} \left(n(\sigma - 4) - \frac{\sigma^2}{2} + 2\sigma + 4 \right) - \sigma(\sigma - 2)\phi_{\sigma + 4} \right.$$

$$\left. + \frac{\phi_\sigma}{1 + \lambda^2}(\sigma^2 - 4\sigma - 4 - n(\sigma - 4)) \right] + \beta_t \left| \phi'_\sigma - \frac{\sigma \lambda}{1 + \lambda^2} \left(\frac{\phi_{\sigma - 4}}{2(1 + \lambda^2)} - \phi_\sigma \right) \right| = 0. \tag{7.4.27}$$

From (7.4.26) we know that ϕ_{σ_0} $\phi'_{\sigma_0} \in L^2(R \times (0, T))$ for some convenient σ_0. Apply (7.4.27) with $\sigma_1 = \sigma_0 + 4$, and note that ϕ_{σ_1} $\phi'_{\sigma_1} \in L^2(R \times (0, T))$ since $\omega_{\sigma_1} \le \omega_{\sigma_0}$. According to the regularity theorem of LSU used in the proof of Theorem 7.2.2, $\phi_{\sigma_1} \in W^{2,1,2}$, hence from the embedding theorem ϕ_{σ_1} $\phi'_{\sigma_1} \in L^6(R \times (0, T))$. Applying (7.4.27) with $\sigma = \sigma_2 = \sigma_1 + 4$ we deduce $\phi_{\sigma_2} \in W^{2,1,6}(R \times (0, T))$ hence ϕ_{σ_2}, ϕ'_{σ_2} are Hölder continuous on $R \times [0, T]$. Applying (7.4.27) again with $\sigma = \sigma_3 = \sigma_4 + 4$, we can assert, turning to z,

$$z \in W^{2,1,q}_{\omega_\sigma}, \quad \forall q, \; 2 \le q < \infty$$

$$\omega_\sigma z, \, \omega_\sigma z' \text{ are Hölder continuous in } R \times [0, T]. \tag{7.4.28}$$

Now consider the linear equation

$$-\frac{\partial \chi}{\partial t} - \gamma_t^2 \frac{\partial^2 \chi}{\partial \lambda^2} - \frac{\partial \chi}{\partial \lambda} \gamma_t^2 \frac{(n - 1)}{\lambda} + \frac{\partial \chi}{\partial \lambda} \beta_t \frac{\lambda}{|\lambda|} = 0$$

$$\chi(\lambda, T) = \lambda^2 + \text{tr } P_T. \tag{7.4.29}$$

As for z we can assert that (7.4.29) has a unique solution in the functional space

$$\chi \in W^{2,1,q}_{\omega_\sigma}(R \times (0, T)), \quad \forall q, \; 2 \le q < \infty, \tag{7.4.30}$$

for some convenient σ. Moreover $\omega_\sigma \chi$, $\omega_\sigma \chi'$ are Hölder continuous.

We have

Lemma 7.4.3 *The function z, the solution of (7.4.21) coincides with χ, the solution of (7.4.29).*

Proof

As in Lemma 7.4.2, it is enough to show that

$$\frac{\partial \chi}{\partial \lambda} \ge 0 \quad \text{for} \quad \lambda \ge 0$$

$$\frac{\partial \chi}{\partial \lambda} \le 0 \quad \text{for} \quad \lambda \le 0. \tag{7.4.31}$$

Again χ is even, so we have to consider only the case $\lambda \geq 0$. Let $\psi = \partial \chi / \partial \lambda \in W_{\omega_o}^{1,0,q}$, then we see from (7.4.29)

$$-\frac{\partial \psi}{\partial t} - \gamma_t^2 \frac{\partial^2 \psi}{\partial \lambda^2} - \frac{\partial \psi}{\partial \lambda} \gamma_t^2 \frac{(n-1)}{\lambda} + \psi \gamma_t^2 \frac{(n-1)}{\lambda^2}$$
$$+ \frac{\partial \psi}{\partial \lambda} \beta_t = 0, \quad \lambda > 0 \tag{7.4.32}$$

and

$$\psi(0,t) = 0$$
$$\psi(\lambda, T) = 2\lambda.$$

The equation (7.4.32) presents an additional singularity corresponding to the zero order term, but the coefficient $\gamma_t^2(n-1)/\lambda^2 > 0$. Therefore the maximum principle applies to assert that $\psi \geq 0$, which proves the desired result. ∎

Therefore we can state the following result concerning the solution of the Bellman equation

Theorem 7.4.2 *We make the assumptions (7.1.1), (7.1.4), (7.2.2), (7.2.16) and (7.4.15), (7.4.16), (7.4.17); then the solution of (7.4.18) is given by the formula*

$$u(x,t) = \chi(|M_t x + r_t|, t)$$

where M_t, r_t are defined by (7.4.20) and χ is the solution of the linear problem (7.4.29).
∎

Here again we can deduce an 'optimal' feedback, although unfortunately it is not continuous.

Define

$$\hat{v}(x,t) = \frac{-G_t^* M_t^* (M_t x + r_t)}{\beta_t |M_t x + r_t|} \tag{7.4.33}$$

then using

$$|\hat{v}|^2 = \hat{v}^* \hat{v} = 1$$

and since

$$Du = \frac{\partial \chi}{\partial \lambda} \frac{M_t^* (M_t x + r_t)}{|M_t x + r_t|}$$

we have

$$-G_t \hat{v} \cdot Du = \frac{\partial \chi}{\partial \lambda}.$$

On the other hand

$$|G^* Du| = \left| \frac{\partial \chi}{\partial \lambda} \right| \frac{|G^* M^* (M x + r)|}{|M x + r|} = \frac{\partial \chi}{\partial \lambda}$$

therefore the feedback (7.4.33) satisfies

$$-G_t \hat{v} \cdot Du = |G^* Du| = - \inf_{|v| \leq 1} G_t v \cdot Du.$$

It is thus optimal.

Defining

$$\rho_t = \int_t^T \Phi(t,s) f_s \mathrm{d}s \qquad (7.4.34)$$

hence

$$M_t \rho_t = r_t$$

we can notice that, using (7.4.17) again,

$$G_t \hat{v}(x,t) = -\beta_t \frac{(x+\rho_t)}{|M_t(x+\rho_t)|}. \qquad (7.4.35)$$

But as in the predicted-miss problem the feedback obtained is not a continuous function of x.

7.5   Solution of the predicted-miss and minimum-distance problems

We shall prove in this section that, although the optimal feedbacks $\hat{v}$ obtained from the Bellman equation in these two cases are not continuous functions of the state, they still define optimal controls for the stochastic control problem with partial information.

Naturally the difficulty stems from solving the equation of the Kalman filter in a strong sense.

7.5.1   The predicted-miss problem

We follow the article of Christopeit and Helmes (1982), who used some ideas of Ruzicka (1977). In fact it is sufficient to prove *strong uniqueness*. This comes from the theory of Yamada and Watanabe (1971) which guarantees that strong uniqueness and weak existence imply strong existence.

Let

$$\phi_N(x) = \begin{cases} 1 & \text{if } x \geq 1/N \\ -1 & \text{if } x \leq -1/N \\ Nx/2(3 - N^2x^2) & \text{if } -1/N \leq x \leq 1/N \end{cases}$$

which is C^1, $|\phi_N| \leq 1$, and

$$\phi'_N(x) = \begin{cases} 0 & \text{if } |x| \geq 1/N \\ 3N/2(1 - N^2x^2) & \text{if } |x| \leq 1/N. \end{cases}$$

The function ϕ_N will be a convenient smoothing of the function sign.

We shall use some notation of § 7.3.2; see (7.3.11) in particular. From the feedback (7.4.13) we can write

$$G_t\hat{v}(x,t) = \rho_t \operatorname{sign}(r_t^*x + \nu_t)$$

where ρ_t is a fixed vector (bounded in t).

The equation (7.3.11) can then be written

$$\frac{d\xi}{dt} = A_t\xi_t + \rho_t \operatorname{sign}(\alpha_t + r_t^*\xi_t) \tag{7.5.1}$$

$$\xi(0) = 0$$

where we have set

$$\alpha_t = r_t^*\mu_t + \nu_t \tag{7.5.2}$$

and the process α_t captures the randomness. Assume that there exist two solutions ξ_t and η_t of (7.5.1). We represent the difference

$$\zeta_t = \xi_t - \eta_t$$

as

$$\zeta_t = \int_0^t \left\{ A_s \zeta_s + \rho_s [\text{sign}(\alpha_s + r_s^* \xi_s) - \text{sign}(\alpha_s + r_s^* \eta_s)] \right\} ds.$$

We are going to show that

$$X(t) = \int_0^t \rho_s [\text{sign}(\alpha_s + r_s^* \xi_s) - \text{sign}(\alpha_s + r_s^* \eta_s)] ds = \int_0^t d\Psi(s)\zeta_s \qquad (7.5.3)$$

where $\Psi(t)$ is a matrix process which is a.s. with bounded variations. That suffices to imply $\zeta_t = 0$. Consider the quantity

$$X^N(t) = \int_0^t \rho_s [\phi_N(\alpha_s + r_s^* \xi_s) - \phi_N(\alpha_s + r_s^* \eta_s)] ds$$

we can assert that

$$\text{a.s. } X^N(t) \to X(t), \quad \forall t \text{ on } [0, T]. \qquad (7.5.4)$$

Indeed since $\alpha_s + r_s^* \xi_s$ and $\alpha_s + r_s^* \eta_s$ are non degenerate diffusions, we have

$$\text{a.s. } \alpha_s + r_s^* \xi_s \neq 0 \quad \text{for almost all } t.$$

But, if $\alpha_s + r_s^* \xi_s \neq 0$,

$$\phi_N(\alpha_s + r_s^* \xi_s) \to \text{sign}(\alpha_s + r_s^* \xi_s)$$

and similarly with η_s instead of ξ_s. This implies (7.5.4).

Now we write

$$X^N(t) = \int_0^t \rho_s \int_0^1 \phi_N'(\alpha_s + \lambda r_s^* \xi_s + (1 - \lambda) r_s^* \eta_s) d\lambda r_s^* \zeta_s ds$$

$$= \int_0^t d\Psi^N(s)\zeta_s$$

where

$$\Psi^N(t) = \int_0^t \rho_s r_s^* \left[\int_0^1 \phi_N'(Z^\lambda(s)) d\lambda \right] ds \qquad (7.5.5)$$

in which we have set

$$Z^\lambda(t) = \alpha_t + \lambda r_t^* \xi_t + (1 - \lambda) r_t^* \eta_t.$$

We shall study the limit of $\Psi_N(t)$, at least for a subsequence. We can write the Ito differential of $Z^\theta(t)$ as

$$dZ^\lambda(t) = \gamma_t^\lambda dt + \sigma_t db_t \qquad (7.5.6)$$

where

$$\sigma_t^2 = r_t^* P_t H_t^* R_t^{-1} H_t P_t r_t$$

and

$$b(t) = \int_0^t \frac{r_s^* P_s H_s^* R_s^{-1}}{\sigma_s} dz.$$

Clearly $\sigma_t^2 \geq c_0$, by our assumptions (see (7.4.5)).

We have emphasized the fact that the diffusion term does not contain λ. Moreover, among other properties

$$E|\gamma_t^\lambda|^2 \le C \text{ independent of } \lambda \text{ and } t \in [0, T]. \qquad (7.5.7)$$

We are going to prove that for a subsequence

$$\text{a.s} \quad \int_0^t \sigma_s^2 \left[\int_0^1 \phi_N'(Z^\lambda(s)) \mathrm{d}\lambda \right] \mathrm{d}s \quad \text{converges uniformly in } t \in [0, T]. \qquad (7.5.8)$$

This fact will imply, for this subsequence,

$$\text{a.s.} \quad \Psi^N(t) \to \Psi(t) \quad \text{converges uniformly in } t \in [0, T]. \qquad (7.5.9)$$

Indeed, if ρ_t and σ_t were differentiable in t, the result would follow from an easy integration by parts in t, between (7.5.5) and (7.5.8). Noting that $\int_0^1 \phi_N' Z^\lambda(s) \mathrm{d}\lambda$ is uniformly bounded in N, s we can approximate ρ_t and σ_t by smooth functions of time and deduce the desired result for our case.

Let us prove (7.5.8). Define

$$\psi_N(x) = \int_0^x \phi_N(y) \mathrm{d}y$$

then

$$\psi_N(x) \to |x|, \quad |\psi_N(x)| \le |x|.$$

From Ito's formula, we can write

$$\psi_N(Z_t^\lambda) - \psi_N(r_0^* x_0) = \int_0^t \phi_N(Z_s^\lambda) \mathrm{d}Z_s^\lambda + \frac{1}{2} \int_0^t \phi_N'(Z_s^\lambda) \sigma_s^2 \mathrm{d}s. \qquad (7.5.10)$$

Clearly

$$\int_0^1 \psi_N(Z_t^\lambda) \mathrm{d}\lambda \to \int_0^1 |Z_t^\lambda| \mathrm{d}\lambda.$$

Let us check that, for a subsequence,

$$\int_0^1 \left[\int_0^1 \phi_N(Z_s^\lambda) \mathrm{d}Z_s^\lambda \right] \mathrm{d}\lambda \to \int_0^1 \left[\int_0^t \text{sign}\,(Z_s^\lambda) \, \mathrm{d}Z_s^\lambda \right] \mathrm{d}\lambda \qquad (7.5.11)$$

uniformly in t.

From (7.5.6) we infer

$$E \sup_{0 \le t \le T} \left| \int_0^1 \left\{ \int_0^t [\phi_N(Z_s^\lambda) - \text{sign}(Z_s^\lambda)] \mathrm{d}Z_s^\lambda \right\} \mathrm{d}\lambda \right|^2$$

$$\le \int_0^1 \left\{ E \sup_{0 \le t \le T} \left| \int_0^t [\phi_N(Z_s^\lambda) - \text{sign}(Z_s^\lambda)] \mathrm{d}Z_s^\lambda \right|^2 \right\} \mathrm{d}\lambda$$

and from (7.5.6), it easily follows that

$$\int_0^1 \left\{ E \sup_{0 \le t \le T} \left| \int_0^t [\phi_N(Z_s^\lambda) - \text{sign}(Z_s^\lambda)] \mathrm{d}Z_s^\lambda \right|^2 \right\} \mathrm{d}\lambda$$

$$\le cE \int_0^1 \int_0^T |\phi_N(Z_t^\lambda) - \text{sign}(Z_t^\lambda)|^2 \mathrm{d}t \mathrm{d}\lambda$$

$$= cI_N$$

and we shall prove that

$$I_N \to 0 \quad \text{as } N \to \infty \tag{7.5.12}$$

which will imply (7.5.11).

We use the estimate

$$I_N \le E \int_0^1 \int_0^T \mathcal{I}_{|Z_t^\lambda| \le 1/N} \mathrm{d}\lambda \mathrm{d}t.$$

Let $\beta(x)$ be a $C_0^\infty(R)$ function, $\beta \ge 0$, $\beta'' \le 0$ and $-\beta'' = 1$ for $x \in [-1, +1]$.

By Ito's calculus, one has

$$E\beta(NZ_T^\lambda) = E\beta(NZ_0^\lambda) + E \int_0^T \left(N\beta'(NZ_s^\lambda)\gamma_s^\lambda + \frac{N^2}{2}\beta''(NZ_s^\lambda)\sigma_s^2 \right) \mathrm{d}s$$

hence

$$E \int_0^T \mathcal{I}_{|Z_s^\lambda| \le 1/N} \mathrm{d}s \le \frac{c}{N^2} E\beta(NZ_0^\lambda) + \frac{1}{N} E \int_0^T |\beta'|(NZ_s^\lambda)\mathrm{d}s$$

$$\le \frac{c}{N}, \quad \forall \lambda$$

which implies (7.5.12).

Therefore we have proven (7.5.8), and thus (7.5.9).

It remains to check that $\psi(t)$ has bounded variations, a.s. But

$$E \int_0^T \|\mathrm{d}\Psi^N(s)\| = E \int_0^T \left| \rho_t \|r_t\| \int_0^1 \phi'_N(Z_t^\lambda)\mathrm{d}\lambda \right| \mathrm{d}t$$

$$\le cN \int_0^T \int_0^1 \mathcal{I}_{|Z_t^\lambda| \le 1/N} \mathrm{d}\lambda \mathrm{d}t$$

$$\le c$$

from the above estimate.

We thus have proved that the equation of the Kalman filter

$$\begin{aligned}
\mathrm{d}\hat{y} &= [F_t \hat{y}_t + f_t + G_t \rho_t \mathrm{sign}(r_t^* \hat{y}_t + \nu_t)]\mathrm{d}t \\
&\quad + P_t H_t^* R_t^{-1}[\mathrm{d}z_t - (H_t \hat{y}_t + h_t)\mathrm{d}t] \\
\hat{y}(0) &= x_0
\end{aligned} \tag{7.5.13}$$

has a unique strong solution.

The existence of a weak solution to (7.5.13) is a standard consequence to the Girsanov transformation.

We will turn to the use of the weak solution to (7.5.13) in the discussion of Section 7.6.

We can state

Theorem 7.5.1 *We make the assumptions of Theorem 7.4.1; then the process*

$$\hat{v}_t = \rho_t \, \mathrm{sign}\,(r_t^* \hat{y}_t + \nu_t) \tag{7.5.14}$$

is an optimal control for the problem of stochastic control with partial information (7.1.6), (7.1.12), *corresponding to the predicted-miss problem* (7.4.1).

7.5.2   The minimum-distance problem

The Kalman filter equation corresponding to this problem is, according to the feedback rule (7.4.35), given by

$$d\hat{y}_t = \left[(F_t - P_tH_t^*R_t^{-1}H_t)\hat{y}_t + f_t - \beta_t\frac{(\hat{y}_t + \rho_t)}{|M_t(\hat{y}_t + \rho_t)|}\right.$$
$$\left. - P_tH_t^{-*}R_t^{-1}h_t\right]dt + P_tH_t^*R_t^{-1}dz_t \qquad (7.5.15)$$

$$\hat{y}(0) = x_0.$$

Again as in § 7.5.1, the weak existence being a standard consequence of Girsanov's theorem, what has to be proven is the strong uniqueness.

As before, we split $\hat{y}_t$ as

$$\hat{y}_t = \mu_t + \xi_t$$

where μ_t is as defined in Proposition 7.3.1 and ξ_t is the solution of

$$\frac{d\xi}{dt} = A_t\xi_t \quad \frac{-\beta_t(\xi_t + \nu_t)}{|M_t(\xi_t + \nu_t)|} \qquad (7.5.16)$$

$$\xi(0) = 0$$

where we have set

$$\nu_t = \mu_t + \rho_t \qquad (7.5.17)$$

and we recall that

$$A_t = F_t - P_tH_t^*R_t^{-1}H_t$$
$$M_t = \Phi(T, t)$$

where $\Phi(t, s)$ is the fundamental matrix corresponding to F_t. In (7.5.16) ν_t captures the randomness of the system. As in § 7.5.1, if there are two solutions ξ, η of (7.5.16), we set

$$\zeta_t = \xi_t - \eta_t$$

and

$$X(t) = -\int_0^t \beta_s\left[\frac{\xi_s + \nu_s}{|M_s(\xi_s + \nu_s)|} - \frac{\eta_s + \nu_s}{|M_s(\eta_s + \nu_s)|}\right]ds.$$

We shall prove again that we can write

$$X(t) = -\int_0^t d\Psi(s)\zeta_s$$

where $\Psi(t)$ is a matrix process, which is a.s. with bounded variations.

Let

$$\phi^N(x) = \left(1/N + |x|^2\right)^{-1/2}x$$

and consider the approximation

$$X^N(t) = -\int_0^t \beta_sM_s^{-1}\left[\phi^N(M_s(\xi_s + \nu_s)) - \phi^N(M_s(\eta_s + \nu_s))\right]ds$$

$$= -\int_0^t \beta_sM_s^{-1}\left[\int_0^1 D\phi^N(Z_s^\lambda)d\lambda\right]M_s\zeta_s ds$$

where

$$\mathrm{D}\phi^N(x) = \left(\frac{1}{N} + |x|^2\right)^{-1/2} \left[I - \frac{xx^*}{\frac{1}{N} + |x|^2}\right]$$

and

$$Z_t^\lambda = M_t(\nu_t + \lambda\xi_t + (1-\lambda)\eta_t).$$

Clearly

$$\text{a.s.} \quad X^N(t) \to X(t) \quad \forall t$$

and

$$X^N(t) = -\int_0^t \mathrm{d}\Psi^N(s)\zeta_s$$

with

$$\Psi^N(t) = \int_0^t \beta_s M_s^{-1} \left[\int_0^1 \mathrm{D}\phi^N(Z_s^\lambda)\mathrm{d}\lambda\right] M_s \mathrm{d}s.$$

We can write for Z_t^λ,

$$\mathrm{d}Z_t^\lambda = \gamma_t^\lambda \mathrm{d}t + \Sigma_t \mathrm{d}z_t$$

where

$$\Sigma_t = M_t P_t H_t^* R_t^{-1}, \quad \Sigma_t R_t^{-1}\Sigma_t^* \geq c_0 I, \quad c_0 > 0$$

and

$$E|\gamma_t^\lambda|^2 \leq C.$$

We shall show that for the dimension $n \geq 2$

$$E\int_0^T \int_0^1 \frac{\mathrm{d}s\mathrm{d}\lambda}{|Z_s^\lambda|} < \infty. \tag{7.5.18}$$

If (7.5.18) is proven, then one has

$$\text{a.s.} \quad \int_0^T \int_0^1 \frac{\mathrm{d}s\mathrm{d}\lambda}{|Z_s^\lambda|} < \infty$$

and, by Lebesgue's theorem,

$$\text{a.s.} \quad \Psi^N(t) \to \Psi(t) = \int_0^t \beta_s M_s^{-1} \left[\int_0^1 \mathrm{D}\phi(Z_s^\lambda)\mathrm{d}\lambda\right] M_s \mathrm{d}s$$

uniformly in $t \in [0, T]$,

where we have set $\phi(t) = |x|^{-1/2}x$.

The fact that $\Psi(t)$ is with bounded variations follows from

$$E\int_0^T \|\mathrm{d}\Psi^N(t)\| \leq cE\int_0^T \frac{\mathrm{d}s\mathrm{d}\lambda}{|Z_s^\lambda|}.$$

Let us prove (7.5.18). We shall use the fact that

$$E \int_0^T \int_0^1 \operatorname{tr} \Sigma_s R_s^{-1} \Sigma_s^* D\phi^N(Z_s^\lambda) ds d\lambda$$

$$= E \int_0^T \int_0^1 \left(\frac{1}{N} + |Z_s^\lambda|^2 \right)^{-1/2} \left(\operatorname{tr} \Sigma_s R_s^{-1} \Sigma_s^* - \frac{(Z_s^\lambda)^* \Sigma_s R_s^{-1} \Sigma_s^* Z_s^\lambda}{1/N + |Z_s^\lambda|^2} \right) ds d\lambda \quad (7.5.19)$$

$$\geq c_0 E \int_0^T \int_0^1 \frac{ds d\lambda}{(1/N + |Z_s^\lambda|^2)^{1/2}}.$$

and thus it is sufficient to prove that the left hand side of (7.5.19) is bounded as $N \to \infty$†.

But

$$\phi^N(x) = D\theta^N(x)$$

where

$$\theta^N(x) = \left(\frac{1}{N} + |x|^2 \right)^{1/2}$$

hence

$$X_N = E \int_0^T \int_0^1 \operatorname{tr} \Sigma_s R_s^{-1} \Sigma_s^* D\phi^N(Z_s^* \lambda) ds d\lambda$$

$$= E \int_0^T \int_0^1 \operatorname{tr} \Sigma_s R_s^{-1} \Sigma_s^* D^2 \theta^N(Z_s^\lambda) ds d\lambda.$$

By Ito's formula

$$\frac{1}{2} X_N = E \int_0^1 \theta^N(Z_T^\lambda) d\lambda - \theta^N(M_0(x_0 + \rho_0)) - E \int_0^T D\theta^N(Z_s^\lambda) \gamma_s^\lambda ds$$

which is easily seen to be bounded. Hence the desired result (7.5.18) has been proven.

We have thus proven strong uniqueness when $n \geq 2$. When $n = 1$, the minimum distance problem and the predicted miss problem coincide; the strong uniqueness results from § 7.5.1.

We thus can assert

Theorem 7.5.2. *We make the assumptions of Theorem 7.4.2; then the process*

$$\hat{v}_t = -\frac{G_t^* M_t^*(M_t \hat{y}_t + r_t)}{\beta_t |M_t \hat{y}_t + r_t|} \quad (7.5.20)$$

is an optimal control for the problem of stochastic control with partial information (7.1.6), (7.1.12) corresponding to the minimum-distance problem (7.4.15).

---

† The fact that the dimension is ≥ 2 is used at this stage.

7.6 An extension of the concept of solution

7.6.1 General comments

We assume here the assumptions of Theorem 7.2.2, i.e., (7.1.1), (7.1.2), (7.1.3), (7.1.4), (7.2.1), (7.2.2), (7.2.16), (7.2.17). Therefore the solution of the Bellman equation (7.2.5) exists and is unique in the space $W^{2,1,q}_{\Pi_s}$, $\forall q$, $2 \leq q < \infty$, and s in Π_s (see (7.2.6)) depends on q. Moreover one has (7.2.22).

In Section 7.3, we have exhibited a feedback $\hat{V}(x,t,p)$ which was, under the assumptions (7.3.1), (7.3.2), a continuous function of x, p. It is clear that (7.3.1), (7.3.2) are very restrictive, especially (7.3.2), which is not satisfied in the predicted-miss or the minimum-distance problem. To obtain a feedback, it is sufficient to assume that

$$g_t(v), \ l_t(x,v) \text{ are continuous in } v. \tag{7.6.1}$$

Indeed in that case the function $l_t(x,v) + p \cdot g_t(v)$ is Borel in x,t,p,v and continuous in v. Since we minimize on a compact set, it follows from classical selection theorems (see Ekeland and Temam 1976 for instance) that there exists a *Borel* map $\hat{V}(x,t,p)$, which satisfies

$$l_t(x,\hat{V}) + p \cdot g_t(\hat{V}) = \min_{v \in U_{ad}} [\tilde{l}_t(x,v) + p \cdot g_t(v)] \quad \forall x,t,p. \tag{7.6.2}$$

Naturally $\hat{V}$ is not continuous in x,p as was the case when (7.3.1), (7.3.2) held.

Setting

$$\hat{v}(x,t) = \hat{V}(x,t,Du) \tag{7.6.3}$$

we obtain a Borel map, such that

$$\tilde{H}(x,t,Du) = \tilde{l}_t(x,\hat{v}(x,t)) + Du(x,t) \cdot g_t(\hat{v}(x,t)), \quad \forall x,t. \tag{7.6.4}$$

This was in particular the situation in the predicted-miss and minimum-distance problems.

In the discussion of § 7.3.1, we have indicated that we could not assert that $\hat{v}(x,t)$ leads to an optimal control for the problem of stochastic control under partial information, unless we could prove that the Kalman filter equation (7.3.6) has a strong solution, i.e., a solution adapted to Z^t. This is somewhat annoying, since such a result may fail to be true, and on the other hand the associated problem with *full* information considered at the end of § 7.3.1 has a solution, which is provided by the feedback $\hat{v}(x,t)$. In order to show that $\hat{v}(x,t)$ yields an optimal control, in some natural sense, we shall extend the concept of solution as we shall see in the next paragraph.

7.6.2   A new definition of the problem of optimal stochastic control with partial information

We shall reformulate the problem described in § 7.1.1, in a slightly different manner, allowing an extended class of admissible controls for the applications.

Let $(\Omega, \mathcal{A}, P, \mathcal{F}^t)$ be the type of system, on which are defined two independent Wiener processes with values in R^n and R^m respectively, denoted by w_t, $\hat{x}_t - x_0$, whose covariance matrices are Q_t, $P_t H_t^* R_t^{-1} H_t P_t$ respectively. Eventually, after some convenient change of probability measure, $\hat{x}_t$ will be the Kalman filter. But we introduce it beforehand, instead of the observation, as was done previously in § 7.1.1.

Consider a process v_t with values in U_{ad} (at this stage it is $\mathcal{F}^t$ measurable).

We construct the process x_t, the solution of

$$
\begin{aligned}
dx_t &= (F_t x_t + f_t + g_t(v_t))dt + dw_t \\
x(0) &= \xi
\end{aligned}
\tag{7.6.5}
$$

where as usual

ξ is $\mathcal{F}^0$ measurable, gaussian, with mean x_0 and covariance matrix P_0.   (7.6.6)

We next define b_t^v by

$$
\begin{aligned}
db_t^v &= (PH^*R^{-1})^{-1}\{d\hat{x}_t - [F_t\hat{x}_t + PH^*R^{-1}H(x_t - \hat{x}_t) + f_t + g_t(v_t)]dt\} \\
b^v(0) &= 0
\end{aligned}
\tag{7.6.7}
$$

and z_t^v by

$$
\begin{aligned}
dz_t^v &= (H_t x_t + h_t)dt + db_t^v \\
z_0^v &= 0.
\end{aligned}
\tag{7.6.8}
$$

We shall now make precise the class of admissible controls:†

v_t is admissible if the processes v_t and w_t are independent and

$\forall t$, v_t is measurable with respect to the σ-algebra generated   (7.6.9)

by v_s, $s < t$ and z_s, $s \le t$.

As a particular case (that is very important subsequently) let us show that any feedback $v(x, t)$ yields an admissible control, in the sense (7.6.9), as follows:

$$
v_t = \hat{v}(\hat{x}_t, t) \quad \text{is admissible.}
\tag{7.6.10}
$$

Indeed, since $\hat{x}_t$ is by construction independent of $\sigma(w_s, s \le t)$, v_t is independent of this σ-algebra. Now from (7.6.7), (7.6.8) we can also view $\hat{x}_t$ as the solution of

$$
\begin{aligned}
d\hat{x}_t &= (F_t\hat{x}_t + f_t + g_t(v_t))dt + P_t H_t^* R_t^1 [dz_t - (H_t\hat{x}_t + h_t)dt] \\
\hat{x}(0) &= 0,
\end{aligned}
\tag{7.6.11}
$$

---

† Similar to the idea of wide sense admissible controls, see Chapter 9.

or also

$$\hat{x}_t = \bar{x}_t + \int_0^t K(t,s)[g_s(v_s)ds + P_s H_s^* R_s^{-1} dz_s] \tag{7.6.12}$$

where $\bar{x}_t$ is deterministic, and $K(t,s)$ is the fundamental matrix associated to $F_t - P_t H_t^* R_t^{-1} H_t$.

From the formula (7.6.12) it appears clearly that $\hat{x}_t$ is measurable with respect to the σ-algebra generated by v_s, $s < t$ and z_s, $s \leq t$, and thus from (7.6.10), it follows that the same is true for the process v_t itself.

Naturally, any process adapted to the σ-algebra generated by $\hat{x}_s$, $s \leq t$, will be admissible in the sense (7.6.9). We shall denote

$$\mathcal{M}^{v,t} = \sigma(v_s, \ s < t, \ z_s \leq t) \tag{7.6.13}$$

as the σ-algebra representing the observations. We perform next a change of probability as follows. Let μ_t^v be the martingale

$$\begin{aligned}
d\mu_t^v &= \mu_t^v \left[F_t \hat{x}_t + P_t H_t^* R_t^{-1} H_t (x_t - \hat{x}_t) + f_t + \theta_t(v_t) \right]^* \\
&\quad (PH^* R^{-1} HP)^{-1} d\hat{x}_t \\
\mu_0^v &= 1.
\end{aligned} \tag{7.6.14}$$

Recalling that v_t is bounded, one has

$$E\mu_t^v = 1 \tag{7.6.15}$$

and thus we may define a new probability by setting

$$\left. \frac{dP^v}{dP} \right|_{\mathcal{F}^t} = \mu_t^v. \tag{7.6.16}$$

For the system $(\Omega, \mathcal{A}, P^v, \mathcal{F}^t)$ the pair w_t, $b_{v,t}$ become a pair of $\mathcal{F}^t$ Wiener processes, independent, with covariance matrices Q_t, R_t respectively.

We first can assert that

$$\hat{x}_t = E^v[x_t | \mu^{v,t}] \tag{7.6.17}$$

for any admissible control. We prove (7.6.17) in a way similar to the proof of Lemma 2.4.3. Indeed, write

$$x_t = \alpha_t + x_{1t}$$

with

$$\begin{aligned}
d\alpha_t &= (F_t \alpha_t)dt + dw_t \\
\alpha(0) &= \xi
\end{aligned} \tag{7.6.18}$$

and

$$\begin{aligned}
\frac{dx_{1t}}{dt} &= F_t x_{1t} + g_t(v_t) \\
x_1(0) &= 0
\end{aligned} \tag{7.6.19}$$

then

$$E^v[x_t | \mathcal{M}^{v,t}] = x_{1t} + E^v[\alpha_t | \mathcal{M}^{v,t}]. \tag{7.6.20}$$

But

$$E^v[\alpha_t|\mathcal{M}^{v,t}] = \frac{E[\mu_t^v\alpha_t|\mathcal{M}^{v,t}]}{E[\mu_t^v|\mathcal{M}^{v,t}]}.$$

From (7.6.14) one has

$$d\mu_t^v = \mu_t^v(\gamma_t + P_tH_t^*R_t^{-1}H_t\alpha_t)^*(P_tH_t^*R_t^1H_tP_t)^{-1}d\hat{x}_t$$

where

$$\gamma_t = F_t\hat{x}_t + P_tH_t^*R_t^{-1}H_t(x_{1t} - \hat{x}_t) + f_t + g_t(v_t)$$

is clearly adapted to $\mathcal{M}^{v,t}$. Therefore, after simplifications, one has

$$E^v[\alpha_t|\mathcal{M}^{v,t}] = \frac{E[\nu_t\alpha_t|\mathcal{M}^{v,t}]}{E[\nu_t|\mathcal{M}^{v,t}]} \qquad (7.6.21)$$

where

$$\begin{aligned}
\nu_t &= \exp\left[\int_0^t \alpha^*P^{-1}d\hat{x} - \frac{1}{2}\int_0^t(\alpha^*H^*R^{-1}H\alpha + 2\gamma^*P^{-1}\alpha)ds\right]\\
&= \exp\left[\int_0^t \alpha_s^*P_s^{-1}(d\hat{x} - \gamma_sds) - \frac{1}{2}\int_0^t \alpha^*H^*R^{-1}H\alpha ds\right]\\
&= \exp\left[\int_0^t \alpha_s^*H_s^*R_s^{-1}d\beta_s^v - \frac{1}{2}\int_0^t \alpha^*H^*R^{-1}H\alpha ds\right]
\end{aligned}$$

where

$$d\beta_t^v = (H_t\alpha_t + h_t)dt + db_t^v, \qquad \beta^v(0) = 0.$$

Now

$$\beta_t^v = \int_0^t (PH^*R^{-1})^{-1}(d\hat{x}_s - \gamma_sds)$$

is measurable with respect to the σ-algebra generated by $v(s)$, $s < t$ and $\hat{x}_s$, $s \le t$. Denote

$$\mathcal{N}^{v,t} = \sigma(v(s), s < t, \hat{x}_s, s \le t).$$

From the relation (7.6.11) and the invertibility of PH^*R^{-1}, it is readily seen that

$$\mathcal{M}^{v,t} = \mathcal{N}^{v,t}.$$

But, from the definition of admissible controls, (7.6.9), it follows that the processes w_t and β_t^v are independent, and thus we compute the numerator and denominator of (7.6.21) by freezing the process β_t^v and taking the corresponding expectations with the remainder, in which the random process is the gaussian process α_t.

Reasoning as in § 2.4.1.2, and noting that

$$\mathcal{B}^{v,t} = \sigma(\beta_s^v, s \le t) \subset \mathcal{M}^{v,t}$$

we can then assert that

$$\begin{aligned}
E^v[\alpha_t|\mathcal{M}^{v,t}] &= E^v[\alpha_t|\mathcal{B}^{v,t}]\\
&= \hat{\alpha}_t
\end{aligned}$$

where $\hat{\alpha}_t$ is the Kalman filter

$$d\hat{\alpha}_t = (F_t\hat{\alpha}_t + f_t)dt + PH^*R^{-1}[d\beta^v_t - (H_t\hat{\alpha}_t + h_t)dt]$$

$$\hat{\alpha}(0) = x_0;$$

it follows that (7.6.17) holds.

We can now pose the control problem. Let us denote by $\mathcal{U}$ the set of admissible controls, i.e., the set of processes satisfying (7.6.9). For $v(\cdot) \in \mathcal{U}$, we define the cost function by

$$J(v(\cdot)) = E^v \left[\int_0^T l_t(x_t, v_t)dt + m(x_T) \right], \tag{7.6.22}$$

where the state x_t is the solution of (7.6.5).

We shall show that there exists an optimal control, which is provided by the feedback (7.6.3).

Considering on $(\Omega, \mathcal{A}, P, \mathcal{F}^t)$ a pair $w_t, \hat{y}_t - x_0$ of independent Wiener processes, with covariance matrices $Q_t, P_t H_t^* R_t^{-1} H_t P_t$, we define

$$\hat{v}_t = \hat{v}(\hat{y}_t, t) \tag{7.6.23}$$

and y_t by

$$dy_t = (F_t y_t + f_t + g_t(\hat{v}_t))dt + dw_t$$

$$y(0) = \xi. \tag{7.6.24}$$

Next we consider b_t and z_t defined by

$$db_t = (PH^*R^{-1})^{-1}\{d\hat{y}_t - [F_t\hat{y}_t + P_t H_t^* R_t^{-1} H_t(y_t - \hat{y}_t) + f_t + g_t(\hat{v}_t)]dt\}$$

$$b(0) = 0 \tag{7.6.25}$$

$$dz_t = (H_t y_t + h_t)dt + db_t$$

$$z(0) = 0. \tag{7.6.26}$$

We then assert the following

Theorem 7.6.1 *We assume (7.1.1), (7.1.2), (7.1.3), (7.1.4), (7.2.1), (7.2.2), (7.2.16), (7.2.17) and (7.6.1). Then the process $\hat{v}_t$ defined by (7.6.23) is an optimal control, i.e.,*

$$J(\hat{v}(\cdot)) = \min_{v(\cdot) \in \mathcal{U}} J(v(\cdot)). \tag{7.6.27}$$

Proof

Consider the innovation

$$I^v_t = z_t - \int_0^t (H_s \hat{x}_s + h_s)ds$$

$$= \beta^v_t - \int_0^t (H_s \hat{\alpha}_s + h_s)ds$$

which is a P^v, $\mathcal{M}^{v,t}$ Wiener process, with covariance matrix R_t. The process $\hat{x}_t$ can be written as follows:

$$d\hat{x}_t = (F_t\hat{x}_t + f_t + g_t(v_t))dt + P_t H_t^* R_t^{-1} dI^v_t$$

$$\hat{y}(0) = x_0. \tag{7.6.28}$$

We then proceed as in the proof of Theorem 7.3.1, to show that

$$u(x_0, 0) = J(\hat{v}(\cdot)) \le J(v(\cdot)), \quad \forall v(\cdot)$$

where u is the solution of the Bellman equation (7.2.5). ∎

7.7   Use of approximations

If, in general, one cannot expect the separation principle to hold, it is reasonable to expect an 'approximate' separation principle to hold in situations where there is a small noise to signal ratio like those described in Chapter 5 (see Section 5.2).

This is the objective of this section. The results presented here are very similar to those of Qing Zhang (1988).

7.7.1   Setting of the problem

The assumptions follow § 5.2.1 with adequate modifications to take account of the control.

We consider $g(x,v)$, $\sigma(x)$, $h(x)$ as follows:†

$g: R^n \times U_{ad} \to R^n$, where U_{ad} is a compact subset of R^k; g is Borel, is C^2 with respect to x, with bounded derivatives. Moreover   (7.7.1)

$g(x,v) = g_0(x) + g_1(x,v)$

where $g_1(x,v)$ is bounded, g_0 is C^2 with bounded derivatives.

$\sigma: R^n \to L(R^n; R^n)$ is C^3; σ and its derivatives are bounded.   (7.7.2)

The function $a = \sigma\sigma^*/2$ is uniformly positive definite.

$h: R^n \to R^n$ is C^3, with bounded derivatives.   (7.7.3)

$g_i\partial/\partial x_i a_{jk}$, $g_i\partial^2 h_k/\partial x_i\partial x_j$ are bounded, $\forall j,k$.   (7.7.4)

Let also

$$p_0: R^n \to R^+ - \{0\}; \quad \int p_0(x)dz = 1;$$

p_0 is a smooth function, $\int \frac{|Dp_0|^2}{p_0}dx < \infty,$ (7.7.5)

$$p_0 \exp[-(1+|x|^2)^{1/2}] \in H^1(R^n); \quad \int |x|^4 p_0(x)dx < \infty.$$

Let $(\Omega, \mathcal{A}, \tilde{P}, \mathcal{F}^t)$ be a probability space equipped with a filtration, and

w_t, z_t are two independent $\mathcal{F}^t$ Wiener processes with values in R^n.

The process w_t is standard (i.e. its covariance matrix is I, the   (7.7.6) identity), whereas z_t has a covariance matrix equal to $\epsilon^2 I$.

---

† To simplify the notation, we omit the dependence in t.

ξ is a random variable with values in R^n, which is $\mathcal{F}^0$ measurable.

The probability distribution of ξ, denoted by Π_0, has a density $\qquad$ (7.7.7)
with respect to Lebesgue measure, equal to p_0.

Denote $\mathcal{Z}^t = \sigma(z_s, \ s \leq t)$. An admissible control is, as usual, a process v_t such that

$$v_t \text{ is adapted to } \mathcal{Z}^t \text{ and } v_t \in U_{ad}, \text{ a.e., a.s.} \qquad (7.7.8)$$

Given an admissible control, we can solve the stochastic differential equation

$$\begin{aligned} \mathrm{d}x_t &= g(x_t, v_t)\mathrm{d}t + \sigma(x_t)\mathrm{d}w_t \\ x(0) &= \xi. \end{aligned} \qquad (7.7.9)$$

In view of the regularity of g and σ with respect to x, we can solve (7.7.9) in the strong sense.

We then define the process b_t by the relation

$$b_t = \frac{1}{\epsilon}\left(z_t - \int_0^t h(x_s)\mathrm{d}s\right)$$

or more conveniently we write

$$\mathrm{d}z_t = h(x_t)\mathrm{d}t + \epsilon\mathrm{d}b_t. \qquad (7.7.10)$$

Of course b_t is not a Wiener process, since z_t is a Wiener process. To obtain the usual model, we perform a Girsanov transformation.

Define

$$\eta_t = \exp\left(\int_0^t \frac{h^*(x_s)}{\epsilon^2}\mathrm{d}z_s - \frac{1}{2\epsilon^2}\int_0^t |h(x_s)|^2\mathrm{d}s\right) \qquad (7.7.11)$$

and the probability $P = P^v$ such that

$$\left.\frac{\mathrm{d}P}{\mathrm{d}\tilde{P}}\right|_{\mathcal{F}^t} = \eta_t. \qquad (7.7.12)$$

For the system $(\Omega, \mathcal{A}, P, \mathcal{F}^t)$, w_t and b_t are independent standard Wiener processes.

Now let $l(x, v)$ and $k(x)$ be such that

l is a Borel map from $R^n \times U_{ad} \to R$, l is C^2 with respect
to x, and $l(0, v)$, $\mathrm{D}l(0, v)$ are bounded, $\mathrm{D}^2 l(x, v)$ is bounded. $\qquad$ (7.7.13)
$k : R^2 \to R$ is C^2 and $\mathrm{D}^2 k$ is bounded. $\qquad$ (7.7.14)

We define the cost function

$$J(v(\cdot)) = E\left[\int_0^T l(x_t, v_t)\mathrm{d}t + k(x_T)\right] \qquad (7.7.15)$$

which we wish to minimize over the set $\mathcal{U}$ of admissible controls (defined by (7.7.8)). Note that through v_t and its dependence with respect to z_t, the functional $J(v(\cdot))$ depends *a priori* on ϵ. We have omitted to mention it explicitly.

7.7.2   The approximate control problem

With the notation of Chapter 5, (5.2.19) to (5.2.25), we may consider the process m_t given by

$$dm_t = g(m_t, v_t)dt + \frac{1}{2}\theta(m_t)[dz_t - h(m_t)dt] \tag{7.7.16}$$

$$m(0) = m_0$$

for the same control v_t, and the cost function

$$\tilde{J}(v(\cdot)) = E\left[\int_0^T l(m_t, v_t)dt + k(m_T)\right] \tag{7.7.17}$$

and we consider $\tilde{J}(v(\cdot))$ as an approximation to $J(v(\cdot))$. We can state

Theorem 7.7.1 *Under the assumptions (7.7.1)–(7.7.7) and (7.7.13), (7.7.14) one has the following estimate*

$$|J(v(\cdot)) - \tilde{J}(v(\cdot))| \leq C\epsilon \tag{7.7.18}$$

where the constant C does not depend on ϵ, nor $v(\cdot)$.

Proof

Using Theorem 5.2.1, we can state

$$E|\hat{x}_t - m_t|^2 \leq C\left(\epsilon^2 + \epsilon e^{-\gamma_0 t/\epsilon} + e^{\frac{-\gamma_1 t}{\epsilon}}\right) \tag{7.7.19}$$

where $\hat{x}_t = E[x_t|Z^t]$, and the constants C, γ_0, γ_1 are clearly independent of the control $v(\cdot)$, and also of T. Moreover from Lemma 7.2.1, one also has

$$E|x_t - m_t|^2 \leq Ce^{-\gamma_0 t/\epsilon} + C\epsilon \tag{7.7.20}$$

where C does not depend on the control or T.

Now

$$\begin{aligned}
J(v(\cdot)) - J(\tilde{v}(\cdot)) = & E\int_0^T Dl(m_t, v_t)(\hat{x}_t - m_t)dt \\
& + E\int_0^T \int_0^1 \int_0^1 D^2 l(m_t + \lambda\mu(x_t - m_t), v_t)(x_t - m_t)^2 \lambda d\lambda d\mu \\
& + EDk(m_T)(\hat{x}_T - m_T) \\
& + E\int_0^1 \int_0^1 D^2 k(m_T + \lambda\mu(x_T - m_T))(x_T - m_T)^2 \lambda d\lambda d\mu.
\end{aligned} \tag{7.7.21}$$

Since

$$E\sup_{0 \leq t \leq T} |x_t|^2 \leq C_T$$

we have

$$|EDl(m_t, v_t)(\hat{x}_t - m_t)| \leq C_T(\epsilon + \epsilon^{1/2}e^{-\gamma_0 t/2\epsilon} + e^{-\gamma_1 t/2\epsilon})$$

hence

$$\left|E\int_0^T Dl(m_t, v_t)(\hat{x}_t - m_t)dt\right| \leq C_T\epsilon.$$

Similar estimates hold for the other terms of the difference, hence (7.7.18) is proved.

∎

It follows from (7.7.18) that

$$\left| \inf_{\mathcal{U}} J(v(\cdot)) - \inf_{\mathcal{U}} \tilde{J}(v(\cdot)) \right| \leq C\epsilon \qquad (7.7.22)$$

and thus we can replace the control problem with partial observation, $\inf_{\mathcal{U}} J(v(\cdot))$, by the control problem with full observation, $\inf_{\mathcal{U}} \tilde{J}(v(\cdot))$, up to an approximation of order ϵ.

Unfortunately the control problem (7.7.16), (7.7.17) is not easily solved since the process z_t which is the external noise of the system is not a Wiener process and has a complicated distribution. We can assert however

$$dz_t = h(\hat{x}_t)dt + d\nu_t \qquad (7.7.23)$$

where ν_t, the innovation process, is a $\mathcal{Z}^t$ Wiener process with covariance matrix $\epsilon^2 I$. But then $\hat{x}_t$ enters in the relation (7.7.23), and it is not easily computable.

7.7.3   The approximate separation principle

The natural question, in view of the general philosophy of the separation principle, is the following. Suppose the problem were fully observable; consider an optimal feedback corresponding to this case and use it to replace the state (which is not accessible) by m_t, which is an approximation of the best estimate $\hat{x}_t$. What is the value of such a rule?

We shall address this question in the following case (cf. Qing Zhang 1988, Ch. 3, Section 2)

$$\begin{aligned} h(x) &= x \\ \sigma(x) &= \sigma \end{aligned} \qquad (7.7.24)$$

$$p_0 \text{ is a gaussian with mean } x_0 \text{ and variance } O(\epsilon). \qquad (7.7.24')$$

In that case

$$\theta(x) = (\sigma\sigma^*)^{1/2}$$

and (7.7.16) yields, choosing $m_0 = x_0$,

$$\begin{aligned} dm_t &= g(m_t, v_t)dt + \frac{1}{\epsilon}(\sigma\sigma^*)^{1/2}[dz_t - h(m_t)dt] \\ m(0) &= x_0. \end{aligned} \qquad (7.7.25)$$

Consider now the process μ_t defined by (cf (7.7.23))

$$\begin{aligned} d\mu_t &= g(\mu_t, v_t)dt + \frac{1}{\epsilon}(\sigma\sigma^*)^{1/2}d\nu_t \\ \mu(0) &= x_0. \end{aligned} \qquad (7.7.26)$$

We shall introduce a new control problem, associated to (7.7.26), with the cost function

$$\tilde{J}(v(\cdot)) = E\left[\int_0^T l(\mu_t, v_t)dt + k(\mu_T) \right]. \qquad (7.7.27)$$

We have the following result.

Theorem 7.7.2 *We make the assumptions of Theorem 7.7.1 and (7.7.24), (7.7.24′).
One has the estimate*

$$|\tilde{J}(v(\cdot)) - \tilde{\tilde{J}}(v(\cdot))| \leq C\sqrt{\epsilon} \tag{7.7.28}$$

where the constant does not depend on ϵ or $v(\cdot)$.

Proof

The assumption (7.7.24) allows a better estimate for (7.7.19). We can rely on the
formulas established in (5.2.46) (see also Remark 7.2.2). Recalling that (see (5.2.36)
and (5.2.46))

$$\rho(t) = \int q u_i a_{ir} u_r dx,$$

we have

$$\frac{d}{dt} E\rho(t) = -2E \int q a_{\lambda\mu} a_{ir} u_{i\lambda} u_{r\mu} dx + 2E \int q u_\lambda u_r (-a_{ir} g_{\lambda,i} - \frac{2}{\epsilon} a_{ri} \phi_{i\mu} a_{\mu\lambda}) dx$$
$$+ \frac{2}{\epsilon} E \int q u_r a_{ir} [(x_j - m_j) \phi_{kj} g_{k,i} + (g_k - g_k^t) \phi_{ki}] dx - 2E \int q a_{ir} u_r g_{\lambda,\lambda i} dx \tag{7.7.29}$$

and

$$\phi = \frac{1}{\sqrt{2}} a^{-1/2}.$$

Therefore

$$\frac{d}{dt} E\rho(t) \leq -\frac{C_0}{\epsilon} E\rho(t) + C_1 + \frac{C_2}{\epsilon} e^{-\gamma_0 t/\epsilon}$$

hence, noting that by virtue of (7.7.24′), $E\rho(0) = O(1/\epsilon)$,

$$E\rho(t) \leq \epsilon C_1 + \frac{C_2}{\epsilon} e^{-C_0 t/\epsilon}.$$

Since (cf. Theorem 5.2.1)

$$E|\hat{x}_t - m_t|^2 \leq C\epsilon^2 E\rho(t)$$

we deduce from (7.7.27)

$$E|\hat{x}_t - m_t|^2 \leq C\epsilon^3 + C\epsilon e^{-C_0 t/\epsilon}.$$

Naturally, all the constants c are independent from the control v_t.

Taking account of (7.7.25), (7.7.26), using the condition $\mu_0 = m_0$ and (7.7.23) we
have

$$\frac{d}{dt}(m_t - \mu_t) = g(m_t, v_t) - g(\mu_t, v_t) + \frac{1}{\epsilon}(\sigma\sigma^*)^{1/2}(\hat{x}_t - m_t)$$
$$(m - \mu)(0) = 0.$$

Therefore we deduce

$$E|m_t - \mu_t| \leq C\epsilon^{1/2}.$$

From this estimate, it follows easily that

$$|\tilde{J}(v(\cdot)) - \tilde{\tilde{J}}(v(\cdot))| \leq C\epsilon^{1/2}$$

which is the desired result.   ∎

The solution of the problem (7.7.26), (7.7.27) is provided by the optimal feedback of the problem with full observation. Indeed, since ν_t is a Wiener process with variance $\epsilon^2 I$, the dynamic programming equation related to (7.7.26), (7.7.27) is

$$-\frac{\partial u}{\partial t} - \text{tr } a\text{D}^2 u = \inf_{v \in U_{ad}} \{l(x,v) + \text{D}u \cdot g(x,v)\}$$
$$u(x,T) = k(x) \tag{7.7.30}$$

and in view of the regularity of l and g with respect to x, the solution of (7.7.30) exists and is unique in the functional space $W^{2,1,p}_{\Pi_s}$, $\forall p$, $2 \leq p < \infty$ and s chosen depending on p. This follows by techniques similar to those of Theorem 7.2.2 (note that the last part of assumption (7.7.1) is essential here; see Remarks 7.2.1, 7.2.2).

To proceed we shall make the following assumptions:

$$g \text{ is } C^1 \text{ in } x, v \text{ with bounded derivatives; } l \text{ is } C^1 \text{ in } x, v,$$
$$|\text{D}_x l|, \ |\text{D}_v l| \leq C(1 + |x|); \tag{7.7.31}$$

There exists $\hat{v}(x,t)$, Borel and uniformly Lipschitz in x, such that

$$l(x, \hat{v}) + \text{D}u \cdot g(x, \hat{v}) = \inf_{v \in U_{ad}} \{l(x,v) + \text{D}u \cdot g(x,v)\}. \tag{7.7.32}$$

We can use the feedback $\hat{v}(x,t)$ in equation (7.7.25), to yield the equation

$$d\hat{m}_t = g(\hat{m}_t, \hat{v}(\hat{m}_t, t))dt + \frac{1}{\epsilon}(\sigma\sigma^*)^{1/2}(dz_t - \hat{m}_t dt)$$
$$\hat{m}(0) = x_0. \tag{7.7.33}$$

In the probability space $(\Omega, A, \tilde{P}, \mathcal{F}^t)$, z_t is a Wiener process and (7.7.33) is an Ito equation in the strong sense. Define next

$$\hat{v}_t = \hat{v}(\hat{m}_t, t)$$

and proceed as in the general construction of § 7.7.1 to define $x_t, \hat{x}_t, b_t$ and the probability $P = P^{\hat{v}}$, as well as the innovation ν_t (a $\mathcal{Z}^t$ Wiener process with covariance $\epsilon^2 I$). For this innovation process ν_t, construct the process

$$d\hat{\mu}_t = g(\hat{\mu}_t, \hat{v}(\hat{\mu}_t, t))dt + \frac{1}{\epsilon}(\sigma\sigma^*)^{1/2}d\nu_t$$
$$\hat{\mu}(0) = x_0. \tag{7.7.34}$$

It is useful to make precise the following possible source of confusion. If we set

$$\hat{\hat{v}} = \hat{v}(\hat{\mu}_t, t)$$

which is an admissible control, and if we proceed with the general construction of the state, ..., innovation as done in the beginning of the section, then the corresponding innovation is different from ν_t and thus $\hat{\mu}_t$ is not the process defined by equation (7.7.26), when v_t is replaced by $\hat{\hat{v}}_t$.

Denote by

$$J^* = E\left[\int_0^T l(\hat{\mu}_t, \hat{v}_t)\mathrm{d}t + k(\hat{\mu}_T)\right]$$

which is not $\tilde{J}(\hat{v})$, for reasons mentioned above. Nevertheless, since ν_t is a Wiener process, it is easy to check that

$$u(x_0, 0) = J^*. \tag{7.7.35}$$

Similarly, from the definition of $\tilde{J}(v(\cdot))$ one can check that

$$\tilde{J}(v(\cdot)) \geq u(x_0, 0), \quad \forall v(\cdot)$$

hence

$$\inf \tilde{J}(v(\cdot)) \geq J^*. \tag{7.7.36}$$

Now comparing (7.7.33), (7.7.34), we notice that

$$\mathrm{d}z_t = \hat{x}_t \mathrm{d}t + \mathrm{d}\nu_t$$

and the general property (7.7.28) yields

$$E|\hat{x}_t - \hat{m}_t| \leq C\epsilon^{3/2} + C\epsilon^{1/2}\mathrm{e}^{-C_0 t/\epsilon}.$$

Furthermore, thanks to (7.7.31), (7.7.32)

$$|g(\hat{m}_t, \hat{v}(\hat{m}_t, t)) - g(\hat{\mu}_t, \hat{v}(\hat{\mu}_t, t))| \leq C|\hat{m}_t - \hat{\mu}_t|.$$

Therefore, since

$$\frac{\mathrm{d}}{\mathrm{d}t}(\hat{m}_t - \hat{\mu}_t) = g(\hat{m}_t, \hat{v}(\hat{m}_t, t)) - g(\hat{\mu}_t, \hat{v}(\hat{\mu}_t, t)) + \frac{1}{\epsilon}(\sigma\sigma^*)^{1/2}(\hat{x}_t - \hat{m}_t)$$

we have

$$E|\hat{m}_t - \hat{\mu}_t| \leq C\epsilon^{1/2}.$$

Therefore also, using (7.7.31),

$$|\tilde{J}(\hat{v}(\cdot)) - J^*| \leq C\epsilon^{1/2}.$$

Hence from (7.7.36)

$$\tilde{J}(\hat{v}(\cdot)) \leq \inf \tilde{J}(v(\cdot)) + C\epsilon^{1/2}. \tag{7.7.37}$$

On the other hand, using (7.7.28) and (7.7.22) we have

$$|\inf J(v(\cdot)) - \inf \tilde{J}(v(\cdot))| \leq C\epsilon^{1/2}$$

$$|J(\hat{v}(\cdot)) - \tilde{J}(\hat{v}(\cdot))| \leq C\epsilon^{1/2}.$$

Using these inequalities in (7.7.37) we get

$$J(\hat{v}(\cdot)) \leq \inf J(v(\cdot)) + C\epsilon^{1/2}$$

and thus we have proved the following

Theorem 7.7.3 *We make the assumptions of Theorem 7.7.1, and (7.7.24), (7.7.24'), (7.7.31), (7.7.32); then the control $\hat{v}_t$ defined by (7.7.33) is 'quasi' optimal for the original problem (7.7.9), (7.7.15) in the sense that*

$$\left|J(\hat{v}(\cdot)) - \inf_{v(\cdot) \in \mathcal{U}} J(v(\cdot))\right| \leq C\epsilon^{1/2}. \tag{7.7.38}$$

■

In practice the assumption (7.7.32) is not easily satisfied. It is thus useful to investigate whether a smoothing process is applicable.

We begin with a technical result:

Lemma 7.7.1 *Let v_t be an admissible control and m_t be the corresponding solution of (7.7.25). Then one has*

$$E|m_t|^\lambda \leq C_\lambda \quad \forall \lambda \geq 2, \tag{7.7.39}$$

where C_λ does not depend on $v(\cdot)$.

Proof

This is just an extension of the general estimates of § 5.2.4 (see Lemma 5.2.1).

Note that

$$\begin{aligned}
\mathrm{d}m_t &= g(m_t, v_t)\mathrm{d}t + \frac{1}{\epsilon}(\sigma\sigma^*)^{1/2}(x_t - m_t)\mathrm{d}t + (\sigma\sigma^*)^{1/2}\mathrm{d}b_t \\
m(0) &= x_0
\end{aligned}$$

hence

$$\begin{aligned}
\mathrm{d}(x_t - m_t) &= (g(x_t, v_t) - g(m_t, v_t))\mathrm{d}t - \frac{1}{\epsilon}(\sigma\sigma^*)^{1/2}(x_t - m_t)\mathrm{d}t \\
&\quad + \sigma\mathrm{d}w_t - (\sigma\sigma^*)^{1/2}\mathrm{d}b_t \\
x(0) - m(0) &= \xi - x_0.
\end{aligned}$$

Operating as in Lemma 5.2.1, we deduce easily the estimate

$$E|x_t - m_t|^\lambda \leq C(\lambda\epsilon)^{\lambda/2} + E|\xi - x_0|^\lambda e^{-\lambda\gamma_0 t/\epsilon}$$

and (7.7.39) follows easily. ∎

Now let $v_\beta(x, t)$ be any admissible feedback (i.e. Borel measurable with values in U_{ad}), which is also Lipschitz with respect to x. We can solve the equation (as in (7.7.33))

$$\begin{aligned}
\mathrm{d}m^\beta &= g(m^\beta, v_\beta(m^\beta, t))\mathrm{d}t + \frac{1}{\epsilon}(\sigma\sigma^*)^{1/2}(\mathrm{d}z_t - m^\beta\mathrm{d}t) \\
m^\beta(0) &= x_0.
\end{aligned} \tag{7.7.40}$$

Note

$$v^\beta(t) = v_\beta(m^\beta, t). \tag{7.7.41}$$

It is convenient to introduce the linear equation

$$\begin{aligned}
-\frac{\partial u^\beta}{\partial t} - \mathrm{tr}\, a\mathrm{D}^2 u^\beta &= l(x, v_\beta) + \mathrm{D}u^\beta \cdot g(x, v_\beta) \\
u^\beta(x, T) &= k(x).
\end{aligned} \tag{7.7.42}$$

We can assert that u^β exists and is unique in the space $W_{\Pi_*}^{2,1,p}$ as for u. Of course, we may claim that $g(x, v_\beta)$ is a Lipschitz function, but this will yield estimates of the norm in $W_{\Pi_*}^{2,1,p}$ depending on β, which is undesirable. However, thanks to the decomposition of g (see the last part of assumption (7.7.1)), the estimates are independent of β. Using Ito's formula, one derives easily that

$$u^\beta(x_0, 0) = \tilde{J}(v^\beta) - \frac{1}{\epsilon}E\int_0^T \mathrm{D}u^\beta(m^\beta(t), t) \cdot \sigma\sigma^*(\hat{x}_t^\beta - m^\beta)\mathrm{d}t \tag{7.7.43}$$

where, naturally, $\hat{x}_t^\beta$ represents the best estimate of the state x^β defined from the control v^β.

Since $u^\beta \in W_{\Pi_s}^{2,1,p}$, for any p and a convenient s depending on p, with norms independent of β, we can claim that

$$|Du^\beta(x)| \le c_0(1 + |x|^2)^{s_0/2}$$

for some s_0, where s_0 and also c_0 do not depend on β. Now, the integral part in the right hand side of (7.7.43) can be estimated by

$$\int_0^T \{E[1 + |m^\beta(t)|^{s_0}]^2\}^{1/2} \frac{(E|\hat{x}^\beta - m^\beta|^2)^{1/2}}{\epsilon} dt$$

and from Lemma 7.7.1 and (7.7.28) it is $O(\epsilon^{1/2})$. Hence we have proved that

$$|u^\beta(x_0, 0) - \tilde{J}(v^\beta)| \le C\epsilon^{1/2} \tag{7.7.44}$$

where C is independent of β. From the general property (7.7.22) we deduce

$$J(v^\beta) \le u^\beta(x_0, 0) + C\epsilon^{1/2}. \tag{7.7.45}$$

Now we know that

$$u(x_0, 0) \le \inf J(v(\cdot)) + C\epsilon^{1/2}$$

and thus from (7.7.45) we can assert that

$$J(v^\beta) \le \inf J(v(\cdot)) + u^\beta(x_0, 0) - u(x_0, 0) + C\epsilon^{1/2}. \tag{7.7.46}$$

Therefore, we are led to estimate the difference $u^\beta(x_0, 0) - u(x_0, 0)$, which can be done from PDE theory.

Note that, if (7.7.31) is satisfied, then the optimal feedback $\hat{v}(x, t)$ exists, although it may fail to satisfy (7.7.32). In that case (7.7.30) becomes

$$-\frac{\partial u}{\partial t} - \text{tr } aD^2 u = l(x, \hat{v}) + Du \cdot g(x, \hat{v})$$
$$u(x, T) = k(x). \tag{7.7.47}$$

Therefore writing $z^\beta = u^\beta - u$, we obtain

$$-\frac{\partial z^\beta}{\partial t} - \text{tr } aD^2 z^\beta = Dz^\beta \cdot g(x, v_\beta) + F^\beta$$
$$z^\beta(x, T) = 0$$

where we have set

$$F^\beta = l(x, v_\beta) - l(x, \hat{v}) + Du \cdot (g(x, v_\beta) - g(x, \hat{v})).$$

Now if

$$v_\beta \to \hat{v} \quad \text{a.e.} \tag{7.7.48}$$

we will have

$$F^\beta \to 0 \text{ in } L_{\Pi_s}^q, \quad \forall q, \ 2 \le q < \infty,$$

and s conveniently chosen depending on q.

But then $z^\beta \to 0$ in $W_{\Pi_s}^{2,1,q}$, and thus since q is arbitrarily large

$$u^\beta(x,0) - u(x,0) \to 0 \quad \text{uniformly in } x.$$

We can gather the results in the following:

Theorem 7.7.4 *We make the assumptions of Theorem 7.7.3, except (7.7.32). Pick any feedback v^β which is Lipschitz with values in U_{ad}, and satisfies (7.7.48). Then one has*

$$J(v^\beta) \leq \inf J(v(\cdot)) + \theta(\beta) + C\epsilon^{1/2} \qquad (7.7.49)$$

where $\theta(\beta) \to 0$ with β, and v^β is the control defined by (7.7.40), (7.7.41). The rate $\theta(\beta)$ can be estimated in terms of the rate of convergence (7.7.48) in $L_{\Pi_s}^q$ spaces. ∎

Remark 7.7.1 In a recent paper Haussmann and Zhang (1988) have considered smoothing feedbacks. They do not use the PDE for u^β, and rely heavily on technical results of Krylov (1980). The proof is intricate and seems limited to approximate feedbacks obtained by a smoothing procedure. We show here that any pointwise approximation by a Lipschitz function is acceptable. This permits approximations by piecewise linear functions. Note however the restriction on g due to PDE techniques. ∎

Stochastic maximum principle and dynamic programming for systems with partial observation

Introduction

The stochastic maximum principle and dynamic programming are among the main methods stochastic control theory. It is also possible to develop such methods for partially observable systems. This is *a priori* expectable since the stochastic control problem with partial observation can be reduced to a stochastic control problem with complete observation, with the reservation that the system with full observation (to be controlled) is not finite dimensional, but infinite dimensional. With this remark in mind, we see that we are led to use the maximum principle or dynamic programming for an infinite dimensional system. The situation is very similar to that for systems governed by partial differential equations.

We shall not attempt to cover all possible cases in one theorem. Our approach will be to reduce the problem to the control with full observation of a stochastic PDE, namely the Zakai equation. This equation will be formulated as a differential equation in a Hilbert space, using variational techniques (see Section 4.7). In this framework, it is convenient to use variational techniques to derive necessary conditions of optimality. One advantage of this approach is that it is mostly analytic. On the other hand, the case of unbounded coefficients (for instance the linear quadratic case) is not easily covered in this formulation, without substantial technical transformations. However, the result is formally applicable without any difficulty.

The present developments improve previous results of mine (see Bensoussan 1983 and simplify the derivation. Besides the application to the LQG problem, we consider the situation of the separation principle and obtain it in a new way, through the stochastic maximum principle. This derivation is new. Other approaches have been given in the literature. Besides the first attempt of Kwakernaak (1981), Haussmann (1987) derived a stochastic maximum principle, using local modifications of the control and an explicit formula for the adjoint process, but did not derive a backward equation for it. Baras, Elliott and Kohlmann (1987) have more recently used the ideas of stochastic flows to compute the change in the cost due to a local modification of the control, which brings some simplifications.

From Section 8.4 to the end, we present a dynamic programming approach to the problem of stochastic control with partial observations. The spirit of dynamic programming is to define an analytic problem and identify its solution with the value function of a control problem. In finite dimensions (for the full observation case for instance) the analytic problem is the Bellman equation, which amounts to a non linear PDE. It has been recognized, since the work of Nisio (1976), that a nice alternative for the Bellman equation, when the latter is out of reach, is provided by the *envelope of semigroup* problem. This approach has a broad range of applicability. A related idea also due to Nisio (1976) is that of the *non linear semigroup* which is more or less equivalent to the optimality principle of dynamic programming.

These ideas can be carried over to the situation of stochastic control with partial information. At this stage, because of the intrinsic infinite dimensionality of the underlying dynamic system (formally the normalized or unnormalized conditional probability), the treatment varies very much with the choice of the functional space on which the conditional probability is defined. Basically there are two possibilities. One is the variational approach, in which the conditional probability has a density with respect to Lebesgue measure in R^n, which is an L^2 function. In this case the state space is the Hilbert space $H = L^2(R^n)$. In the other one, the conditional probability is a positive finite measure on R^n, and this is the state space.

I have chosen the variational approach in order to keep the treatment as analytic as possible, relying on papers of Bensoussan (1988) and Bensoussan and Nisio (1989). The reader interested in the other approach can consult the papers of Borkar (1982), Davis (1984), Davis and Kohlmann (1988), Fleming (1982).

The study of the Bellman equation as a partial differential equation with a space variable which is infinite dimensional has been undertaken by Cannarsa and Daprato (1988). The specific case of controlling the Zakai equation has been considered by Lions (1988), using the concept of a viscosity solution.

8.1   Setting of the problem

8.1.1   Assumptions notation

We consider the functions

$$g(x,v,t): \ R^n \times R^k \times (0,\infty) \to R^n, \quad \text{Borel bounded}$$
$$\sigma(x,t): \ R^n \times (0,\infty) \to L(R^n; R^n), \quad \text{Borel bounded} \tag{8.1.1}$$

where

$$|g(x_1,v,t) - g(x_2,v,t)| \le k|x_1 - x_2|$$
$$\|\sigma(x_1,t) - \sigma(x_2,t)\| \le k|x_1 - x_2| \tag{8.1.2}$$

and

$$h(x,t): \ R^n \times (0,\infty) \to R^m, \quad \text{Borel bounded.} \tag{8.1.3}$$

Now let (Ω, A, P) be a probability space equipped with a filtration $\mathcal{F}^t$ on which are given *independent, standard* Wiener processes $w(t)$ and $z(t)$. Let also

$$\xi \text{ be a random variable, } \mathcal{F}^0 \text{ measurable, independent of } w, z,$$
$$\text{with values in } R^n. \text{ The probability law of } \xi \text{ is denoted by } \Pi_0. \tag{8.1.4}$$

Let $\mathcal{U}_{ad}$ be a non empty subset of R^k, and let $\mathcal{Z}^t = \sigma(z_s, \ s \le t)$.
An *admissible* control is a process which is $\mathcal{Z}^t$ adapted, with values in U_{ad}.
If v_t is an admissible control, we can solve the stochastic differential equation

$$\mathrm{d}x_t = g(x_t, v_t)\mathrm{d}t + \sigma(x_t)\mathrm{d}w_t\dagger, \quad x(0) = \xi \tag{8.1.5}$$

in the strong sense.

By virtue of (8.1.2), this is possible and there exists one and only one process x_t which is a solution of (8.1.5) such that

$$x(\cdot) \in L^2(\Omega, A, P; C(0, T; R^n)). \tag{8.1.6}$$

Consider next the process η_t defined by

$$\eta_t = \exp\left[\int_0^t h^*(x_s)\mathrm{d}z_s - \frac{1}{2}\int_0^t |h(x_s)|^2\mathrm{d}s\right] \tag{8.1.7}$$

which is an $\mathcal{F}^t$ martingale, and

$$E\eta_t = 1, \quad \forall t. \tag{8.1.8}$$

---

† To simplify the notation, we omit to write the dependence of g and σ on the variable t.

We perform a change of probability by setting

$$\frac{\mathrm{d}P^{v(\cdot)}}{\mathrm{d}P}\bigg|_{\mathcal{F}^t} = \eta_t. \tag{8.1.9}$$

Define $b_t = b_t^{v(\cdot)}$ by the formula

$$b_t = z_t - \int_0^t h(x_s)\mathrm{d}s \tag{8.1.10}$$

which becomes for the system $(\Omega, \mathcal{A}, P, \mathcal{F}^t)$ a standard Wiener process, independent of w_T.

Remark 8.1.1 In the presentation of Chapter 4, we have begun by considering *a priori* b_t as a Wiener process, instead of z_t. This is more natural from the modelling point of view. When there is a control involved, which must be adapted to the observations, it is much more convenient to consider z_t as a *priori* given. This is why we take this approach in the present context. ∎

8.1.2 Unnormalized conditional probablity

We shall define the unnormalized conditional probability by

$$q_t^{v(\cdot)}(\psi) = E[\psi(x_t)\eta_t|\mathcal{Z}^t] \tag{8.1.11}$$

where ψ is a test function (Borel bounded to fix the ideas).

Because of the control, the expression (8.1.11) does not follow immediately from the results of Chapter 4. For what follows, we shall be particularly interested in the stochastic PDE approach to the unnormalized conditional probability (the Zakai equation). We shall concentrate on this approach.

We recall the notation (§ 4.7.2)

$$H = L^2(R^n), \quad V = H^1(R^n)$$

and define

$$A_t\phi = -\sum_{i,j} \frac{\partial}{\partial x_i}\left(a_{ij}\frac{\partial\phi}{\partial x_j}\right), \quad \text{for } \phi \in V \tag{8.1.12}$$

where

$$a_{ij} = \frac{1}{2}(\sigma\sigma^*)_{ij}.$$

We also define

$$G_t(v)\phi = \sum_i \frac{\partial}{\partial x_i}(a_i(x,v)\phi), \quad \forall\phi \in H \tag{8.1.13}$$

where

$$a_i(x,v) = -g_i(x,v) + \sum_j \frac{\partial}{\partial x_j}a_{ij}$$

and

$$B_t \phi = \phi h(x,t), \quad \forall \phi \in H. \tag{8.1.14}$$

We now consider the following stochastic PDE:

$$dq + A_t q dt = G_t(v_t) q_t dt + B_t q_t dz$$
$$q(0) = p_0 \tag{8.1.15}$$

in the functional space

$$q \in L^2_Z(0,T;V) \cap L^2(\Omega, \mathcal{A}, P; C(0,T;H)) \tag{8.1.16}$$

where in (8.1.15) q_0 is related to Π_0 as follows:

Π_0 has a density with respect to Lebesgue measure, denoted p_0,

which belongs to H. $\tag{8.1.17}$

Moreover in (8.1.15) v_t is any admissible control, hence it is a process adapted to $\mathcal{Z}^t$. Naturally the very definition of a_i presupposes

$$\frac{\partial}{\partial x_j} a_{ij} \quad \text{is bounded.} \tag{8.1.18}$$

Note that $G_t(v)$ maps H into V'. Thus if q_t belongs to (8.1.6), $G_t(v_t)q_t \in L^2_Z(0,T;V')$, and

$$B_t q_t \in L^2_Z(0,T;\mathcal{L}(R^m;H)).$$

Remark 8.1.2 We have slightly modified the notation, with respect to Chapter 4 and (8.1.12). This is to stress better the role of the control. Moreover, the notation $G_t(q,v)$, $B_t(q)$ could also fit with non linear operators. The problem (8.1.15) belongs to a class of controlled stochastic PDEs (see Bensoussan (1982) for further details in this direction). ∎

We can state the following:

Theorem 8.1.1 *We assume (8.1.1), (8.1.2), (8.1.3), (8.1.17), (8.1.18); then for any v_t given (admissible control), there exists one and only one solution of (8.1.15).*

Proof

It is a variant of the proof of Theorem 4.7.1. We first notice that for any $q \in L^2_Z(0,T;V)$, the following condition holds (equivalent to (4.7.15))

$$2E \int_0^T \langle A_t, q, q \rangle dt + \lambda_0 E \int_0^T |q|^2 dt$$
$$\geq 2E \int_0^T \langle G_t(v_t)q, q \rangle dt + E \sum_{i=1}^m \int_0^T |B_t^* q \cdot e_i|^2 dt + \alpha_0 E \int_0^T \|q\|^2 dt \tag{8.1.19}$$

where $\alpha_0 > 0$, $\lambda_0 \geq 0$, and $e_1, \ldots, e_m$ is an orthonormal basis of R^m.

In view of (8.1.13), (8.1.14), the property (8.1.19) follows from the relation

$$2 \int_{R^n} a_{ij} \frac{\partial q}{\partial x_j} \frac{\partial q}{\partial x_i} dx + \lambda_0 \int_{R^n} q^2 dx \geq -2 \int_{R^n} a_i(x,v) q \frac{\partial q}{\partial x_i} dx$$
$$+ \int_{R^n} |h|^2 q^2 dx + \alpha_0 \int_{R^n} \|q\|^2 dx, \quad \forall v, \tag{8.1.20}$$

for convenient choices of α_0, λ_0.

We perform in (8.1.15) the change of unknown function $q \to qe^{kt}$, and see that (8.1.15) is equivalent to

$$dq + (A_t q + kq)dt = G_t(v_t)q dt + B_t q dz$$
$$q(0) = p_0. \qquad (8.1.21)$$

We define a map $\eta \to \zeta$ from $L^2_Z(0, T; V)$ into itself, by setting

$$d\zeta + (A_t \zeta + k\zeta)dt = G_t(v_t)\eta dt + B_t \eta dz$$
$$\zeta(0) = p_0. \qquad (8.1.22)$$

We then proceed as in the proof of Theorem 4.7.1, to check that, for a convenient choice of k, the preceding map is a contraction.

The proof has been completed. ∎

We next relate the solution of (8.1.15) to the unnormalized conditional probability, by the following result.

Theorem 8.1.2 *We make the assumptions of Theorem 8.1.1. Let $\phi \in C_0^\infty(R^n)$, then one has*

$$\int_{R^n} \phi(x)q(x,t)dx = q_t^{v(\cdot)}(\phi) \quad \text{a.s. } \forall t. \qquad (8.1.23)$$

where q is the solution of (8.1.15), and the right hand side of (8.1.23) is defined by (8.1.11).

Proof

We shall assume that

$$\frac{\partial h}{\partial t}, \quad \frac{\partial h}{\partial x_i}, \quad \frac{\partial^2 h}{\partial x_j \partial x_k} \quad \text{are bounded,} \qquad (8.1.24)$$

and prove that (8.1.23) holds in this case. By a regularization procedure, one can check that (8.1.23) holds without this particular assumption.

Now, if (8.1.24) holds, we can write from (8.1.11)

$$q_t^{v(\cdot)}(\phi) = E\Bigg[\phi(x_t)\exp(z_t \cdot h(x_t))$$
$$\exp\Bigg(-z_t \cdot h(x_t) + \int_0^t h \cdot dz - \frac{1}{2}\int_0^t |h|^2 ds\Bigg)\Bigg|Z^t\Bigg] \qquad (8.1.25)$$
$$= E[\phi(x_t)\exp(z_t \cdot h(x_t))\exp \xi_t | Z^t]$$

where

$$\xi_t = \int_0^t \Bigg[\Bigg(-z\frac{\partial h}{\partial s} + a_k\frac{\partial}{\partial x_k}(hz) + A(hz) - \frac{1}{2}|h|^2\Bigg)ds - D(hz)\cdot\sigma dw\Bigg].$$

We note that

$$d\exp \xi_t = \exp \xi_t\Bigg[\Bigg(-z\frac{\partial h}{\partial t} + a_k\frac{\partial}{\partial x_k}(hz) + aD(hz)\cdot D(hz) + A(hz) - \frac{1}{2}|h|^2\Bigg)dt$$
$$- D(hz)\cdot\sigma dw\Bigg].$$
$$\qquad (8.1.26)$$

In the calculation (8.1.25), we notice that, because of the independence of z_t and w_t, it suffices to freeze z_t as if it were deterministic, and compute the expectation with respect to the remaining source of uncertainty w_t.

We can then proceed as follows. Consider the function $\mu(x,t)$, depending parametrically on z_t, defined by

$$-\frac{\partial \mu}{\partial t} + A_t\mu + a_k\frac{\partial}{\partial x_k}\mu + 2D\mu \cdot aD(hz)$$

$$+\mu\left(z\frac{\partial h}{\partial t} - a_k\frac{\partial}{\partial x_k}(hz) - aD(hz)\cdot D(hz) - A(hz) + \frac{1}{2}|h|^2\right) = 0 \qquad (8.1.27)$$

$$\mu(x,T) = \phi(x)\exp(z_T\cdot h(x)).$$

Note that from (8.1.25)

$$q_T^{v(\cdot)}(\phi) = E[\mu(x_T,T)\exp\xi_T|\mathcal{Z}^T]. \qquad (8.1.28)$$

Using (8.1.26) and (8.1.27), we see that

$$d(\mu(x_t,t)\exp\xi_t) = \exp\xi_t(D\mu - \mu D(hz))\cdot\sigma dw$$

and thus it follows that

$$q_T^{v(\cdot)}(\phi) = \int \mu(x,0)p_0(x)dx. \qquad (8.1.29)$$

On the other hand, one can assert that

$$\int_{R^n} \phi(x)q(x,T)dx = \int_{R^n} \mu(x,T)q(x,T)\exp(-z_T\cdot h(x))dx. \qquad (8.1.30)$$

Setting

$$\nu(x,t) = q(x,t)\exp(-z_t\cdot h(x))$$

we deduce easily from (8.1.15) that ν satisfies the PDE

$$\frac{\partial \nu}{\partial t} + A\nu - \frac{\partial}{\partial x_k}(a_k\nu) - 2D\nu \cdot aD(hz)$$

$$+\nu\left(A(hz) - D(hz)aD(hz) + z\frac{\partial h}{\partial t} + \frac{1}{2}|h|^2 - a_k\frac{\partial}{\partial x_k}(zh)\right) = 0 \qquad (8.1.31)$$

$$\nu(x,0) = p_0(x)$$

and the right hand side of (8.1.30) reads $\int_{R^n} \mu(x,T)\nu(x,T)dx$. Using (8.1.31) and (8.1.27), an integration by parts yields

$$\int_{R^n} \mu(x,T)\nu(x,T)dx = \int_{R^n} \mu(x,0)p_0(x)dx$$

and comparing with (8.1.29), (8.1.30) we obtain the result (8.1.23) .   ∎

8.1.3   Control problem

Let us consider functions

$$l_t(x,v) \quad \text{Borel bounded;}$$
$$k(x) \quad \text{Borel bounded.} \tag{8.1.32}$$

Let v_t be an *admissible* control; we define the cost function

$$J(v(\cdot)) = E^{v(\cdot)}\left[\int_0^T l(x_t, v_t)\mathrm{d}t + k(x_T)\right]. \tag{8.1.33}$$

One objective is to minimize $J(v(\cdot))$.

From the definition of $P^{v(\cdot)}$ one has

$$J(v(\cdot)) = E\eta_T\left[\int_0^T l(x_t, v_t)\mathrm{d}t + k(x_T)\right]$$

and, from the martingale property of η_t,

$$J(v(\cdot)) = E\left[\int_0^T \eta_t l(x_t, v_t)\mathrm{d}t + k(x_T)\eta_T\right]$$
$$= E\left[\int_0^T q_t^{v(\cdot)}(l(\cdot, v_t))\mathrm{d}t + q_T^{v(\cdot)}(k)\right]$$

and, from Theorem 8.1.2, one has

$$J(v(\cdot)) = E\int_0^T \int_{R^n} q(x,T)l(x,v_t)\mathrm{d}x\mathrm{d}t + E\int_{R^n} q(x,t)k(x)\mathrm{d}x \tag{8.1.34}$$

We can write (8.1.34) as a functional defined on the solution of (8.1.15) provided the following assumptions hold:

$$l(\cdot, v) \in H, \ |l(\cdot, v)|_H \leq C, \text{ independent of } v. \tag{8.1.35}$$

$$k \in H. \tag{8.1.36}$$

Then the functional $J(v(\cdot))$ can be written as

$$J(v(\cdot)) = E\left[\int_0^T (l_t(v_t), q_t)\mathrm{d}t + (k, q_T)\right]. \tag{8.1.37}$$

8.2   Stochastic maximum principle

8.2.1   Preliminary results

It is convenient to consider that $J(v(\cdot))$ is a functional on $L_Z^2(0, T; R^k)$ (although we shall be interested only in the subset of admissible controls).

We shall need the additional assumptions

$$\partial g_i/\partial v_j \text{ is bounded and continuous in } v \qquad (8.2.1)$$

$$\partial l/\partial v(x, v) \text{ is continuous in } v \text{ and } |\partial l/\partial v(x, v)| \leq \beta(x), \text{ where } \beta \in H \qquad (8.2.2)$$

$$U_{ad} \text{ is convex compact.} \qquad (8.2.3)$$

We shall use the notation

$$
G_v(v)\phi = G_{v,t}(v)\phi = \sum_i \frac{\partial}{\partial x_i} \left(\frac{\partial a_i}{\partial v}(x, v)\phi \right)
$$
$$
= -\sum_i \frac{\partial}{\partial x_i} \left(\frac{\partial g_i}{\partial v}(x, v)\phi \right) \qquad (8.2.4)
$$
$$
G_v(v)\phi \in \mathcal{L}(R^k; \mathcal{L}(H; V'))
$$

$$
(l_v(v), \phi) = (l_{v,t}(v), \phi) = \int_{R^n} \frac{\partial l}{\partial v}(x, v)\phi(x)\mathrm{d}x \qquad (8.2.5)
$$
$$
(l_v(v), \phi) \in \mathcal{L}(R^k; \mathcal{L}(H; R))
$$

We can now state

Lemma 8.2.1 *The functional $J(v(\cdot))$ is Gateaux differentiable and one has (for $u(\cdot)$, $v(\cdot)$ admissible)*

$$
\frac{d}{d\theta} J(u(\cdot) + \theta v(\cdot))\Big|_{\theta=0} = E\left\{ \int_0^T [(l(u_t), \zeta_t) + (l_v(u_t), p_t) \cdot v_t]dt + (k, \zeta_T) \right\} \qquad (8.2.6)
$$

where p_t is the solution of (8.1.21) corresponding to the control u_t, i.e.

$$
dp + A_t p dt = G_t(u_t)p_t dt + B_t p_t \cdot dz
$$
$$
p(0) = p_0 \qquad (8.2.7)
$$

and ζ is the solution of

$$
d\zeta_t + A_t \zeta_t dt = (G_t(u_t)\zeta_t + G_{v,t}(u_t)p_t \cdot v_t)dt + B_t \zeta_t dz
$$
$$
\zeta(0) = 0 \qquad (8.2.8)
$$

and

$$\zeta \in L_Z^2(0, T; V) \cap L^2(\Omega, \mathcal{A}, P; C(0, T; H)). \qquad (8.2.9)$$

Proof

Note that U_{ad} compact is used to assert that $G_{v,t}(u_t)p_t \cdot v_t$ belongs to $L_Z^2(0, T; V')$.

Let p_θ be the trajectory corresponding to the control $u(\cdot) + \theta v(\cdot)$:

$$\tilde{p}_\theta = \frac{p_\theta - p}{\theta} - \zeta$$

we have

$$\begin{aligned}
\mathrm{d}\tilde{p}_\theta + A\tilde{p}_\theta \mathrm{d}t =& G(u + \theta v)\tilde{p}_\theta \mathrm{d}t + \int_0^1 \mathrm{d}\lambda(G_v(u + \lambda\theta v) - G_v(u))p \cdot v \mathrm{d}t \\
&+ \theta \int_0^1 \mathrm{d}\lambda G_v(u + \lambda\theta v)\zeta \cdot v \mathrm{d}t + B\tilde{p}_\theta \cdot \mathrm{d}z.
\end{aligned}$$

We can write the energy equality (cf. (4.3.3))

$$\begin{aligned}
E|\tilde{p}_\theta(t)|^2 &+ 2E \int_0^t \langle A\tilde{p}_\theta, \tilde{p}_\theta \rangle \mathrm{d}s \\
=& 2E \int_0^t \langle \tilde{p}_\theta, G(u + \theta v)\tilde{p}_\theta \rangle \mathrm{d}s \\
&+ 2E \int_0^t \left\langle \tilde{p}_\theta, \int_0^1 \mathrm{d}\lambda(G_v(u + \lambda\theta v) - G_v(u))p \cdot v \right\rangle \mathrm{d}s \\
&+ 2\theta E \int_0^t \left\langle \tilde{p}_\theta, \int_0^1 \mathrm{d}\lambda G_v(u + \lambda\theta v)\zeta \cdot v \right\rangle \mathrm{d}s + E \int_0^t |B(\tilde{p}_\theta)|^2 \mathrm{d}s.
\end{aligned}$$

Making the above relation explicit, we get

$$\begin{aligned}
E|\tilde{p}_\theta(t)|^2 &+ 2E \int_0^t \int_{R^n} a_{ij} \frac{\partial \tilde{p}_\theta}{\partial x_j} \frac{\partial \tilde{p}_\theta}{\partial x_i} \mathrm{d}x \mathrm{d}s + 2E \int_0^t \int_{R^n} \frac{\partial \tilde{p}_\theta}{\partial x_i} a_i(x, u + \theta v)\tilde{p}_\theta \mathrm{d}x \mathrm{d}s \\
=& 2E \int_0^t \int_{R^n} \frac{\partial \tilde{p}_\theta}{\partial x_i} \int_0^1 \mathrm{d}\lambda \left(\frac{\partial g_i}{\partial v}(x, u + \lambda\theta v) - \frac{\partial g_i}{\partial v}(x, u) \right) vp \mathrm{d}x \mathrm{d}s \\
&+ 2\theta E \int_0^t \int_{R^n} \frac{\partial \tilde{p}_\theta}{\partial x_i} \int_0^1 \mathrm{d}\lambda \frac{\partial g_i}{\partial v}(x, u + \lambda\theta v) v\zeta \mathrm{d}x \mathrm{d}s + E \int_0^t \int_{R^n} \tilde{p}_\theta^2 |h|^2 \mathrm{d}x \mathrm{d}s
\end{aligned}$$

from which we deduce

$$E|\tilde{p}_\theta(t)|^2 + \alpha E \int_0^t \|\tilde{p}_\theta(s)\|^2 \mathrm{d}s \leq K \left(E \int_0^t |\tilde{p}_\theta(s)|^2 \mathrm{d}s + \rho_\theta \right) \qquad (8.2.10)$$

in which the constant K does not depend on θ, and

$$\begin{aligned}
\rho_\theta =& \sum_i E \int_0^T \int_{R^n} \int_0^1 \mathrm{d}\lambda \left| \left(\frac{\partial g_i}{\partial v}(x, u + \lambda\theta v) - \frac{\partial g_i}{\partial v}(x, u) \right) \right|^2 |v|^2 |p|^2 \mathrm{d}x \mathrm{d}s \\
&+ \theta E \int_0^T \int_{R^n} |v|^2 |\zeta|^2 \mathrm{d}x \mathrm{d}s.
\end{aligned}$$

Clearly $\rho_\theta \to 0$ as $\theta \to 0$. Therefore

$$\sup_{0 \le t \le T} E|\tilde{p}_\theta(t)|^2 \to 0 \text{ as } \theta \to 0. \tag{8.2.11}$$

Now we may write

$$\frac{J(u(\cdot) + \theta v(\cdot)) - J(u(\cdot))}{\theta}$$

$$= E\left\{ \int_0^T [(l(u_t), \zeta_t) + (l_v(u_t), p_t) \cdot v_t] dt + (k, \zeta_T) \right\}$$

$$+ E \int_0^T \int_{R^n} \int_0^1 d\lambda \left(\frac{\partial l}{\partial v}(x, u + \lambda\theta v) - \frac{\partial l}{\partial v}(x, u) \right) vp dx ds$$

$$+ \theta E \int_0^T \int_{R^n} \int_0^1 d\lambda \frac{\partial l}{\partial v}(x, u + \lambda\theta v)v\zeta dx ds + E \int_0^T \int_{R^n} l(x, u + \theta v)\tilde{p}_\theta dx ds$$

and from (8.2.11) and the assumption (8.2.2) the desired result (8.2.6) follows. ∎

8.2.2   Abstract definition of the adjoint process

Let us consider processes $\phi \in L_Z^2(0, T; V')$ and $\psi \in (L_Z^2(0, T; H))^m$. We then define an equation of the type (8.2.8), with data ϕ and ψ as follows:

$$d\rho_t + A_t\rho_t dt = (G_t(u_t)\rho_t + \phi_t)dt + (B_t\rho_t + \psi_t) \cdot dz$$

$$\rho(0) = 0 \tag{8.2.12}$$

$$\rho \in L_Z^2(0, T; V) \cap L^2(\Omega, \mathcal{A}, P; C(0, T; H)).$$

We can write the energy equality corresponding to (8.2.12) as

$$|\rho_t|^2 + 2\int_0^t \langle A\rho_s, \rho_s \rangle ds = 2\int_0^t \langle \rho_s, G_s(u_s)\rho_s + \phi_s \rangle ds$$

$$+ 2\int_0^t (\rho_s, B_s\rho_s + \psi_s) \cdot dz \tag{8.2.13}$$

$$+ \sum_{i=1}^m \int_0^t |h_i\rho + \psi_i|_H^2 ds.$$

It is easy to check that the map $\phi, \psi \to \rho$ is linear continuous for the spaces of the definition.

Therefore there exist uniquely defined processes $\lambda_t \in L_Z^2(0, T; V)$ and $r_{it} \in L_Z^2(0, T; H)$, $i = 1, \ldots, m$ such that the following relation holds

$$E \int_0^T \left(\langle \lambda_t, \phi_t \rangle + \sum_{i=1}^m (r_{it}, \psi_{it}) \right) dt = E\left[\int_0^T (l(u_t), \rho_t)dt + (k, \rho(T)) \right]. \tag{8.2.14}$$

Note that, if we compare (8.2.8) with (8.2.12), the solution of (8.2.12) corresponding to

$$\phi_t = G_{v,t}(u_t)p_t \cdot v_t \text{ and } \psi_t = 0$$

is ζ_t, the solution of (8.2.8). Therefore from (8.2.14) it follows that

$$E \int_0^T \langle \lambda_t, G_{v,t}(u_t)p_t \cdot v_t \rangle \mathrm{d}t = E \left[\int_0^T (l(u_t), \zeta_t) \mathrm{d}t + (k, \zeta_T) \right]$$

hence from (8.2.6) we can assert

$$\begin{aligned}
\frac{\mathrm{d}}{\mathrm{d}\theta} J(u(\cdot) + \theta v(\cdot)) \bigg|_{\theta=0} &= E \int_0^T [\langle \lambda, G_{v,t}(u_t)p_t \cdot v_t \rangle + l_v(u_t), p_t) \cdot v_t] \mathrm{d}t \\
&= E \int_0^T \left[\int_{R^n} \left(D\lambda(x,t) \frac{\partial g}{\partial v}(x, u_t) + \frac{\partial l}{\partial v}(x, u_t) \right) \right. \qquad (8.2.15) \\
&\qquad\qquad p(x,t)\mathrm{d}x \bigg] v_t \mathrm{d}t.
\end{aligned}$$

We can now state the following.

Theorem 8.2.1 *We assume (8.1.1), (8.1.2), (8.1.3), (8.1.4), (8.1.17), (8.1.18), (8.1.32), (8.1.35), (8.1.36), (8.2.1), (8.2.2), (8.2.3); then, if u_t is an optimal control for (8.1.5), (8.1.33) or the equivalent form (8.1.15), (8.1.40), and if p denotes the unnormalized conditional probability corresponding to u_t, the solution of (8.2.7), then there exists $\lambda_t \in L_Z^2(0, T; V)$ such that the following condition holds:*

$$\left[\int_{R^n} \left(D\lambda(x,t) \frac{\partial g}{\partial v}(x, u_t) + \frac{\partial l}{\partial v}(x, u_t) \right) p(x,t)\mathrm{d}x \right] \cdot (v - u_t) \geq 0 \qquad (8.2.16)$$

$$\forall v \in U_{ad}, \quad \text{a.e., a.s.}$$

Proof

Since U_{ad} is convex, if $v(\cdot)$ is an admissible control, $u(\cdot) + \theta(v(\cdot) - u(\cdot))$ is also admissible, hence from

$$J(u(\cdot) + \theta(v(\cdot) - u(\cdot))) \geq J(u(\cdot))$$

it follows that

$$\frac{\mathrm{d}}{\mathrm{d}\theta} J(u(\cdot) + \theta(v(\cdot) - u(\cdot))) \bigg|_{\theta=0} \geq 0$$

and from the formula (8.2.15) we have

$$E \int_0^T \left[\int_{R^n} \left(D\lambda(x,t) \frac{\partial g}{\partial v}(x, u_t) + \frac{\partial l}{\partial v}(x, u_t) \right) p(x,t)\mathrm{d}x \right] (v_t - u_t) \mathrm{d}t \geq 0. \qquad (8.2.17)$$

Let

$$\xi_t = \int_{R^n} \left(D\lambda(x,t) \frac{\partial g}{\partial v}(x, u_t) + \frac{\partial l}{\partial v}(x, u_t) \right) p(x,t)\mathrm{d}x$$

we write (8.2.17) as

$$E \int_0^T \xi_t \cdot (v_t - u_t) \mathrm{d}t \geq 0. \qquad (8.2.18)$$

Note that $\xi_t \in L_Z^1(0, T; R^k)$.

Let N be given as well as $t_0 \in (0, T)$, and ϵ such that $t_0 + \epsilon < T$. Let v_{t_0} be $\mathcal{Z}^{t_0}$ measurable, with values in U_{ad}. We choose

$$v_t = \begin{cases} u_t & \text{if } |\xi_t| \geq N \text{ and } t \notin (t_0, t_0 + \epsilon) \\ v_{t_0} & \text{if } |\xi_t| < N \text{ and } t \in (t_0, t_0 + \epsilon) \end{cases}$$

which is admissible. We deduce from (8.2.18) that

$$E \int_{t_0}^{t_0+\epsilon} \xi_t (v_{t_0} - u_t) \mathcal{I}_{|\xi_t|<N} dt \geq 0. \tag{8.2.19}$$

But

$$\forall N \quad \frac{1}{\epsilon} \int_{t_0}^{t_0+\epsilon} \xi_t \mathcal{I}_{|\xi_{t_0}|<N} \to \xi_{t_0} \mathcal{I}_{|\xi_t|<N}$$

for almost all t_0 as $\epsilon \to 0$, in $L^2(\Omega, \mathcal{A}, P)$. Since also

$$E \xi_t u_t \mathcal{I}_{|\xi_t|<N} \to E \xi_{t_0} u_{t_0} \mathcal{I}_{|\xi_{t_0}|<N}$$

for almost all t_0, we deduce from (8.2.19) that

$$\text{for all } N, \text{ almost all } t_0, \quad E \xi_{t_0} (v_{t_0} - u_{t_0}) \mathcal{I}_{|\xi_{t_0}|<N} \geq 0, \quad \forall v_{t_0}.$$

Therefore also

$$\text{for almost all } t_0 \quad E \xi_{t_0} (v_{t_0} - u_{t_0}) \mathcal{I}_{|\xi_{t_0}|<N} \geq 0, \quad \forall v_{t_0}, \forall N$$

and letting N tend to $+\infty$,

$$\text{for almost all } t_0 \quad E \xi_{t_0} (v_{t_0} - u_{t_0}) \geq 0, \quad \forall v_{t_0}, \, \mathcal{Z}^{t_0} \text{ measurable,}$$
$$\text{with values in } U_{ad}.$$

Let now v be deterministic in U_{ad}, and set

$$\mu_{t_0} = \xi_{t_0} (v - u_{t_0}), \quad \text{which is } \mathcal{Z}^{t_0} \text{ measurable.}$$

Let $A_0 = \{\omega | \mu_{t_0} < 0\}$ and take $v_{t_0} = u_{t_0}$ in $\Omega - A_0$. Therefore necessarily

$$E \mu_{t_0} \mathcal{I}_{A_0} \geq 0$$

which is a contradiction unless A_0 has probability 0. Therefore the desired result (8.2.16) has been proven. ∎

Remark 8.2.1 Defining the Hamiltonian $H_t(\phi, \lambda, v)$ for $\phi \in H$, $\lambda \in V$, $v \in R^k$ by

$$H_t(\phi, \lambda, v) = \int_{R^n} (\mathrm{D}\lambda(x) \cdot g(x, v) + l(x, v)) \phi(x) \mathrm{d}x \tag{8.2.20}$$

the condition (8.2.16) becomes

$$\frac{\partial H_t}{\partial v}(p_t, \lambda_t, u_t)(v - u_t) \geq 0, \quad \forall v \in U_{ad}, \text{for almost all } t, \text{ a.s.} \tag{8.2.21}$$

∎

8.2.3   Approximation procedure

We shall define in this paragraph an approximation procedure for equation (8.2.12).

We consider a basis of V (not necessarily orthogonal), denoted by $\eta_1, \ldots, \eta_j, \ldots$ Let

$$V_\mu = [\eta_1, \ldots, \eta_\mu]$$

be the finite dimensional subspace generated by $\eta_1, \ldots, \eta_\mu$. We look for a process $\rho_t^\mu \in L_Z^2(0, T; V_\mu)$ defined by

$$\rho_t^\mu = \sum_{i=1}^\mu \theta_{i,t}^\mu \eta_i$$

such that

$$(d\rho_t^\mu, \eta_j) + \langle A\rho_t^\mu, \eta_j \rangle dt = \langle G_t(u_t)\rho_t^\mu + \phi_t, \eta_j \rangle dt + (B_t\rho_t^\mu + \psi_t, \eta_j) \cdot dz,$$
$$\forall j = 1, \ldots, \mu \tag{8.2.22}$$

and

$$\rho^\mu(0) = 0.$$

Our objective is to prove

Theorem 8.2.2 *The assumptions are those of Theorem 8.2.1. One has*

$$\rho^\mu \to \rho \text{ in } L_Z^2(0, T; V) \cap L^2(\Omega, \mathcal{A}, P; C(0, T; H)). \tag{8.2.23}$$

We shall need several steps for the proof.

Lemma 8.2.2 *We have the estimates*

$$E \int_0^T \|\rho_t^\mu\|^2 dt \le C \tag{8.2.24}$$

$$\sup_{t \in [0,T]} E|\rho_t^\mu|^4 \le C \tag{8.2.25}$$

$$E \sup_{t \in [0,T]} |\rho_t^\mu|^2 \le C. \tag{8.2.26}$$

Proof

Let us denote by Π_μ the orthogonal projection operator on V_μ, considered as a subspace of H. We use Ito's formula to derive the relation

$$d|\rho_t^\mu|^2 + 2\langle A\rho_t^\mu, \rho_t^\mu \rangle dt = \Big[2\langle G(u_t)\rho_t^\mu + \phi_t, \rho_t^\mu \rangle + (\Pi_\mu(h\rho_t^\mu + \psi_t), h\rho_t^\mu + \psi_t) \Big] dt$$
$$+ 2(\rho_t^\mu, h\rho_t^\mu + \psi_t) \cdot dz \tag{8.2.27}$$

hence

$$E|\rho_t^\mu|^2 + 2E \int_0^t \langle A\rho_s^\mu, \rho_s^\mu \rangle ds$$
$$= E \int_0^t [2\langle G(u_s)\rho_s^\mu + \phi_s, \rho_s^\mu \rangle + (\Pi_\mu(h\rho_s^\mu + \psi_s), h\rho_s^\mu + \psi_s)] ds$$

(where the notation $(\Pi_\mu(h\rho_s^\mu+\psi_s), h\rho_s^\mu+\psi_s)$ means $\sum_{i=1}^m(\Pi_\mu(h_i\rho_s^\mu+\psi_{is}), h_i\rho_{is}^\mu+\psi_{is})$) from which (8.2.24) follows as well as

$$\sup_{t\in[0,T]} E|\rho_t^\mu|^2 \leq C.$$

We next derive

$$
\begin{aligned}
\mathrm{d}|\rho_t^\mu|^4 + 4|\rho_t^\mu|^2\langle A\rho_t^\mu, \rho_t^\mu\rangle\mathrm{d}t =& 2|\rho_t^\mu|^2\Big[2\langle G(\mu_t)\rho_t^\mu + \phi_t, \rho_t^\mu\rangle \\
& + (\Pi_\mu(h\rho_t^\mu + \psi_t), h\rho_t^\mu + \psi_t)\Big]\mathrm{d}t + 4(\rho_t^\mu, h\rho_t^\mu + \psi_t) \\
& (\rho_t^\mu, h\rho_t^\mu + \psi_t)\mathrm{d}t + 4|\rho_t^\mu|^2(\rho_t^\mu, h\rho_t^\mu + \psi_t) \cdot \mathrm{d}z
\end{aligned}
$$

hence

$$
\begin{aligned}
E|\rho_t^\mu|^4 + & 4E\int_0^t |\rho_s^\mu|^2\langle A\rho_s^\mu, \rho_s^\mu\rangle\mathrm{d}s \\
=& 2E\int_0^t |\rho_s^\mu|^2[2\langle G(\mu_s)\rho_s^\mu + \rho_s, \rho_s^\mu\rangle + (\Pi_\mu(h\rho_s^\mu + \psi_s), h\rho_s^\mu + \psi_s)]\mathrm{d}s \\
& + 4E\int_0^t (\rho_s^\mu, h\rho_s^\mu + \psi_s)(\rho_s^\mu, h\rho_s^\mu + \psi_s)\mathrm{d}s \\
\leq& CE\int_0^t |\rho_s^\mu|^2(|\rho_s^\mu| + \|\phi_s\|)\|\rho_s^\mu\|\mathrm{d}s + CE\int_0^t |\rho_s^\mu|^4\mathrm{d}s + CE\int_0^t |\psi_s|^2\mathrm{d}s
\end{aligned}
$$

and (8.2.25) follows easily from this estimate.

Now (8.2.26) follows from (8.2.25) combined with (8.2.27). ∎

Lemma 8.2.3 *The convergence (8.2.23) holds in a weak sense.*
Proof
We extract a subsequence of ρ_t^μ, still denoted ρ_t^μ such that $\rho_t^\mu \to \tilde\rho_t$ in $L_Z^2(0,T;V)$ weakly and $L^2(\Omega, \mathcal{A}, P; C(0,T;H))$ in the weak star topology.

Let $\beta \in L^\infty(0,T;R^m)$ be deterministic and consider the martingale

$$\gamma_t = \exp\left(\mathrm{i}\int_0^t \beta_s \cdot \mathrm{d}z_s + \frac{1}{2}\int_0^t |\beta_s|^2\mathrm{d}s\right).$$

From (8.2.22) we deduce

$$
\begin{aligned}
E\gamma_t(\rho_t^\mu, \eta_j) + & E\int_0^t \gamma_s\langle A\rho_s^\mu, \eta_j\rangle\mathrm{d}s \\
=& E\int_0^t \gamma_s\langle G_s(u_s)\rho_s^\mu + \phi_s, \eta_j\rangle\mathrm{d}s + E\int_0^t \mathrm{i}\gamma_s\beta_s \cdot (h\rho_s^\mu + \psi_s, \eta_j)\mathrm{d}s, \ \forall t.
\end{aligned}
$$

We can take the limit in μ, to deduce

$$
\begin{aligned}
E\gamma_t(\tilde\rho_t, \eta_j) + & E\int_0^t \gamma_s\langle A\tilde\rho_s, \eta_j\rangle\mathrm{d}s \\
=& E\int_0^t \gamma_s\langle G_s(u_s)\tilde\rho_s + \phi_s, \eta_j\rangle\mathrm{d}s \\
& + E\int_0^t \mathrm{i}\gamma_s\beta_s \cdot (h\tilde\rho_s + \psi_s, \eta_j)\mathrm{d}s, \ \forall t.
\end{aligned}
$$

which can be rewritten as

$$E\gamma_t\left[(\tilde{\rho}_t, \eta_j) + \int_0^t \langle A\tilde{\rho}_s, \eta_j\rangle \mathrm{d}s - \int_0^t \langle G_s(u_s)\tilde{\rho}_s + \phi_s, \eta_j\rangle \mathrm{d}s \right.$$
$$\left. - \int_0^t (h\tilde{\rho}_s + \psi_s, \eta_j)\cdot \mathrm{d}z\right] = 0.$$

Since β is arbitrary, this relation implies

$$(\tilde{\rho}_t, \eta_j) + \int_0^t \langle A\tilde{\rho}_s, \eta_j\rangle \mathrm{d}s = \int_0^t \langle G_s(u_s)\tilde{\rho}_s + \phi_s, \eta_j\rangle \mathrm{d}s + \int_0^t (h\tilde{\rho}_s + \psi_s, \eta_j)\cdot \mathrm{d}z.$$

Since η_j is any element of the basis of V, it follows that $\tilde{\rho}$ satisfies the relation (8.2.12), and thus necessarily

$$\tilde{\rho} = \rho.$$

The proof has been completed. ∎

Proof of Theorem 8.2.2

It remains to prove the strong convergence. Now from (8.2.22) we deduce

$$\mathrm{d}|\rho_t^\mu|^2 + 2\langle A\rho_t^\mu, \rho_t^\mu\rangle \mathrm{d}t = [2\langle G(u_t)\rho_t^\mu + \phi_t, \rho_t^\mu\rangle + (\Pi_\mu(h\rho_t^\mu + \psi_t), h\rho_t^\mu + \psi_t)]\mathrm{d}t$$
$$+ 2(\rho_t^\mu, h\rho_t^\mu + \psi_t)\cdot \mathrm{d}z \tag{8.2.28}$$

hence, for any scalar $k > 0$,

$$E|\rho_t^\mu|^2 e^{-kt} + E\int_0^t \left\{2\langle A\rho_s^\mu, \rho_s^\mu\rangle + k|\rho_s^\mu|^2 - 2\langle G(u_s)\rho_s^\mu, \rho_s^\mu\rangle - (\Pi_\mu(h\rho_s^\mu), h\rho_s^\mu)\right\} e^{-ks}\mathrm{d}s$$

$$= 2\int_0^t E[\langle\phi_s, \rho_s^\mu\rangle + (\Pi_\mu(\psi_s), h\rho_s^\mu)]e^{-ks}\mathrm{d}s + E\int_0^t (\Pi_\mu(\psi_s), \psi_s)e^{-ks}\mathrm{d}s$$

$$\to 2\int_0^t E[\langle\phi_s, \rho_s\rangle + (\psi_s, h\rho_s)]e^{-ks}\mathrm{d}s + E\int_0^t |\psi_s|^2 e^{-ks}\mathrm{d}s.$$

$$= E|\rho_t|^2 e^{-ks} + E\int_0^t \left\{2\langle A\rho_s, \rho_s\rangle + k|\rho_s|^2 - 2\langle G(u_s)\rho_s, \rho_s\rangle - |h\rho_s|^2\right\} e^{-ks}\mathrm{d}s. \tag{8.2.29}$$

We deduce that

$$E|\rho_t^\mu - \rho_t|^2 e^{-kt} + E\int_0^t \left\{2\langle A(\rho_s^\mu - \rho_s), \rho_s^\mu - \rho_s\rangle + k|\rho_s^\mu - \rho_s|^2 \right.$$
$$\left. -2\langle G(u_s)(\rho_s^\mu - \rho_s), \rho_s^\mu - \rho_s\rangle - (\Pi_\mu(h\rho_s^\mu - h\rho_s), h\rho_s^\mu - h\rho_s)\right\} e^{-ks}\mathrm{d}s$$

$$= 2E\int_0^t [\langle\phi_s, \rho_s^\mu\rangle + (\Pi_\mu(\psi_s), h\rho_s^\mu)]e^{-ks}\mathrm{d}s$$

$$+ E\int_0^t (\Pi_\mu(\psi_s), \psi_s)e^{-ks}\mathrm{d}s + 2E\int_0^t [\langle\phi_s, \rho_s\rangle + (\psi_s, h\rho_s)]e^{-ks}\mathrm{d}s$$

$$+ E\int_0^t |\psi_s|^2 e^{-ks}\mathrm{d}s + E\int_0^t [|h\rho_s|^2 - (\Pi_\mu(h\rho_s), h\rho_s)]e^{-ks}\mathrm{d}s$$

$$- 2E(\rho_t^\mu, \rho_t)e^{-kt} - 2E\int_0^t \Big[2\langle A\rho_s^\mu, \rho_s\rangle + k(\rho_s^\mu, \rho_s)$$

$$- \langle G(u_s)\rho_s^\mu, \rho_s\rangle - \langle G(u_s)\rho_s, \rho_s^\mu\rangle - (h\rho_s^\mu, \Pi_\mu(h\rho_s))\Big]e^{-ks}\mathrm{d}s$$

$$\to 0 \quad \text{as} \quad \mu \to \infty.$$

But, for k sufficiently large, the left hand side of the preceding equality is larger than or equal to

$$E|\rho_t^\mu - \rho_t|^2 e^{-kt} + c_0 E \int_0^t \|\rho_s^\mu - \rho_s\|^2 e^{-ks} ds$$

and thus strong convergence in $L_Z^2(0, T; V)$ follows, as well as in $C(0, T; L^2(\Omega, \mathcal{A}, P; H))$.

One can then compute a quantity like $E|\rho_t^\mu - \rho_t|^4 e^{-kt}$, as in the proof of Lemma 8.2.2, and show successively that $\rho_t^\mu \to \rho_t$ in $C(0, T; L^4(\Omega, \mathcal{A}, P; H))$ and then in $L^2(\Omega, \mathcal{A}, P; C(0, T; H))$. The proof of Theorem 8.2.2 has been completed. ∎

We may next define in a unique way

$$\lambda_t^\mu \in L_Z^2(0, T; V^\mu), \quad r_{it}^\mu \in L_Z^2(0, T; V^\mu), \quad i = 1, \ldots, m$$

such that

$$E \int_0^T \left[\langle \lambda_t^\mu, \phi_t \rangle dt + \sum_{i=1}^m (r_{it}^\mu, \psi_{it}) \right] dt = E \left[\int_0^T (l(u_t), \rho_t^\mu) dt + (k, \rho^\mu(T)) \right], \quad (8.2.30)$$
$$\forall \phi \in L_Z^2(0, T; V'), \quad \forall \psi \in (L_Z^2(0, T; H))^m.$$

Note that (8.2.30) defines λ^μ, r^μ in a unique way, since in fact, whatever ϕ, ψ are, only the components $\langle \eta_j, \phi_t \rangle$, (η_j, ψ_{it}) for $j = 1, \ldots, \mu$ play a role.

We have

Lemma 8.2.4 $\lambda^\mu \to \lambda$ in $L_Z^2(0, T; V)$ and $r^\mu \to r$ in $(L_Z^2(0, T; H))^m$, as $\mu \to \infty$.

Proof

From (8.2.30), and the estimates of Lemma 8.2.2, it is clear that λ^μ remains in a bounded subset of $L_Z^2(0, T; V)$ and r^μ in a bounded subset of $(L_Z^2(0, T; H))^m$. If we extract a weakly converging subsequence, it follows from the results of Theorem 8.2.2, and the definitions (8.2.14) of λ_t, r_{it}, that the convergence stated in the lemma holds in a weak sense.

It thus remains to prove the strong convergence. For that we take in (8.2.30)

$$\phi_t = J\lambda_t^\mu, \quad \psi_{it} = r_{it}^\mu$$

where J denotes the isomorphism from $V \to V'$, $J = -\Delta + I$. The corresponding solution of (8.2.12) is denoted by $\tilde{\rho}^\mu$, hence

$$d\tilde{\rho}_t^\mu + A_t \tilde{\rho}_t^\mu dt = (G_t(u_t)\tilde{\rho}_t^\mu + J\lambda_t^\mu)dt + (B_t \tilde{\rho}_t^\mu + r_t^\mu) \cdot dz$$
$$\tilde{\rho}^\mu(0) = 0 \qquad\qquad (8.2.31)$$

and, to save notation, we still denote by ρ^μ the approximation of $\tilde{\rho}^\mu$, the solution of (8.2.22), namely

$$(d\rho_t^\mu, \eta_j) + \langle A_t \rho_t^\mu, \eta_j \rangle dt = \langle G_t(u_t)\rho_t^\mu + J\lambda_t^\mu, \eta_j \rangle dt$$
$$+ (B_t \rho_t^\mu + r_t^\mu, \eta_j) \cdot dz, \quad \forall j = 1, \ldots, \mu. \qquad (8.2.32)$$

By construction, we have from (8.2.30)

$$E \int_0^T \left(\|\lambda_t^\mu\|^2 + \sum_{i=1}^m |r_{it}^\mu|^2 \right) dt = E \left[\int_0^T (l(u_t), \rho_t^\mu) dt + (k, \rho^\mu(T)) \right]. \tag{8.2.33}$$

Still to save notation, we denote by ρ_t the solution of (8.2.12) corresponding to

$$\phi_t = J\lambda_t, \quad \psi_{it} = r_{it}.$$

Since $\lambda^\mu \to \lambda$ in $L_Z^2(0, T; V)$ weakly and $r_{it}^\mu \to r_{it}$ in $L_Z^2(0, T; H)$ weakly, we can argue as in Lemma 8.2.3 to assert that

$$\rho^\mu \to \rho \text{ in } L_Z^2(0, T; V) \text{ weakly and } L^2(\Omega, \mathcal{A}, P; C(0, T, H)) \text{ weak star.}$$

Therefore the right hand side of (8.2.33) tends to $E[\int_0^T (l(u_t), \rho_t) dt + (k, \rho(T))]$, which is, by virtue of (8.2.14), equal to $E \int_0^T (\|\lambda_t\|^2 + \sum_{i=1}^m |r_{it}|^2) dt$. This result and the weak convergence implies the desired result. ∎

8.2.4 Equation of the adjoint process

In this paragraph we shall derive a backward equation for the adjoint process λ_t. It will be useful to begin with an equation for λ_t^μ, which is finite dimensional valued. We shall prove

Lemma 8.2.5 *The processes* λ_t^μ *and* r_t^μ *satisfy*

$$-(d\lambda_t^\mu, \eta_j) + \left[\langle A\lambda_t^\mu, \eta_j \rangle + \sum_{i=1}^m \left(a_i(\cdot, u_t) \frac{\partial \lambda_t^\mu}{\partial x_i}, \eta_j \right) \right] dt$$
$$= [(h \cdot r_t^\mu, \eta_j) + (l(u_t), \eta_j)] dt - (r_t^\mu, \eta_j) \cdot dz \tag{8.2.34}$$
$$(\lambda_T^\mu - k, \eta_j) = 0, \quad \forall j = 1, \ldots, \mu.$$

Proof

From (8.2.34) $\lambda^\mu \in L^2(\Omega, \mathcal{A}, P; C(0, T; V_\mu))$. Let us first check that if λ^μ, r^μ satisfy (8.2.34) then necessarily (8.2.30) holds for any ϕ, ψ, which will imply the uniqueness of the pair λ^μ, r^μ, the solution of (8.2.34) (belonging to $L_Z^2(0, T; V_\mu)$ and $(L_Z^2(0, T; V_\mu))^m$ respectively).

Indeed one has

$$d(\rho_t^\mu, \lambda_t^\mu) = (d\rho_t^\mu, \lambda_t^\mu) + (\rho_t^\mu, d\lambda_t^\mu) + (d\rho_t^\mu, d\lambda_t^\mu)$$

and from (8.2.22) and (8.2.34)

$$(d\rho_t^\mu, \lambda_t^\mu) = [-\langle A\rho_t^\mu t, \lambda_t^\mu \rangle + \langle G_t(u_t)\rho_t^\mu + \phi, \lambda_t^\mu \rangle] dt$$
$$+ (h\rho_t^\mu + \psi_t, \lambda_t^\mu) \cdot dz$$
$$(d\lambda_t^\mu, \rho_t^\mu) = \left[\langle A\lambda_t^\mu, \rho_t^\mu \rangle + \sum_{i=1}^m \left(a_i(\cdot, u(t)) \frac{\partial \lambda_t^\mu}{\partial x_i}, \rho_t^\mu \right) \right.$$
$$\left. - (l(u_t), \rho_t^\mu) - (h \cdot r_t^\mu, \rho_t^\mu) \right] dt + (r_t^\mu, \rho_t^\mu) \cdot dz$$

$$(d\lambda_t^\mu, d\rho_t^\mu) = \left[(h \cdot r_t^\mu, \rho_t^\mu) + \sum_i (\psi_{it}, r_{it}^\mu)\right] dt.$$

Making use of these relations and also

$$(\lambda_T^\mu, \rho_T^\mu) = (\rho_T^\mu, k)$$

we deduce easily that (8.2.30) holds for any ϕ, ψ. Therefore there exists at most one pair λ^μ, r^μ, as a solution of (8.2.34).

Let us now prove the existence of a pair λ^μ, r^μ, the solution of (8.2.34).

We consider the fundamental matrix related to (8.2.22). It is a process

$$\Phi^\mu \in L_Z^2(0, T; \mathcal{L}(V_\mu; V_\mu)) \cap L^2(\Omega, \mathcal{A}, P; C(0, T; \mathcal{L}(V_\mu; V_\mu)))$$

such that

$$d(\Phi_t^\mu \eta_i, \eta_j) + \langle A\Phi_t^\mu \eta_i, \eta_j \rangle dt = \langle G_t(u_t)\Phi_t \eta_i, \eta_j \rangle dt + (h\Phi_t^\mu \eta_i, \eta_j) \cdot dz$$
$$\Phi^\mu(0) = I_\mu \quad \forall i, j = 1, \dots, \mu, \tag{8.2.35}$$

and I_μ represents the identity in V_μ (as a subset of H).

The matrix Φ^μ is invertible. Indeed let

$$\Psi^\mu \in L_Z^2(0, T; \mathcal{L}(V_\mu; V_\mu)) \cap L^2(\Omega, \mathcal{A}, P; C(0, T; \mathcal{L}(V_\mu; V_\mu)))$$

be such that

$$d(\Psi_t^\mu \eta_i, \eta_j) = \left[\langle A\Psi_t^\mu \eta_i, \eta_j \rangle + \sum_{\beta=1}^m \left(a_\beta(\cdot, u_t)\frac{\partial}{\partial x_\beta}(\Psi_t^\mu \eta_i), \eta_j\right) + (h\Psi_t^\mu \eta_i, \Pi_\mu(h\eta_j))\right] dt$$
$$- (h\Psi_t^\mu \eta_i, \eta_j) \cdot dz$$
$$\Psi^\mu(0) = I_\mu. \tag{8.2.36}$$

Then we have

$$(\Psi_t^\mu)^* \Phi_t^\mu = I_\mu \tag{8.2.37}$$

where $(\Psi_t^\mu)^*$ is the adjoint of Ψ_t^μ in the sense

$$((\Psi_t^\mu)^* \eta_i, \eta_j) = (\eta_i, \Psi_t^\mu \eta_j), \quad \forall i, j = 1, \dots, \mu.$$

To prove (8.2.37) one has to prove that

$$(\Phi_t^\mu \eta_i, \Psi_t^\mu \eta_j) = (\eta_i, \eta_j), \quad \forall i, j = 1, \dots, \mu, \forall t. \tag{8.2.38}$$

Since (8.2.38) is true at $t = 0$, it suffices to show that

$$d(\Phi_t^\mu \eta_i, \Psi_t^\mu \eta_j) = 0$$

which follows from (8.2.35), (8.2.36).

Consider the random variable with values in V_μ

$$X^\mu = (\Phi_T^\mu)^* \Pi_\mu k + \int_0^T (\Phi_t^\mu)^* \Pi_\mu l(u_t) dt$$

and set

$$\xi_t^\mu = E^{Z^t} X^\mu - \int_0^t (\Phi_s^\mu)^* \Pi_\mu l(u_s) \mathrm{d}s.$$

From the Kunita–Watanabe representation theorem we can also write

$$E^{Z^t} X^\mu = EX^\mu + \int_0^t \Gamma_s^\mu \cdot \mathrm{d}z_s$$

where

$$\Gamma^\mu \in (L_Z^2(0,T;V_\mu))^m.$$

Let us set

$$\lambda_t^\mu = \Psi_t^\mu \xi_t^\mu$$
$$r_t^\mu = \Psi_t^\mu \Gamma_t^\mu - \Pi_\mu(h\Psi_t^\mu \xi_t^\mu).$$

We shall check that this pair satisfies (8.2.34).

Indeed

$$\lambda_T^\mu = \Psi_T^\mu (\Phi_T^\mu)^* \Pi_\mu k = \Pi_\mu k.$$

Next one has

$$(\mathrm{d}\lambda_t^\mu, \eta_j) = \mathrm{d}(\xi_t^\mu, (\Psi_t^\mu)^* \eta_j)$$
$$= (\mathrm{d}\xi_t^\mu, (\Psi_t^\mu)^* \eta_j) + (\xi_t^\mu, \mathrm{d}(\Psi_t^\mu)^* \eta_j) + (\mathrm{d}\xi_t^\mu, \mathrm{d}(\Psi_t^\mu)^* \eta_j).$$

But

$$(\mathrm{d}\xi_t^\mu, (\Psi_t^\mu)^* \eta_j) = (\eta_j, \Psi_t^\mu \Gamma_t^\mu) \cdot \mathrm{d}z_t - (\eta_j, l(u_t))\mathrm{d}t$$

$$(\xi_t^\mu, \mathrm{d}(\Psi_t^\mu)^* \eta_j) = \left[\langle A\Psi_t^\mu \xi_t^\mu, \eta_j \rangle + \sum_\beta \left(a_\beta(\cdot, u_t) \frac{\partial}{\partial x_\beta} (\Psi_t^\mu \xi_t^\mu), \eta_j \right) \right.$$
$$\left. + (\Pi_\mu(h\Psi_t^\mu \xi_t^\mu), h\eta_j) \right] \mathrm{d}t - (h\Psi_t^\mu \xi_t^\mu, \eta_j) \cdot \mathrm{d}z$$

$$(\mathrm{d}\xi_t^\mu, \mathrm{d}(\Psi_t^\mu)^* \eta_j) = -(h \cdot \Psi_t^\mu \Gamma_t^\mu, \eta_j)\mathrm{d}t.$$

Collecting results, we see that (8.2.34) is satisfied. The proof of the Lemma has been completed. ∎

We can now state the following.

Theorem 8.2.3 *The assumptions are those of Theorem 8.2.1. The adjoint process λ_t satisfies the relation*

$$\lambda_t \in L_Z^2(0,T;V) \cap L^2(\Omega, \mathcal{A}, P; C(0,T;H)) \qquad (8.2.39)$$

$$-\mathrm{d}\lambda_t + \left[A\lambda_t + \sum_i a_i(\cdot, u_t) \frac{\partial \lambda_t}{\partial x_i} \right] \mathrm{d}t = (l(u_t) + h \cdot r_t)\mathrm{d}t - r_t \cdot \mathrm{d}z \qquad (8.2.40)$$

$$\lambda_T = k$$

in which

$$r_t \in (L_Z^2(0,T;H))^m.$$

Moreover there exists only one pair λ_t, r_t in the preceding functional spaces such that (8.2.40) holds.

Proof

From Lemma 8.2.4, we know $\lambda^\mu \to \lambda$ in $L^2_Z(0, T; V)$ and $r^\mu \to r$ in $(L^2_Z(0, T; H))^m$. From this information, it is easy to take the limit in (8.2.34), and to check that (8.2.40) holds.

It remains to check the uniqueness. Suppose the pair is the λ_t, r_t solution of (8.2.40), then considering the equation (8.2.12) for arbitrary ϕ and ψ, one can compute $d(\lambda_t, \rho_t)$ and check easily that (8.2.14) holds. But (8.2.14) characterizes λ_t, r_t in a unique way, hence the desired result has been proven. ∎

8.3 Applications of the stochastic maximum principle

8.3.1 Linear quadratic case

We are going to apply Theorem 8.2.1 and Theorem 8.2.3 to the linear quadratic gaussian case, although the assumptions are *not satisfied*. The treatment is thus formal.

We take

$$
\begin{aligned}
g(x,v) &= Fx + f + Gv \\
\sigma(x) &= I \\
h(x,t) &= Hx + h
\end{aligned} \tag{8.3.1}
$$

$$
\begin{aligned}
l(x,v) &= \frac{1}{2}(Qx \cdot x + Nv \cdot v) + q \cdot x + n \cdot v \\
k(x) &= \frac{1}{2}Mx \cdot x + m \cdot x.
\end{aligned} \tag{8.3.2}
$$

The Kalman filter is defined by the relation

$$
\begin{aligned}
d\hat{x}_t &= (F\hat{x}_t + f + Gu_t)dt + PH^*(dz - (H\hat{x} + h)dt) \\
\hat{x}(0) &= x_0
\end{aligned} \tag{8.3.3}
$$

and P_t is the solution of the Riccati equation

$$
\begin{aligned}
P + PH^*HP - FP - PF^* - I &= 0 \\
P(0) &= P_0
\end{aligned} \tag{8.3.4}
$$

where x_0, P_0 are the mean and the covariance matrix of the initial state ξ, which is a gaussian, i.e.

$$
d\Pi_0(x) = \frac{\exp[-\frac{1}{2}P_0^{-1}(x - x_0) \cdot (x - x_0)]}{(2\Pi)^{n/2}|P_0|^{1/2}}dx = p_0(x)dx \tag{8.3.5}
$$

Note that

$$
a_i(x,v) = -(F_{ij}x_j + f_i + G_{ij}v_j).
$$

The Zakai equation (8.2.7) becomes

$$
\begin{aligned}
dp - \frac{1}{2}\Delta p dt &= -\frac{\partial}{\partial x_i}[(F_{ij}x_j + f_i + G_{ij}u_{jt})p]dt + p(Hx + h) \cdot dz \\
p(x,0) &= p_0(x)
\end{aligned} \tag{8.3.6}
$$

and it has an explicit solution

$$p(x,t) = \frac{\exp[-\frac{1}{2}(P_t^{-1}(x - \hat{x}_t) \cdot (x - \hat{x}_t))]}{(2\Pi)^{n/2}|P_t|^{1/2}} \nu_t \qquad (8.3.7)$$

where

$$\nu_t = \exp\left[-\frac{1}{2}\int_0^t |H\hat{x} + h|^2 \mathrm{d}s + \int_0^t (H\hat{x} + h) \cdot \mathrm{d}z\right]. \qquad (8.3.8)$$

Let us use the notation (already used for the statement of the separation principle, see Section 7.1)

$$\tilde{\phi}_t(x) = \int \phi(x + P_t^{1/2}\xi, t)\frac{\exp(-\frac{1}{2}|\xi|^2)}{(2n)^{n/2}}\mathrm{d}\xi$$

if $\phi(x,t) = \phi_t(x)$ is a given function.

The conditions (8.2.16) and (8.2.40) become

$$G^*\mathrm{D}\tilde{\lambda}(\hat{x}_t, t) + Nu_t + n_t = 0 \qquad (8.3.9)$$

$$- \mathrm{d}\lambda_t + \left[-\frac{1}{2}\Delta\lambda_t - (Fx + f) \cdot \mathrm{D}\lambda_t - Gu_t \cdot \mathrm{D}\lambda_t\right]\mathrm{d}t$$

$$= \left(\frac{1}{2}(Qx \cdot x + Nu_t \cdot u_t) + q \cdot x + n \cdot u_t + r_t \cdot (Hx + h)\right)\mathrm{d}t - r_t \cdot \mathrm{d}z, \qquad (8.3.10)$$

$$\lambda(x, T) = \frac{1}{2}Mx \cdot x + m \cdot x.$$

We look for a process $\lambda(x, t)$ of the form

$$\lambda(x,t) = \frac{1}{2}\Sigma_t x \cdot x + \frac{1}{2}\Gamma_t(x - \hat{x}_t) \cdot (x - \hat{x}_t) + \rho_t x + \beta_t \qquad (8.3.11)$$

where Σ_t, Γ_t, ρ_t, β_t are deterministic functions.

The initial condition yields

$$\Sigma_T = M, \quad \Gamma_T = 0, \quad \rho_T = m, \quad \beta_T = 0. \qquad (8.3.12)$$

Now

$$\mathrm{D}\lambda_t = \Sigma_t x + \Gamma_t(x - \hat{x}_t) + \rho_t$$
$$\mathrm{D}^2\lambda_t = \Sigma_t + \Gamma_t \qquad (8.3.13)$$

and (8.3.9) yields

$$G^*(\Sigma_t\hat{x}_t + \rho_t) + Nu_t + n_t = 0.$$

i.e.

$$u_t = -N_t^{-1}[G_t^*(\Sigma_t\hat{x}_t + \rho_t) + n_t]. \qquad (8.3.14)$$

We now insert (8.3.11) in (8.3.10), and identify the martingale terms and the remainders. We obtain

$$\mathrm{d}\lambda = \left[\frac{1}{2}\dot{\Sigma}_t x \cdot x + \frac{1}{2}\dot{\Gamma}_t(x - \hat{x}_t) \cdot (x - \hat{x}_t) + \dot{\rho}_t x + \dot{\beta}_t + \frac{1}{2}\mathrm{tr}\,\Gamma_t PH^*HP\right]\mathrm{d}t$$
$$- \Gamma_t(x - \hat{x}_t)\mathrm{d}\hat{x}_t$$

hence

$$r_t = -HP\Gamma_t(x - \hat{x}_t) \tag{8.3.15}$$

and

$$
\begin{aligned}
&-\frac{1}{2}\dot{\Sigma}_t x \cdot x - \frac{1}{2}\dot{\Gamma}_t(x - \hat{x}_t)(x - \hat{x}_t) - \dot{\rho}_t x - \dot{\beta}_t - \frac{1}{2}\mathrm{tr}\,(\Sigma_t + \Gamma_t) \\
&- (Fx + f)\cdot(\Sigma_t x + \Gamma(x - \hat{x}_t) + \rho_t) + GN^{-1}(G^*(\Sigma_t \hat{x} + \rho_t) + n_t)(\Sigma_t x + \rho_t) \\
&= \frac{1}{2}Qx \cdot x + q \cdot x + \frac{1}{2}N^{-1}(G^*(\Sigma_t \hat{x}_t + \rho_t) + n_t)\cdot(G^*(\Sigma_t \hat{x}_t + \rho_t) + n_t) \\
&\quad - n_t N^{-1}[G_t^*(\Sigma_t \hat{x}_t + \rho_t) + n_t] - (Hx + h)\cdot HP\Gamma_t(x - \hat{x}_t).
\end{aligned}
\tag{8.3.16}
$$

We identify the coefficients of x. We get

$$\dot{\Sigma} + F^*\Sigma + \Sigma F + Q + \dot{\Gamma} + (F^* - H^*HP)\Gamma + \Gamma(F - PH^*H) = 0 \tag{8.3.17}$$

$$
\begin{aligned}
&[\dot{\Gamma} + \Gamma(F - PH^*H) + (F^* - H^*HP)\Gamma + \Sigma GN^{-1}G^*\Sigma]\hat{x}_t \\
&= \dot{\rho} + (F^* - \Sigma GN^{-1}G^*)\rho + q + \Sigma f - \Sigma GN^{-1}n
\end{aligned}
\tag{8.3.18}
$$

$$
\begin{aligned}
&-\frac{1}{2}\Big[\dot{\Gamma} + \Gamma(F - PH^*H) + (F^* - H^*HP)\Gamma + \Sigma GN^{-1}G^*\Sigma\Big]\hat{x}_t \cdot \hat{x}_t \\
&- \dot{\beta}_t - \frac{1}{2}\mathrm{tr}\,(\Gamma PH^*HP + \Sigma + \Gamma) - \rho f + \frac{1}{2}N^{-1}(G^*\rho + n)\cdot(G^*\rho + n) = 0.
\end{aligned}
\tag{8.3.19}
$$

We then define Σ, Γ, ρ by the relation

$$\dot{\Sigma} + F^*\Sigma + \Sigma F - \Sigma GN^{-1}G^*\Sigma + Q = 0; \quad \Sigma_T = M \tag{8.3.20}$$

$$\dot{\Gamma} + (F^* - H^*HP)\Gamma + \Gamma(F - PH^*H) + \Sigma GN^{-1}G^*\Sigma = 0; \quad \Gamma_T = 0 \tag{8.3.21}$$

$$\dot{\rho} + (F^* - \Sigma GN^{-1}G^*)\rho + q - \Sigma GN^{-1}n + \Sigma f = 0; \quad \rho_T = m \tag{8.3.22}$$

and β_t by

$$
\begin{aligned}
&-\dot{\beta}_t - \frac{1}{2}\mathrm{tr}\,(\Sigma + \Gamma + \Gamma PH^*HP) - \rho f + \frac{1}{2}N^{-1}(G^*\rho + n)\cdot(G^*\rho + n) = 0; \\
&\beta_T = 0.
\end{aligned}
\tag{8.3.23}
$$

The function $\tilde{\lambda}(x, t)$ is given by

$$\tilde{\lambda}(x, t) = \frac{1}{2}\Sigma_t x \cdot x + \frac{1}{2}\Gamma_t(x - \hat{x}_t)(x - \hat{x}_t) + \rho_t x + \delta_t \tag{8.3.24}$$

where

$$\delta_t = \beta_t + \frac{1}{2}\mathrm{tr}\,(\Sigma + \Gamma)$$

and it is easy to check that

$$
\begin{aligned}
&\dot{\delta}_t = -\rho f + \frac{1}{2}N^{-1}(G^*\rho + n)(G^*\rho + n) - \frac{1}{2}\mathrm{tr}\,QP - \frac{1}{2}\mathrm{tr}\,\Sigma PH^*HP \\
&\delta_T = \frac{1}{2}\mathrm{tr}\,M_T P_T.
\end{aligned}
\tag{8.3.25}
$$

8.3.2 Stochastic maximum principle and separation principle

Let us consider a situation where the separation principle holds. We shall remain slightly formal to shorten the presentation.

We assume

$$g(x, v) = Fx + f + g(v)$$
$$\sigma(x) = I \qquad (8.3.26)$$
$$h(x, t) = Hx + h.$$

Consider the Kalman filter

$$d\hat{x} = (F\hat{x} + f + g(u_t))dt + PH^*[dz - (H\hat{x} + h)dt]$$
$$\hat{x}(0) = x_0 \qquad (8.3.27)$$

and P_t is the solution of (8.3.4). The unnormalized conditional probability $p(x, t)$ is again given by (8.3.7).

The condition (8.2.16) reduces to

$$\left[D\tilde{\lambda}(\hat{x}_t, t)\frac{\partial g}{\partial v}(u_t) + \frac{\partial \tilde{l}}{\partial v}(\hat{x}_t, u_t) \right] \cdot (v - u_t) \geq 0 \qquad (8.3.28)$$
$$\forall v \in U_{ad}, \qquad \text{a.e., a.s.}$$

where $\tilde{\lambda}(x, t)$ is deduced from $\lambda(x, t)$ as in (8.3.9). The adjoint process $\lambda(x, t)$ satisfies (see (8.2.40))

$$-d\lambda_t + \left[-\frac{1}{2}\Delta\lambda_t - (Fx + f)D\lambda - g(u_t) \cdot D\lambda \right] dt$$
$$= (l(x, u_t) + (Hx + h) \cdot r_t(x))dt - r_t(x)dz \qquad (8.3.29)$$
$$\lambda_T = k(x).$$

From the formula

$$\tilde{\lambda}(x, t) = \int \lambda(x + P_t^{1/2}\xi, t)\frac{\exp(-\frac{1}{2}|\xi|^2)}{(2n)^{n/2}}d\xi \qquad (8.3.30)$$

we can derive the equation of which $\tilde{\lambda}$ is a solution. It is the following:

$$-d\tilde{\lambda} + \left[-\frac{1}{2}\text{tr } D^2\tilde{\lambda}PH^*HP - (Fx + f) \cdot D\tilde{\lambda} - g(u_t)D\tilde{\lambda} \right] dt$$
$$= (\tilde{l}(x, u_t) + (Hx + h)\tilde{r}_t + \text{tr } D\tilde{r}PH^*)dt - \tilde{r}_t \cdot dz \qquad (8.3.31)$$
$$\tilde{\lambda}(x, T) = \tilde{k}(x).$$

At this stage we search for a process $\tilde{\lambda}(x, t)$ of the form

$$\tilde{\lambda}(x, t) = \chi(x, \hat{x}_t, t)$$

where $\chi(x, y, t)$ is a deterministic function of the arguments. In other words, we assume that the randomness of $\tilde{\lambda}$ is due to the Kalman filter $\hat{x}_t$.

We have

$$d\tilde{\lambda} = \left\{ \frac{\partial \chi}{\partial t} + D_y\chi(F\hat{x}_t + f + g(u_t) - PH^*[H\hat{x}_t + h]) + \frac{1}{2}\text{tr } D_y^2\chi PH^*HP \right\} dt$$
$$+ HPD_y\chi \cdot dz.$$

Comparing with (8.3.31) we deduce that

$$\tilde{r}_t = HPD_y\chi(x, \hat{x}_t, t). \tag{8.3.32}$$

Moreover χ must satisfy

$$-\frac{\partial \chi}{\partial t} - \frac{1}{2}\text{tr } (D_x^2\chi + D_y^2\chi + 2D_{yx}\chi)PH^*HP$$
$$- (Fx + f + g(u_t))D_x\chi - (F\hat{x}_t + f + g(u_t))D_y\chi \tag{8.3.33}$$
$$- PH^*H(x - \hat{x}_t)D_y\chi = \tilde{l}(x, u_t)$$

where the second argument of χ is evaluated at $\hat{x}_t$.

To obtain a PDE for χ we make use of (8.3.28) as follows. Consider the variational inequality

$$\left(p\frac{\partial g}{\partial v}(V) + \frac{\partial \tilde{l}}{\partial v}(x, V) \right)(v - V) \geq 0 \tag{8.3.34}$$
$$\forall v \in U_{ad}, \quad \forall x, p, \quad V \in U_{ad}$$

and we assume that (8.3.34) has a solution, written as $V(x, t, p)$, which minimizes the function $p \cdot g(v) + \tilde{l}(x, v)$ over U_{ad}. The condition (8.3.28) amounts to

$$u_t = V(\hat{x}_t, t, D_x\chi(\hat{x}_t, \hat{x}_t, t)). \tag{8.3.35}$$

We shall guess the equation for χ and check afterwards that it is the right one.

Let

$$\Phi(x, t) = \chi(x, x, t) \tag{8.3.36}$$

and

$$v(x, t) = V(x, t, D\Phi(x, t)). \tag{8.3.37}$$

At this stage we must be careful that u_t is a priori different from $v(\hat{x}_t, t)$, since

$$D\Phi = D_x\chi + D_y\chi.$$

Later we shall check that

$$D_y\chi(x, x, t) = 0 \tag{8.3.38}$$

and thus (8.3.35) will coincide with $v(\hat{x}_t, t)$. The equation for χ is now written as

$$-\frac{\partial \chi}{\partial t} - \frac{1}{2}\text{tr } (D_x^2\chi + D_y^2\chi + 2D_{yx}\chi)PH^*HP$$
$$- (Fx + f)D_x\chi - (Fy + f)D_y\chi - PH^*H(x - y)D_y\chi \tag{8.3.39}$$
$$= \tilde{l}(x, v(y)) + (D_x\chi + D_y\chi)g(v(y))$$
$$\chi(x, y, T) = \tilde{k}(x).$$

At this stage, we can just guarantee that, if χ satisfies (8.3.39) and (8.3.38), the conditions (8.3.33) and (8.3.35) will be satisfied, with the definition of v.

From (8.3.39) we derive the equation for Φ, recalling (8.3.36):

$$-\frac{\partial \Phi}{\partial t} - \frac{1}{2}\operatorname{tr} D^2\Phi PH^*HP - (Fx+f)D\Phi = \tilde{l}(x,v(x)) + D\Phi g(v(x)) \qquad (8.3.40)$$
$$\Phi(x,T) = \tilde{k}(x).$$

Since

$$\tilde{l}(x,v(x)) + D\Phi \cdot g(v(x)) = \inf_v [\tilde{l}(x,v) + D\Phi \cdot g(v)]$$

the function Φ is the solution of the Bellman equation (7.2.5). Define next

$$\Psi(x,y,t) = \chi(x,y,t) - \Phi(x,t).$$

we deduce for Ψ the equation

$$-\frac{\partial \Psi}{\partial t} - \frac{1}{2}\operatorname{tr}(D_x^2\Psi + D_y^2\Psi + 2D_{yx}\Psi)PH^*HP - (Fx+f)D_x\Psi$$
$$- (Fy+f)D_y\Psi - PH^*H(x-y)D_y\Psi - (D_x\Psi + D_y\Psi)g(v(y)) \qquad (8.3.41)$$
$$= \tilde{l}(x,v(y)) - \tilde{l}(x,v(x)) + D\Phi(x) \cdot [g(v(y)) - g(v(x))]$$
$$\Psi(x,y,T) = 0.$$

It is a linear equation and thus defines $\Psi(x,y,t)$ uniquely. Let us verify (8.3.38). We shall in fact check that

$$D_x\Psi(x,x,t) = 0$$

which will be sufficient. Let

$$\Psi_k = \frac{\partial \Psi}{\partial x_k}(x,y,t).$$

Then from (8.3.41) we deduce, differentiating with respect to x_k,

$$-\frac{\partial \Psi_k}{\partial t} - \frac{1}{2}\operatorname{tr}(D_x^2\Psi_k + D_y^2\Psi_k + 2D_{yx}\Psi_k)PH^*HP$$
$$- (Fx+f)D_x\Psi_k - (Fy+f)D_y\Psi_k - PH^*H(x-y)D_y\Psi_k$$
$$- \Psi_l F_{lk} - \frac{\partial \Psi}{\partial y_l}(PH^*H)_{lk} - (D_x\Psi_k + D_y\Psi_k)g(v(y))$$
$$= \frac{\partial}{\partial x_k}\tilde{l}(x,v(y)) + D\frac{\partial \Phi}{\partial x_k}\cdot(g(v(y)) - \frac{\partial}{\partial x_k}\left[\tilde{l}(x,v(x)) + D\Phi(x)\cdot g(v(x))\right]$$
$$\Psi_k(x,y,T) = 0.$$

Next let

$$\zeta_k(x,t) = \Psi_k(x,x,t)$$

and note that

$$\frac{\partial \Psi}{\partial y_l}(x,x,t) = -\frac{\partial \Psi}{\partial x_l}(x,x,t).$$

Moreover by the optimality of $v(x)$, one has also

$$\frac{\partial}{\partial x_k}[\tilde{l}(x,v(x)) + D\Phi(x)\cdot g(v(x))] = \frac{\partial}{\partial x_k}\tilde{l}(x,v(x)) + D\frac{\partial \Phi}{\partial x_k}\cdot g(v(x))$$

and thus we see that ζ_k satisfies

$$-\frac{\partial}{\partial t}\zeta_k - \frac{1}{2}\text{tr } D^2\zeta_k PH^*HP - [Fx + f + g(v(x))]D\zeta_k - \zeta_l(F - PH^*H)_{lk} = 0$$
$$\zeta_k(x, T) = 0.$$

This linear system, whose solution is $\zeta_1, \ldots, \zeta_n$ necessarily has 0 as solution. Hence (8.3.38) has been verified.

Let us make (8.3.41) more precise in special cases. For instance assume that

$$g(v) = Gv$$
$$l(x, v) = l(x) + \frac{1}{2}Nv \cdot v + n \cdot v \tag{8.3.42}$$

and $U_{ad} = R^k$ in which case the function $V(x, t, p)$ is given by

$$V(x, t, p) = -N^{-1}(n + G^*p)$$

hence

$$v(x, t) = -N^{-1}(n + G^*D\Phi).$$

The function Φ is the solution of

$$-\frac{\partial\phi}{\partial t} - \frac{1}{2}\text{tr } D^2\Phi PH^*HP - (Fx + f)D\Phi$$
$$= \tilde{l}(x) - \frac{1}{2}N^{-1}(n + G^*D\Phi) \cdot (n + G^*D\Phi) \tag{8.3.43}$$
$$\Phi(x, T) = \tilde{k}(x).$$

and Ψ is then given by

$$-\frac{\partial\Psi}{\partial t} - \frac{1}{2}\text{tr } (D_x^2\Psi + D_y^2\Psi + 2D_{yx}\Psi)PH^*HP$$
$$- (Fx + f)D_x\Psi - (Fy + f)D_y\Psi - PH^*H(x - y)D_y\Psi$$
$$+ (D_x\Psi + D_y\Psi)GN^{-1}(n + G^*D\Phi(y)) \tag{8.3.44}$$
$$= \frac{1}{2}GN^{-1}G^*(D\Phi(x) - D\Phi(y)) \cdot (D\Phi(x) - D\Phi(y))$$
$$\Psi(x, y, T) = 0.$$

If

$$l(x) = \frac{1}{2}Qx \cdot x + q \cdot x,$$
$$k(x) = \frac{1}{2}Mx \cdot x + m \cdot x$$

then

$$\tilde{l}(x) = l(x) + \frac{1}{2}\text{tr } Q_t P_t$$
$$\tilde{k}(x) = k(x) + \frac{1}{2}\text{tr } M_t P_T$$

and

$$\Phi(x, t) = \frac{1}{2}\Sigma_t x \cdot x + \rho_t x + \delta_t$$

(see (8.3.24)). Moreover

$$\Psi(x, y, t) = \frac{1}{2}\Gamma_t(x - y)(x - y)$$

where Γ_t is the solution of (8.3.21).

Remark 8.3.1 It seems likely that with the above techniques, one can derive necessary conditions of optimality for systems of the type considered in Chapter 6. This is left as an open problem.∎

8.4   Preliminaries to dynamic programming

In this section and the following, we shall develop a dynamic programming approach to the control of diffusions with partial observations.

8.4.1   Setting of the problem

We shall mainly consider the framework of Section 8.1. Consider $g(x, v)$ and $a(x) \equiv a_{ij}(x)$ such that

$$g \text{ is bounded}, \quad \frac{\partial g_i}{\partial x_i} \text{ is bounded} \tag{8.4.1}$$

$$a_{ij} = a_{ji}, \quad a_{ij} \text{ bounded}, \quad \frac{\partial a_{ij}}{\partial x_j}, \frac{\partial^2 a_{ij}}{\partial x_i \partial x_j} \text{ are bounded} \tag{8.4.2}$$

$$a_{ij}\xi_i\xi_j \geq \alpha|\xi|^2, \quad \forall \xi \in R^n, \quad \alpha > 0.$$

Let

$$A\phi = -\sum_{ij} \frac{\partial}{\partial x_i}\left(a_{ij}\frac{\partial \phi}{\partial x_j}\right) \tag{8.4.3}$$

and

$$a_i(x, v) = -g_i(x, v) + \sum_j \frac{\partial}{\partial x_j}a_{ij}$$

$$G(v)\phi = \sum_i \frac{\partial}{\partial x_i}(a_i(x, v)\phi). \tag{8.4.4}$$

Next let

$$h(x): R^n \to R^m \quad \text{be Borel bounded}. \tag{8.4.5}$$

The 'control' v belongs to R^k, and in fact to a subset, which will be made precise later. At the moment, it is just a parameter. We shall consider the Hilbert spaces

$$H = L^2(R^n), \quad V = H^1(R^n), \quad V' = H^{-1}(R^n).$$

Consider a probability space $(\Omega, \mathcal{A}, P)$, equipped with a filtration $\mathcal{F}^t$, on which is given a standard Wiener process z_t with values in R^m. We denote $\mathcal{Z}^t = \sigma(z_s, s \leq t)$ as usual, which represents the filtration generated by the observation process. We consider stochastic processes $v(\cdot) \in L^2_{\mathcal{Z}}(0, T; R^k)$, for any finite T, and the Zakai equation (see (8.1.15))

$$dq + Aq\,dt = G(v_t)q\,dt + qh \cdot dz$$

$$q(0) = p \tag{8.4.6}$$

where $p \in H$ is given.

The equation (8.4.6) has a unique solution (see Theorem 8.1.1)

$$q \in L^2_Z(0, T; V) \cap L^2(\Omega, \mathcal{A}, P; C(0, T; H)), \quad \forall T \text{ finite}. \tag{8.4.7}$$

Note that the coefficients do not depend on t, except of course through v_t. This will allow us to study various semigroups.

8.4.2 Linear semigroup

Consider the equation (8.4.6) with a control v_t which is frozen to a fixed value v and write $q^v_p(t)$ for the solution at time t. In fact we shall treat this process as a Markov process with values in H. To that extent, we shall embed (8.4.6) in a family of equations, as follows.

For $t > s$, let $z_s(t) = z(s + t) - z(s)$, and $\mathcal{Z}^t_s = \sigma(z_s(\lambda), s \leq \lambda \leq t)$. Consider p_s to be a random variable with values in H, which is $\mathcal{Z}^s$ measurable, and $q_{s,p_s}(t)$ to be the solution of

$$dq + Aqdt = G(v)qdt + qh \cdot dz_s(t) \tag{8.4.8}$$
$$q(s) = p_s.$$

The Markov property is viewed as the property

$$q_p(s + t) = q_{s, q_p(s)}(t), \quad \text{a.s.} \tag{8.4.9}$$

which can be checked by a discretization in time.

As usual, a linear semigroup is attached to a Markov process. Consider the Banach spaces

B = space of Borel bounded functionals in H

C = space of uniformly continuous bounded functionals in H.

We provide B, C with the sup norm, and C is a closed subspace of B. To any $F \in B$ or C, we associate

$$\Phi^v(t)(F)(p) = E[F(q^v_p(t))]. \tag{8.4.10}$$

We have

Proposition 8.4.1 $\Phi^v(t)$ *is a Markov semigroup on* B *or* C, *i.e.*

$$\Phi^v(0) = I$$
$$\Phi^v(t + s) = \Phi^v(t)\Phi^v(s) \tag{8.4.11}$$
$$\|\Phi^v(t)\| \leq 1.$$

Proof

The map $p \to q^v_p(t)$ is linear. Moreover from Ito's calculus

$$d|q(t)|^2 + 2\langle Aq, q \rangle dt = \left[2\langle q, G(v)q \rangle + \int_{R^n} q^2 |h|^2 dx \right] dt + 2(q, qh) \cdot dz$$

and thus one has

$$\frac{\mathrm{d}}{\mathrm{d}t} E|q(t)|^2 \le cE|q(t)|^2.$$

Therefore we have proved the estimate

$$E|q_p^v(t)|^2 \le |p|^2 e^{ct} \tag{8.4.12}$$

which shows, among other things, that

$$p \to q_p^v(t) \in \mathcal{L}(H; L^2(\Omega, \mathcal{A}, P; H)).$$

If $F \in C$ and $\rho_F(\delta)$ is the modulus of continuity of F, defined by

$$\rho_F(\delta) = \sup_{|p-p'| \le \delta} |F(p) - F(p')|$$

we have

$$|\Phi^v(t)(F)(p) - \Phi^v(t)(F)(p')|$$
$$= E\left[(F(q_p^v(t)) - F(q_{p'}^v(t)))(\mathcal{I}_{|q_p^v(t)-q_{p'}^v| \le |p-p'|^{1/2}} + \mathcal{I}_{|q_p^v(t)-q_{p'}^v| > |p-p'|^{1/2}})\right]$$
$$\le \rho_F(|p-p'|^{1/2}) + 2\|F\| |p-p'|^{1/2} e^{ct/2}$$

and thus we can assert that

$$\rho_{\Phi^v(t)(F)}(\delta) \le \rho_F(\delta^{1/2}) + 2\|F\|\delta^{1/2} e^{ct/2} \tag{8.4.13}$$

and thus $\Phi^v(t)(F)$ maps B into B and C into C. The first and third of properties (8.4.11) are obvious.

Next

$$\Phi^v(s+t)(F)(p) = E[F(q_p^v(s+t))]$$

and from (8.4.9)

$$E[F(q_p^v(s+t))] = E[F(q_{s,q_p(s)}^v(t))].$$

Since $z_s(t)$ is independent of $\mathcal{Z}^s$, for $t \ge s$, we have

$$E[F(q_{s,q_p(s)}^v(t))|\mathcal{Z}^s] = \Phi^v(t)(F)(q_p(s))$$

and thus the second property (8.4.11) is satisfied. ∎

For the applications to stochastic control the spaces B and C are restrictive as far as growth conditions are concerned. It is useful to introduce the spaces

$$B_1 = \text{space of Borel functionals on } H, \text{ which have linear growth.}$$

The norm in B_1 is defined by

$$\|F\|_1 = \sup \frac{|F(p)|}{1+|p|}$$

and B_1 is a Banach space. Similarly, one defines

$$C_1 = \text{subspace of } B_1 \text{ of functionals } F \text{ such that}$$

$$\frac{F(p)}{1+|p|} \in C.$$

From (8.4.12) one gets immediately

$$\|\Phi^v(t)\|_{\mathcal{L}(B_1;B_1)} \le e^{ct/2}. \tag{8.4.14}$$

We can also estimate the modulus of continuity of $\Phi^v(t)(F)$ in C_1 with respect to that of F in C_1, as done in (8.4.13). Define

$$F_1(p) = \frac{F(p)}{1+|p|}$$

and

$$\rho_{1,F}(\delta) = \rho_{F_1}(\delta).$$

We can write

$$
\begin{aligned}
\frac{\Phi^v(t)(F)(p)}{1+|p|} - \frac{\Phi^v(t)(F)(p')}{1+|p'|} &= E\frac{F(q_p^v(t))}{1+|p|} - E\frac{F(q_{p'}^v(t))}{1+|p'|} \\
&= E\left[\left(\frac{F(q_p^v(t))}{1+|q_p^v(t)|} - \frac{F(q_{p'}^v(t))}{1+|q_{p'}^v(t)|}\right)\frac{1+|q_p^v(t)|}{1+|p|}\right] \\
&\quad + E\frac{F(q_{p'}^v(t))}{1+|q_{p'}^v(t)|}\left(\frac{1+|q_p^v(t)|}{1+|p|} - \frac{1+|q_{p'}^v(t)|}{1+|p'|}\right) \\
&= X_1 + X_2.
\end{aligned}
$$

But

$$
X_1 = E\left[\left(\frac{F(q_p^v(t))}{1+|q_p^v(t)|} - \frac{F(q_{p'}^v(t))}{1+|q_{p'}^v(t)|}\right)\right.
$$
$$
\left.\left(\mathcal{I}_{q_p^v(t)-q_{p'}^v(t)|\le|p-p'|^{1/2}} + \mathcal{I}_{q_p^v(t)-q_{p'}^v(t)|>|p-p'|^{1/2}}\right)\frac{1+|q_p^v(t)|}{1+|p|}\right]
$$

hence

$$|X_1| \le \rho_{1,F}(|p-p'|^{1/2})e^{ct/2} + 2\|F\|_1 e^{ct}|p-p'|^{1/2}$$

and

$$|X_2| \le 2\|F\|_1 e^{ct/2}|p-p'|.$$

Therefore we can assert that

$$\rho_{1,\Phi^v(t)(F)}(\delta) \le \rho_{1,F}(\delta^{1/2})e^{ct/2} + 2\|F\|_1(e^{ct}\delta^{1/2} + e^{ct/2}\delta). \tag{8.4.15}$$

8.4.3   Additional properties

We shall next prove some additional properties.

Proposition 8.4.2 *The solution of (8.4.8) satisfies*

$$t \mapsto q_p^v(t) \quad \text{is continuous from } [0,\infty) \text{ to } L^2(\Omega, \mathcal{A}, P; H). \tag{8.4.16}$$

Proof

We can write from (8.4.6), for $s > t$

$$q(s) - q(t) = -\int_t^s Aq \, d\lambda + \int_t^s G(v)q \, d\lambda + \int_t^s qh \cdot dz$$

which is an equality between random variables with values in V'. We deduce

$$E\|q(s) - q(t)\|_{V'}^2 \leq 2M^2(s-t)E\int_t^s \|q(\lambda)\|_V^2 d\lambda + 2E\int_t^s \int_{R^n} |q|^2 |h|^2 dx d\lambda.$$

where M is a constant (independent of v) such that

$$\|A - G(v)\|_{\mathcal{L}(V;V')} \leq M.$$

Now from (8.4.6) and the energy equality (see Proposition 8.4.1), one can also estimate

$$E\int_0^T \|q(\lambda)\|_V^2 d\lambda, \quad \sup_{0 \leq t \leq T} E|q(t)|^2 \leq C_T |p|^2$$

where C_T depends only on T. Therefore we have

$$E\|q(s) - q(t)\|_{V'}^2 \leq C_T |s - t|. \tag{8.4.17}$$

On the other hand we have for $s \geq t$

$$|E|q(s)|^2 - E|q(t)|^2| \leq C\int_t^s E\|q(\lambda)\|_V^2 d\lambda + (s-t)C_T|p|^2. \tag{8.4.18}$$

The estimates (8.4.17), (8.4.18) imply (8.4.16). Indeed if $t_n \to t$, from (8.4.17) we deduce that

$$q(t_n) \to q(t) \text{ in } L^2(\Omega, \mathcal{A}, P; V') \quad \text{strongly.}$$

$$E|q(t_n)|^2 \to E|q(t)|^2.$$

Since $q(t_n)$ is bounded in $L^2(\Omega, \mathcal{A}, P; H)$, we can pick subsequences which converge weakly, necessarily towards $q(t)$. The limit is unique, hence $q(t_n) \to q(t)$ weakly in $L^2(\Omega, \mathcal{A}, P; H)$. In view of the convergence of the norm, we also have $q(t_n) \to q(t)$ in $L^2(\Omega, \mathcal{A}, P; H)$ strongly. ∎

Proposition 8.4.3. *The function*

$$t \to \Phi^v(t)(F)(p) \in C([0, \infty)), \quad \forall p \in H, \forall F \in C_1. \tag{8.4.19}$$

Proof

One has

$$|\Phi^v(t)(F)(p) - \Phi^v(s)(F)(p)|$$

$$= |EF(q_p^v(t)) - EF(q_p^v(s))|$$

$$= \left| E\left[\left(\frac{F(q_p^v(t))}{1 + |q_p^v(t)|} - \frac{F(q_p^v(s))}{1 + |q_p^v(s)|}\right)(1 + |q_p^v(t)|)\right] + E\frac{F(q_p^v(s))}{1 + |q_p^v(s)|}(|q_p^v(t)| - |q_p^v(s)|)\right|.$$

In the first term we decompose with the indicator function of the set

$$|q_p^v(t) - q_p^v(s)| > (E|q_p^v(t) - q_p^v(s)|^2)^{1/4}.$$

As for the derivation of (8.4.15) we deduce

$$
\begin{aligned}
&|\Phi^v(t)(F)(p) - \Phi^v(s)(F)(p)| \\
&\leq \rho_{1,F}(E|q_p^v(t) - q_p^v(s)|^2)^{1/4}(1 + |p|e^{ct/2}) \\
&\quad + 2\|F\|_1(1 + |p|e^{ct/2})(E|q_p^v(t) - q_p^v(s)|^2)^{1/4} \\
&\quad + \|F\|_1(E|q_p^v(t) - q_p^v(s)|^2)^{1/2}
\end{aligned}
\tag{8.4.20}
$$

and the desired result follows from Proposition 8.4.2.   ∎

We shall finally study some dependence with respect to v. We assume

$$g(x, v) \text{ is continuous with respect to } v. \tag{8.4.21}$$

In fact, we shall prove the following general result.

Proposition 8.4.4 *Assume (8.4.21). Then denoting by $q^{v(\cdot)}(t)$ the solution of (8.4.6) corresponding to a control $v(\cdot)$ in $L_Z^2(0, T; R^k)$, for a fixed initial condition p, one has the property*

$$
\begin{aligned}
&v(\cdot) \to q^{v(\cdot)}(\cdot) \text{ is continuous from } L_Z^2(0, T; R^k) \\
&\text{to } L_Z^2(0, T; V) \cap C(0, T; L^2(\Omega, \mathcal{A}, P; H)).
\end{aligned}
\tag{8.4.22}
$$

In particular for a frozen control $v(\cdot) \equiv v$, one has

$$
\begin{aligned}
&v \to q^v(\cdot) \text{ is continuous from } R^k \\
&\text{to } L_Z^2(0, T; V) \cap C(0, T; L^2(\Omega, \mathcal{A}, P; H)).
\end{aligned}
\tag{8.4.23}
$$

and

$$v \to \Phi^v(t)(F)(p) \text{ is continuous}, \quad \forall F \in C_1, \forall p, \forall t \geq 0. \tag{8.4.24}$$

Proof

Consider a sequence $v^\mu(\cdot)$ converging to $v(\cdot)$ in $L_Z^2(0, T; R^k)$ and let $q^\mu(\cdot)$ be the corresponding solution of (8.4.6). Next let $q(\cdot)$ be the solution corresponding to the limit $v(\cdot)$. One has

$$
\begin{aligned}
\mathrm{d}(q^\mu - q) - A(q^\mu - q)\mathrm{d}t &= (G(v_t^\mu)q^\mu - G(v_t)q)\mathrm{d}t + (q^\mu - q)h \cdot \mathrm{d}z \\
q^\mu(0) - q(0) &= 0.
\end{aligned}
$$

For any constant $\gamma > 0$, we deduce the energy equality

$$
\begin{aligned}
&\frac{\mathrm{d}}{\mathrm{d}t}E|q^\mu(t) - q(t)|^2 e^{-\gamma t} + \gamma e^{-\gamma t}E|q^\mu(t) - q(t)|^2 \\
&\quad + 2e^{-\gamma t}E\int_{R^n} a_{ij}\frac{\partial}{\partial x_j}(q^\mu(x, t) - q(x, t))\frac{\partial}{\partial x_i}(q^\mu(x, t) - q(x, t))\mathrm{d}x \\
&= e^{-\gamma t}E\int_{R^n} \frac{\partial a_i}{\partial x_i}(x, v^\mu)|(q^\mu - q)(x, t)|^2\mathrm{d}x \\
&\quad - 2e^{-\gamma t}E\int_{R^n} \frac{\partial}{\partial x_i}(q^\mu - q)q[a_i(x, v^\mu(t)) - a_i(x, v(t))]\mathrm{d}x \\
&\quad + e^{-\gamma t}E\int |h|^2|q^\mu(x, t) - q(x, t)|^2\mathrm{d}x
\end{aligned}
$$

and by a convenient choice of γ, we deduce the inequality

$$E|q^{\mu}(t) - q(t)|^2 e^{-\gamma t} + c_0 \int_0^t e^{-\gamma s} E\|q^{\mu}(s) - a(s)\|^2 ds$$

$$\leq C \int_0^t e^{-\gamma s} E \Sigma_i \int_{R^n} q^2 [a_i(x, v^{\mu}(s)) - a_i(x, v(s))]^2 dx ds.$$

By the continuity with respect to v and the boundedness of a_i, the right hand side tends to 0 as $\mu \to \infty$. Therefore the desired result (8.4.22) has been proven. The result (8.4.23) follows immediately. As for (8.4.24), we use the estimate, by analogy with (8.4.20):

$$|\Phi^{v_1}(t)(F)(p) - \Phi^{v_2}(t)(F)(p)| \leq \rho_{1,F}[(E|q_p^{v_1}(t) - q_p^{v_2}(t)|^2)^{1/4}](1 + |p|e^{ct/2})$$
$$+ 2\|F\|_1(1 + |p|e^{ct/2})(E|q_p^{v_1}(t) - q_p^{v_2}(t)|^2)^{1/4}$$
$$+ \|F\|_1(E|q_p^{v_1}(t) - q_p^{v_2}(s)|^2)^{1/2}$$

and from (8.4.23), the property (8.4.24) follows. ∎

Proposition 8.4.5 *One has the property*

$$E|q_p^v(t) - p|^2 \leq \gamma(t; p) \tag{8.4.25}$$

where $\gamma(t; p) \to 0$ as $t \to 0$, for any fixed p, independent of v. Also

$$|\Phi^v(t)(F)(p) - F(p)| \leq \delta_F(t; p) \tag{8.4.26}$$

where $\delta_F(t; p) \to 0$ as $t \to 0$, $\forall F, \forall p$; and $\delta_F(t; p)$ is independent of v.

Proof

We know from Propositions 8.4.2 and 8.4.3 that $E|q_p^v(t) - p|^2$ and $|\Phi^v(t)(F)(p) - F(p)|$ tend to 0 as $t \to 0$. But this result holds for a fixed v. Here, we have in addition uniformity with respect to v.

From the energy equality, we have

$$E|q(t)|^2 - |p|^2 + 2E \int_0^t \int_{R^n} a_{ij} \frac{\partial q}{\partial x_j} \frac{\partial q}{\partial x_i} dx ds$$

$$= 2E \int_0^t \int_{R^n} q \frac{\partial}{\partial x_i}(a_i q) dx ds + E \int_0^t \int_{R^n} q^2 |h|^2 dx ds \tag{8.4.27}$$

$$= E \int_0^t \int_{R^n} q^2 a_0(x, v) dx ds$$

where we have set

$$a_0(x, v) = |h|^2 + \sum_i \frac{\partial a_i}{\partial x_i}(x, v) = |h|^2 - \frac{\partial g_i}{\partial x_i}(x, v) + \frac{\partial^2 a_{ij}}{\partial x_i \partial x_j}. \tag{8.4.28}$$

Therefore

$$E|q(t) - p|^2 + 2E \int_0^t \int_{R^n} a_{ij} \frac{\partial q}{\partial x_j} \frac{\partial q}{\partial x_i} dx ds$$

$$= E \int_0^t \int_{R^n} q^2 a_0(x, v) dx ds - 2(p, Eq(t) - p). \tag{8.4.29}$$

Define

$$\beta^v(t) = Eq(t)$$

then $\beta^v(t)$ is the solution of

$$\frac{\partial \beta}{\partial t} + A\beta = G(v)\beta \tag{8.4.30}$$
$$\beta(x, 0) = p.$$

We shall use the property

$$|\beta^v(t) - p| \to 0, \quad \text{as } t \text{ tends to 0, uniformly in } v. \tag{8.4.31}$$

Indeed let $p_n \in V$, $|p_n - p| \to 0$ in H. Let β_n^v be the solution of (8.4.30) corresponding to p_n. We first have

$$|\beta_n^v(t) - \beta^v(t)| \leq |p_n - p|e^{c_0 t}$$

and

$$|\beta_n^v(t) - p_n| \leq C t^{1/2}\|p_n\|_V$$

where c_0, C do not depend on v. Therefore, we deduce that

$$|\beta^v(t) - p| \leq C_T|p_n - p| + C t^{1/2}\|p_n\|_V.$$

Since the right hand side does not depend on v, we deduce

$$\sup_v |\beta^v(t) - p| \to 0 \text{ as } t \to 0.$$

From (8.4.29), the result (8.4.25) follows.

Next, from (8.4.20) we may write

$$\begin{aligned}
|\Phi^v(t)(F)(p) - F(p)| \leq &\rho_{1,F}(E|q_p^v(t) - p|^2)^{1/4}(1 + |p|e^{ct/2}) \\
&+ 2\|F\|_1(1 + |p|e^{ct/2})(E|q_p^v(t) - p|^2)^{1/4} \\
&+ \|F\|_1(E|q_p^v(t) - p|^2)^{1/2}
\end{aligned}$$

which, combined with the estimate (8.4.25), yields the desired result (8.4.26). ∎

8.5   Stationary dynamic programming

We now return to the control problem (8.1.33). Using the formulation (8.1.37), we shall consider $l(x, v)$, a scalar function of x, v such that

$$l_v(x) = l(x, v) \quad \text{belongs to } H, \ \forall v$$
$$|l_v|_H \leq C, \quad \forall v. \tag{8.5.1}$$

We may consider as in (8.1.37) a final term, but in fact we shall generalize it, for the needs of dynamic programming. We shall be interested in the stationary as well as the nonstationary case, which will be treated in the next section. To simplify the behaviour at infinity, we shall use a discounted functional. We shall consider more precisely

$$J_p(v(\cdot)) = E \int_0^\infty e^{-\beta t}(l_{v_t}, q_p^{v(\cdot)}(t)) \mathrm{d}t \tag{8.5.2}$$

where $q_p^{v(\cdot)}$ is the solution of (8.4.6), corresponding to a control $v(\cdot) \in L_Z^2(0, T; R^k)$, $\forall T$ finite, and naturally

$$l_{v_t}(x) = l(x, v_t).$$

We shall also consider in the next section a finite horizon cost functional

$$J_{p,t}(v(\cdot)) = E \left[\int_0^t e^{-\beta s}(l_{v_s}, q_p^{v(\cdot)}(s)) \mathrm{d}s + e^{-\beta t} G(q_p^{v(\cdot)}(t)) \right] \tag{8.5.3}$$

where G belongs to C_1. In particular if $k \in H$, as in (8.1.36), we may pick

$$G(q) = (k, q).$$

8.5.1   Semigroup envelope

The analytic problem related to dynamic programming in our context is the *semigroup envelope* problem. This problem was introduced by Nisio (1976) in stochastic control. A general treatment in the spirit of what follows has been given by Bensoussan and Robin (1982). The adaptation of this technique in the case of stochastic control with partial information is due to Bensoussan (1982).

We introduce the set of functions $S \equiv S(p) \in C_1$ such that

$$S \in C_1$$
$$S \leq \int_0^t e^{-\beta s} \Phi^v(s) l_v \mathrm{d}s + e^{-\beta t} \Phi^v(t) S, \tag{8.5.4}$$
$$\forall t \geq 0, \quad \forall v \in \mathcal{U}_{ad}.$$

We shall study the structure of the set (8.5.4) and prove the existence of a maximum element. It is known, in finite dimensional cases, that this is an equivalent characterization of the solution of the Bellman equation. To keep the analogy, we shall say that finding a maximum element to the set (8.5.4) is the dynamic programming approach, or following Nisio's terminology the semigroup envelope problem. In fact, it will be the analytic characterization of the functional of p, $\inf_{v(\cdot)} J_p(v(\cdot))$.

Now we assume

$$\mathcal{U}_{ad} \text{ is a compact subset of } R^k \tag{8.5.5}$$

$$v \to l_v \quad \text{is continuous from } \mathcal{U}_{ad} \text{ to } H \tag{8.5.6}$$

$$\beta > \frac{c}{2}, \quad \text{where } c \text{ is the constant in (8.4.12).} \tag{8.5.7}$$

The main result is the following:

Theorem 8.5.1 *We assume (8.4.1), (8.4.2), (8.4.5), (8.4.21), (8.5.1), (8.5.5), (8.5.6), (8.5.7). Then the set (8.5.4) is not empty and has a maximum element. This maximum element is uniformly Lipschitz.*

The proof of Theorem 8.5.1 will be given after the introduction of a useful non linear operator.

8.5.2   A non linear operator

Define the affine operator on C_1

$$L_\tau^v(F)(p) = \int_0^\tau e^{-\beta s} \Phi^v(s) l_v(p) ds + e^{-\beta \tau} \Phi^v(\tau)(F)(p). \tag{8.5.8}$$

In view of (8.5.6), (8.5.1) and (8.4.24), we can assert that

$$v \to L_\tau^v(F)(p) \quad \text{is continuous, for fixed } F, p, \tau.$$

Since $\mathcal{U}_{ad}$ is compact, we can define

$$\Gamma_\tau(F)(p) = \min_{v \in \mathcal{U}_{ad}} L_\tau^v(F)(p) \tag{8.5.9}$$

since the minimum is attained. The non linear operator Γ_τ defined on C_1 will play a fundamental role in our treatment. We shall prove several properties of this operator.

Lemma 8.5.1 *The operator Γ_τ maps C_1 into itself:*

Proof

We have

$$L_\tau^v(F)(p) - L_\tau^v(F)(p') = e^{-\beta \tau} [\Phi^v(\tau)(F)(p) - \Phi^v(\tau)(F)(p')]$$
$$+ \int_0^\tau e^{-\beta s} \Phi^v(s) l_v(p - p') ds$$

hence using (8.4.15)

$$\left| \frac{L_\tau^v(F)(p)}{1 + |p|} - \frac{L_\tau^v(F)(p')}{1 + |p'|} \right| \leq 2|l_v||p - p'| \frac{(1 - e^{-(\beta - c/2)\tau})}{\beta - c/2}$$
$$+ e^{-(\beta - c/2)\tau} \left[\rho_{1,F}(|p - p'|^{1/2}) \right.$$
$$\left. + 2\|F\|_1 (|p - p'| + |p - p'|^{1/2} e^{c\tau/2}) \right].$$

Since $\sup_v |l_v|$ is finite we get

$$\begin{aligned}
\rho_{1,\Gamma_\tau(F)}(\delta) \leq &2\sup_v |l_v|\delta\frac{(1-e^{-(\beta-c/2)\tau})}{\beta-c/2} \\
&+ e^{-(\beta-c/2)\tau}[\rho_{1,F}(\delta^{1/2}) + 2\|F\|_1(\delta + \delta^{1/2}e^{c\tau/2})]
\end{aligned} \tag{8.5.10}$$

and the desired result is proven. ∎

Lemma 8.5.2 *One has*

$$\Gamma_\tau(\Gamma_{\tau'}(F)) \leq \Gamma_{\tau+\tau'}(F). \tag{8.5.11}$$

Proof

From the definition

$$\Gamma_\tau(\Gamma_{\tau'}(F))(p) \leq \int_0^\tau e^{-\beta s}\Phi^v(s)l_v(p)\mathrm{d}s + e^{\beta\tau}\Phi^v(\tau)(\Gamma_{\tau'}(F))(p)$$

and

$$\Gamma_{\tau'}(F)(p) \leq \int_0^{\tau'} e^{-\beta s}\Phi^v(s)l_v(p)\mathrm{d}s + e^{-\beta\tau'}\Phi^v(\tau')(F)(p).$$

By the linearity and positivity of $\Phi^v(\tau)$, we have

$$\begin{aligned}
e^{-\beta\tau}\Phi^v(\tau)(\Gamma_{\tau'}(F))(p) \leq &\int_0^{\tau'} e^{-\beta(s+\tau)}\Phi^v(\tau+s)l_v(p)\mathrm{d}s \\
&+ e^{-\beta(\tau+\tau')}\Phi^v(\tau+\tau')(F)(p) \\
= &\int_\tau^{\tau+\tau'} e^{-\beta t}\Phi^v(t)l_v(p)\mathrm{d}t \\
&+ e^{-\beta(\tau+\tau')}\Phi^v(\tau+\tau')(F)(p).
\end{aligned}$$

Therefore

$$\Gamma_\tau(\Gamma_{\tau'}(F))(p) \leq L_{\tau+\tau'}^v(F)(p), \quad \forall v$$

which implies (8.5.11). ∎

Lemma 8.5.3 *The following property holds:*

$$\Gamma_\tau(F)(p) \to F(p), \quad \text{as } \tau \to 0, \ \forall p \in H, \ \forall F \in C_1. \tag{8.5.12}$$

Proof

We have

$$\begin{aligned}
|L_\tau^v(F)(p) - F(p)| \leq &|l_v||p|\frac{1-e^{-(\beta-c/2)\tau}}{\beta-c/2} + (1-e^{-\beta\tau})|F(p)| \\
&+ e^{-\beta\tau}|\Phi^v(\tau)(F)(p) - F(p)|.
\end{aligned}$$

Using (8.4.26) we can assert that

$$|L_\tau^v(F)(p) - F(p)| \leq \gamma_F(\tau, p)$$

independent of v, and $\gamma_F(\tau, p)$ tends to 0 as τ tends to 0, for fixed F and p.

The result (8.5.12) follows. ∎

Lemma 8.5.4. *We have*

$$\|\Gamma_\tau(F)\|_1 \le \frac{\sup_v |l_v|}{\beta - c/2}(1 - e^{-(\beta-c/2)\tau}) + \|F\|_1 e^{-(\beta-c/2)\tau}. \qquad (8.5.13)$$

Proof

This is an easy consequence of (8.4.14) and simple calculations. ∎

Lemma 8.5.5 *The operator Γ_τ is a contraction in C_1, for $\tau > 0$.*

Proof

If $F_1, F_2 \in C_1$

$$L_\tau^v(F_1)(p) - L_\tau^v(F_2)(p) = e^{-\beta\tau}\Phi^v(\tau)(F_1 - F_2)(p)$$

hence

$$|L_\tau^v(F_1)(p) - L_\tau^v(F_2)(p)| \le e^{-(\beta-c/2)\tau}\|F_1 - F_2\|_1(1 + |p|)$$

and thus also

$$\|\Gamma_\tau(F_1) - \Gamma_\tau(F_2)\|_1 \le e^{-(\beta-c/2)\tau}\|F_1 - F_2\|_1$$

which proves the desired result. ∎

8.5.3   Proof of Theorem 8.5.1

We solve the fixed point equation

$$S_\tau = \Gamma_\tau(S_\tau), \quad S_\tau \in C_1, \qquad (8.5.14)$$

which has one and only one solution, by virtue of Lemma 8.5.5. We have first

Lemma 8.5.6. *The functional S_τ is uniformly Lipschitz, and*

$$|S_\tau(p) - S_\tau(p')| \le \sup_v |l_v|\frac{|p - p'|}{\beta - c/2}. \qquad (8.5.15)$$

Proof

Let

$$S_\tau^n = \Gamma_\tau^{(n)}(0).$$

We shall check recursively that S_τ^n is uniformly Lipschitz. Let us define for a uniformly Lipschitz functional on H, $F(p)$ the Lipschitz norm

$$\|\!|F|\!\| = \sup_{p,p'} \frac{|F(p) - F(p')|}{|p - p'|}.$$

Since

$$|\Phi^v(\tau)(F)(p) - \Phi^v(\tau)(F)(p')| \le e^{c\tau/2}\|\!|F|\!\| \, |p - p'|$$

we have also

$$|\Gamma_\tau(F)(p) - \Gamma_\tau(F)(p')| \le \left[\sup_v |l_v|\frac{1 - e^{-(\beta-c/2)\tau}}{\beta - c/2} + e^{-(\beta-c/2)\tau}\|\!|F|\!\|\right]|p - p'|$$

i.e.

$$\|\!|\Gamma_\tau(F)|\!\| \le \sup_v |l_v|\frac{1 - e^{-(\beta-c/2)\tau}}{\beta - c/2} + e^{-(\beta-c/2)\tau}\|\!|F|\!\|. \qquad (8.5.16)$$

It follows that

$$\|\!|S_\tau^n|\!\| \le \sup_v |l_v|\frac{(1 - e^{-(\beta-c/2)\tau})}{\beta - c/2}(1 + e^{-(\beta-c/2)\tau} + \cdots + e^{-(n-1)(\beta-c/2)\tau})$$

and, letting n tend to $+\infty$, we deduce (8.5.15). ∎

We can then give the proof of Theorem 8.5.1.

We first have

$$S_\tau \leq S_{2\tau}. \tag{8.5.17}$$

Indeed, notice that if F satisfies

$$F \leq \Gamma_\tau(F)$$

then necessarily

$$F \leq S_\tau.$$

But

$$\begin{aligned} S_\tau &= \Gamma_\tau(\Gamma_\tau(S_\tau)) \\ &\leq \Gamma_{2\tau}(S_\tau) \end{aligned}$$

therefore (8.5.17) obtains.

On the other hand, since

$$\Phi^v(\tau)(F)(0) = 0, \quad \forall F$$

we have $L_\tau^v(F)(0) = 0, \ \forall \Gamma$, hence $\Gamma_\tau(F)(0) = 0$.

Therefore

$$S_\tau(0) = 0. \tag{8.5.18}$$

From (8.5.15) it follows that

$$|S_\tau(p)| \leq \sup_v |l_v| \frac{|p|}{\beta - c/2}. \tag{8.5.19}$$

Let us define the sequence

$$S_\nu = S_{1/2^\nu}, \quad \nu \text{ an integer}$$

then from (8.5.18)

$$S_\nu(p) \downarrow \bar{S}(p), \quad \text{as} \quad \nu \uparrow +\infty. \tag{8.5.20}$$

From (8.5.15) it follows that $\bar{S}$ is Lipschitz and more precisely

$$\begin{aligned} |\bar{S}(p) - \bar{S}(p')| &\leq \sup_v |l_v| \frac{|p - p'|}{\beta - c/2} \\ \bar{S}(0) &= 0. \end{aligned} \tag{8.5.21}$$

In particular $\bar{S} \in C_1$.

Now using (8.5.11) we can assert that

$$S_\tau \leq \int_0^{m\tau} e^{-\beta t} \Phi^v(t) l_v \mathrm{d}t + e^{-m\tau\beta} \Phi^v(m\tau) S_\tau, \quad \text{for any integer } m. \tag{8.5.22}$$

Taking $\tau = 1/2^\nu$, $m = j2^\nu/2^i$, with $i \leq \nu$, we deduce

$$S_\nu \leq \int_0^{j/2^i} e^{-\beta t} \Phi^v(t) l_v \mathrm{d}t + e^{-\beta j/2^i} \Phi^v\left(\frac{j}{2^i}\right) S_\nu \quad \forall i \leq \nu. \tag{8.5.23}$$

We may let ν tend to $+\infty$, with j, i fixed. Since

$$\Phi^v\left(\frac{j}{2^i}\right) S_\nu(p) = ES_\nu(q_p^v(j/2^i)) \tag{8.5.24}$$

we can use Lebesgue's theorem to assert that the left hand side of (8.5.24) tends to $\Phi^v(j/2^i)\bar{S}(p)$. Therefore we have

$$\bar{S} \le \int_0^{j/2^i} e^{-\beta s}\Phi^v(s)l_v ds + e^{-\beta j/2^i}\Phi^v(j/2^i)\bar{S}. \tag{8.5.25}$$

For fixed t, apply (8.5.25) with $j = t2^i + 1$, and let i tend to $+\infty$. Clearly $j/2^i \to t$. Since $\bar{S} \in C_1$, it follows from Proposition 8.4.3 that $\Phi^v(j/2^i)\bar{S}(p) \to \Phi^v(t)\bar{S}(p)$. Therefore we can take the limit in (8.5.25) pointwise for a fixed value of the argument p. We deduce that $\bar{S}$ is an element of the set (8.5.4).

Moreover $\bar{S}$ is the maximum element of this set. Indeed if $\tilde{S}$ is an element of the set (8.5.4) then by definition

$$\tilde{S} \le L_\tau^v(\tilde{S})$$

hence

$$\tilde{S} \le \Gamma_\tau(\tilde{S}).$$

Therefore

$$\tilde{S} \le SS_\tau$$

hence

$$\tilde{S} \le \bar{S}.$$

The proof of Theorem 8.5.1 has been completed.   ∎

8.5.4   Interpretation of the maximum element

Let us consider the class W of step processes adapted to $\mathcal{Z}^t$, with values in $\mathcal{U}_{ad}$. More precisely, if $v(\cdot) \in W$, there exists a sequence

$$\tau_0 = 0 < \tau_1 \cdots < \tau_n < \cdots$$

which is deterministic, increasing and convergent to $+\infty$, such that

$$v(t; \omega) = v_n(\omega), \quad t \in [\tau_n, \tau_{n+1}]$$

and v_n is $\mathcal{Z}^{\tau_n}$ measurable.

Consider next the subset of W,

$$W_\tau = \{v(\cdot) \in W | \tau_n = n\tau\}$$

then we have the following.

Lemma 8.5.7 *We have the property*

$$S_\tau(p) = \inf_{v(\cdot) \in W_\tau} J_p(v(\cdot)). \tag{8.5.26}$$

Proof

If $v(\cdot) \in W$, let us set

$$q_n = q_n^{v(\cdot)} = q_p^{v(\cdot)}(\tau_n). \tag{8.5.27}$$

Let us also use the notation

$$l_{\tau,v} = \int_0^\tau e^{-\beta t} \Phi^v(t) l_v dt. \tag{8.5.28}$$

Let $t \in [\tau_n, \tau_{n+1}]$, then, as in (8.4.9),

$$(l_{v_t}, q_p^{v(\cdot)}(t)) = (l_{v_n}, q_{\tau_n, q_n}^{v_n}(t - \tau_n)) \quad \text{a.s.}$$

hence

$$E(l_{v_t}, q_p^{v(\cdot)}(t)) = E\Phi^{v_n}(t - \tau_n) l_{v_n}(q_n)$$

therefore for $v(\cdot) \in W$

$$J_p(v(\cdot)) = \sum_{n=0}^\infty \int_{\tau_n}^{\tau_{n+1}} e^{-\beta t} E\Phi^{v_n}(t - \tau_n) l_{v_n}(q_n) dt$$

$$= E \sum_{n=0}^\infty e^{-\beta \tau_n} (l_{\tau_{n+1}-\tau_n, v_n}, q_n) \tag{8.5.29}$$

and for $v(\cdot) \in W_\tau$,

$$J_p(v(\cdot)) = E \sum_{n=0}^\infty e^{-\beta n \tau} (l_{\tau, v_n}, q_n). \tag{8.5.30}$$

But using (see (8.5.22))

$$S_\tau(p) \le \int_0^\tau e^{-\beta t} \Phi^v(t) l_v(p) dt + e^{-\beta \tau} \Phi^v(\tau) S_\tau(p).$$

We apply this inequality, with $p = q_n$, $v = v_n$, and note that

$$\Phi^{v_n}(\tau) S_\tau(q_n) = E[S_\tau(q_{n+1}) | \mathcal{Z}^{n\tau}]$$

therefore we obtain

$$e^{-\beta n \tau} E S_\tau(q_n) \le e^{-\beta n \tau} E(l_{\tau, v_n}, q_n) + e^{-\beta (n+1)\tau} E S_\tau(q_{n+1}).$$

Adding up, when n runs from 0 to $N - 1$, we obtain

$$S_\tau(p) \le E \sum_{n=0}^{N-1} e^{-\beta n \tau} (l_{\tau, v_n}, q_n) + E e^{-\beta N \tau} S_\tau(q_N).$$

But

$$E S_\tau(q_N) \le \|S_\tau\|_1 (1 + E|q_N|) \le \|S_\tau\|_1 (1 + |p|) e^{cN\tau/2},$$

therefore

$$e^{-\beta N \tau} E S_\tau(q_N) \to 0, \quad \text{as} \quad N \to \infty.$$

It follows that

$$S_\tau(p) \le J_p(v(\cdot)), \quad \forall v(\cdot) \in W_\tau. \tag{8.5.31}$$

Turning back to (8.5.9), we may find a Borel map $\hat{v}(p)$, such that

$$\Gamma_\tau(F)(p) = L_\tau^{\hat{v}(p)}(F)(p).$$

Inductively we define the sequences $\hat{q}_n$ and $\hat{v}_n = \hat{v}(\hat{q}_n)$, starting at $n = 0$, with p, which is the fixed element occurring in (8.5.31). The sequence $\hat{v}_n$ defines an element $\hat{v}(\cdot)$ of W_τ. Applying the same argument as above we obtain

$$S_\tau(p) = J_p(\hat{v}(\cdot))$$

which together with (8.5.31) yields the desired result (8.5.26).   ∎

To complete the interpretation, we need to let τ tend to 0 in (8.5.26). For that purpose, we need the additional assumption

$$\mathcal{U}_{ad} \quad \text{is convex.} \tag{8.5.32}$$

We can then state

Theorem 8.5.2 *We make the assumptions of Theorem 8.5.1 and (8.5.32). Then one has*

$$\bar{S}(p) = \inf_{v(\cdot)} J_p(v(\cdot)), \quad \forall p, \tag{8.5.33}$$

where $\bar{S}$ denotes the maximum element of the set (8.5.4).

Proof

We have from (8.5.26)

$$\inf_{v(\cdot)} J_p(v(\cdot)) \le \bar{S}(p). \tag{8.5.34}$$

On the other hand let $v(\cdot)$ be any control.

We approximate $v(\cdot)$ by $v^\tau(\cdot)$, setting

$$v^\tau(t) = \begin{cases} v_0 \in \mathcal{U}_{ad}, & \text{if } 0 \le t < \tau \\ \frac{1}{\tau} \int_{(n-1)\tau}^{n\tau} v(s) ds, & \text{if } t \in [n\tau, (n+1)\tau]. \end{cases}$$

By the convexity of $\mathcal{U}_{ad}$, $v^\tau(t) \in \mathcal{U}_{ad}$, and thus $v^\tau(\cdot) \in W_\tau$. On the other hand

$$v^\tau(\cdot) \to v(\cdot) \text{ in } L_Z^2(0, T; R^k), \quad \forall T \text{ fixed, as } \tau \to 0. \tag{8.5.35}$$

But

$$q^\tau(t) = q^{v^\tau(\cdot)}(t).$$

By Proposition 8.4.4, we have

$$q^\tau(\cdot) \to q^{v(\cdot)}(\cdot) \text{ in } L_Z^2(0, T; V) \cap C(0, T; L^2(\Omega, \mathcal{A}, P; H)), \quad \forall T \text{ fixed.} \tag{8.5.36}$$

By extracting a subsequence we can assume that

$$v^\tau(t) \to v(t) \quad \text{a.e. } t, \text{ a.s.,} \quad t < T$$

hence

$$l_{v^\tau(t)} \to l_{v(t)} \quad \text{a.e. } t, \text{ a.s.}$$

This and (8.5.36) imply

$$E \int_0^T e^{-\beta t}(l_{v^\tau(t)}, q^\tau(t)) \mathrm{d}t \to E \int_0^T e^{-\beta t}(l_{v(t)}, q^{v(\cdot)}(t)) \mathrm{d}t.$$

We also notice that

$$\left| E \int_T^\infty e^{-\beta t}(l_{v_t}, q_p^{v(\cdot)}(t)) \mathrm{d}t \right| \leq \sup_v |l_v| \frac{|p| e^{-(\beta - c/2)T}}{\beta - c/2}$$
$$= K_p e^{-(\beta - c/2)T} \quad \text{for any } v(\cdot),$$

hence

$$S_\tau(p) \leq E \int_0^T e^{-\beta t}(l_{v^\tau(t)}, q^\tau(t)) \mathrm{d}t + K_p e^{-(\beta - c/2)T},$$

for any T, τ. Therefore letting τ tend to 0 with $\tau = 1/2^\nu$, $\nu \to \infty$, we deduce

$$\bar{S}(p) \leq E \int_0^T e^{-\beta t}(l_{v(t)}, q^{v(\cdot)}(t)) \mathrm{d}t + K_p e^{-(\beta - c/2)T}.$$

It follows, letting T tend to $+\infty$, that

$$\bar{S}(p) \to J_p(v(\cdot)).$$

Since $v(\cdot)$ is arbitrary, we deduce, taking (8.5.34) into account, that (8.5.33) holds. ∎

8.6   Non stationary dynamic programming

8.6.1   Statement of the main result

In a way similar to (8.5.4), we shall consider the following set of functions:

$$S(t) \in C_1, \quad \forall t \in [0, T], \quad \text{where } t \mapsto S(t)(p) \text{ is upper semi-continuous } \forall p,$$
$$\text{and } \in C^0([0, T]), \quad \forall p \in V$$

$$S(0) = G \tag{8.6.1}$$

$$S(t) \leq \int_0^{t-s} e^{-\beta\sigma} \Phi^v(\sigma) l_v d\sigma + e^{-\beta(t-s)\Phi^v(t-s)S(s)},$$
$$\forall s, t, 0 \leq s \leq t \leq T, \quad \forall v.$$

The element G entering into the definition of the set (8.6.1) is a given element of C_1.
We shall need the assumption

$$\frac{\partial h_i}{\partial x_j}, \quad \frac{\partial a_{ij}}{\partial x_k}, \quad \frac{\partial^2 a_{ij}}{\partial x_j \partial x_k}, \quad \frac{\partial g_i}{\partial x_k} \quad \text{are bounded.} \tag{8.6.1'}$$

Theorem 8.6.1 *We make the assumptions of Theorem 8.5.1, and (8.5.32), (8.6.1').*
Then the set $\mathcal{K}$ of functionals defined by the conditions (8.6.1) is not empty and has
a maximum element. This element $\bar{S}(t)$ is given by the value function

$$\bar{S}(t)(p) = \inf_{v(\cdot)} J_{p,t}(v(\cdot)), \tag{8.6.2}$$

where $J_{p,t}(v(\cdot))$ has been defined in (8.5.3).

Because of the dependence on t, the proof of Theorem 8.6.1 will be harder than
that of Theorem 8.5.1. Moreover, we shall not be able to give a proof which is purely
analytical, as done for Theorem 8.5.1. In other words, the probabilistic interpretation
will play a role in the proof. Thus the result (8.5.38) is mixed with the characterization
of the maximum element of the set $\mathcal{K}$. We follow the presentation of Bensoussan and
Nisio (1990). To simplify the notation we give the proof with $T = 1$.

8.6.2   Approximation scheme

We shall consider the approximation scheme

$$S_\mu(t) = \Gamma_{2-\mu}^{[t2^\mu]}(G), \quad t \in [0, 1]. \tag{8.6.3}$$

where μ is an integer and $[t2^\mu]$ represents the integer part of $t2^\mu$. Clearly we have

$$S_\mu(t) = \Gamma^j_{2^{-\mu}}(G), \quad \text{for} \quad \frac{j}{2^\mu} \leq t < \frac{j+1}{2^\mu}, \; j = 0, \ldots, 2^\mu - 1. \tag{8.6.4}$$

We study first some properties of this approximation scheme.

Lemma 8.6.1 *The sequence* $S_\mu(j/2^l)$, *where* $0 \leq j < 2^l$, *is decreasing in* μ *for* $\mu \geq l$.

Proof

We have to check that

$$S_{\mu+1}(j/2^l) \leq S_\mu \frac{j}{2^l} \quad \text{if} \quad \mu \leq l. \tag{8.6.5}$$

This amounts to checking that

$$\Gamma^{2j2^{\mu-l}}_{2^{-(\mu+1)}}(G) \leq \Gamma^{j2^{\mu-l}}_{2^{-\mu}}(G)$$

but from (8.5.11)

$$\Gamma^2_{2^{-(\mu+1)}}(G) \leq \Gamma_{2^{-\mu}}(G).$$

Since $\Gamma^2_{2^{-(\mu+1)}}(G)$ and $\Gamma_{2^{-\mu}}$ are both monotone increasing, we obtain the result by iterating $j2^{\mu-l}$ times. ∎

Lemma 8.6.2 *We have the estimate*

$$\|S_\mu(t)\|_1 \leq \frac{\sup |l_v|}{\beta - c/2}(1 - e^{-(\beta-c/2)[t2^\mu]2^{-\mu}}) + \|G\|_1 e^{-(\beta-c/2)[t2^\mu]2^{-\mu}}. \tag{8.6.6}$$

Proof

Apply (8.5.13) with $\tau = 2^{-\mu}$, and iterate $[t2^\mu]$ times. ∎

Using Lemmas 8.6.1 and 8.6.2, we can assert that for any p, j, l,

$$S_\mu\left(\frac{j}{2^l}\right)(p) \downarrow \Sigma\left(\frac{j}{2^l}\right)(p)$$

as $\mu \uparrow +\infty$.

We can state some immediate properties of the sequence $\Sigma(j/2^l)(p)$.

Lemma 8.6.3 *We have* $\Sigma(j/2^l) \in B_1$, *and*

$$\Sigma(0) = G \tag{8.6.7}$$

$$\forall k, j, 0 \leq k \leq j < 2^l, \quad \forall v,$$

$$\Sigma\left(\frac{j}{2^l}\right) \leq \int_0^{(j-k)2^{-l}} e^{-\beta s}\Phi^v(s)l_v ds + e^{-\beta(j-k)2^{-l}}\Phi^v\left(\frac{j-k}{2^l}\right)\Sigma\left(\frac{k}{2^l}\right). \tag{8.6.8}$$

Proof

By construction $\Sigma(j/2^l)(p)$ is a Borel measurable function of p. From the estimate (8.6.6), it follows that $\Sigma(j/2^l)(p)$ has linear growth. Hence $\Sigma(j/2^l) \in B_1$. Since $S_\mu(0) = G$, the result (8.6.7) is verified.

Next we have for $\mu \geq l$

$$S_\mu\left(\frac{j}{2^l}\right) = \Gamma^{(j-k)2^{\mu-l}}_{2^{-\mu}}\left(S_\mu\left(\frac{k}{2^l}\right)\right) \leq \Gamma_{(j-k)2^{-l}}\left(S_\mu\left(\frac{k}{2^l}\right)\right)$$

therefore

$$S_\mu\left(\frac{j}{2^l}\right) \leq \int_0^{(j-k)2^{-l}} e^{-\beta s}\Phi^v(s)l_v ds + e^{-\beta(j-k)2^{-l}}\Phi^v\left(\frac{j-k}{2^l}\right)S_\mu\left(\frac{k}{2^l}\right).$$

Letting $\mu \uparrow +\infty$, and noting that

$$\Phi^v\left(\frac{j-k}{2^l}\right)\left(S_\mu\left(\frac{k}{2^l}\right)\right)(p) \downarrow \Phi^v\left(\frac{j-k}{2^l}\right)\left(\Sigma\left(\frac{k}{2^l}\right)\right)(p)$$

the result (8.6.8) follows.   ∎

To be able to interpolate and define $\Sigma(t)(p)$ for any t, we shall first give the probabilistic interpretation of $\Sigma(j/2^l)(p)$.

8.6.3   Probabilistic interpretation of the approximation scheme

Let us consider a space of step controls similar to W_r. More precisely, we set

$$W_\mu = \left\{ v(\cdot) \middle| v_t = v_i^\mu, \quad \frac{i}{2^\mu} \leq t < \frac{i+1}{2^\mu}, \quad i = 0,\ldots,2^\mu-1 \right\}$$

and v_i^μ is $\mathcal{Z}^{i2^{-\mu}}$ measurable, with values in U_{ad}.

We shall prove the following:

Lemma 8.6.4. *One has the interpretation*

$$S_\mu(j/2^l)(p) = \inf_{v(\cdot)\in W_\mu} J_{p,j2^{-l}}(v(\cdot)). \tag{8.6.9}$$

Proof

It is similar to that of Lemma 8.5.7. If $v(\cdot) \in W_\mu$, we write

$$q_i = q_i^{v(\cdot)} = q_p^{v(\cdot)}(i2^{-\mu}), \quad v_i = v_i^\mu$$

and

$$l_{\mu,v} = \int_0^{2^{-\mu}} e^{-\beta s}\Phi^v(s)l_v ds.$$

Then (cf. (8.5.29)) we have

$$J_{p,j2^{-l}}(v(\cdot)) = E\left[\sum_{i=0}^{j2^{\mu-l}-1} e^{-\beta i2^{-\mu}}(l_{\mu,v_i}, q_i) + e^{-\beta j2^{-l}}G(q_{j2^{\mu-l}})\right]. \tag{8.6.10}$$

Next, for any argument Π, we may write

$$\Gamma_{2^{-\mu}}^{j2^{\mu-l}-i}(G)(\Pi) = \Gamma_{2^{-\mu}}\left(\Gamma_{2^{-\mu}}^{j2^{\mu-l}-(i+1)}(G)\right)(\Pi)$$

$$\leq \int_0^{2^{-\mu}} e^{-\beta s}\Phi^{v_i}(s)l_{v_i}(\Pi)ds + e^{-\beta 2^{-\mu}}\Phi^{v_i}(2^{-\mu})(\Gamma_{2^{-\mu}}^{j2^{\mu-l}-(i+1)}(G))(\Pi)$$

$$= (l_{\mu,v_i}, \Pi) + e^{-\beta 2^{-\mu}}\Phi^{v_i}(2^{-\mu})(\Gamma_{2^{-\mu}}^{j2^{\mu-l}-(i+1)}(G))(\Pi).$$

We apply this inequality with $\Pi = q_i$, multiply by $e^{-\beta i 2^{-\mu}}$ and take the mathematical expectation. We note that

$$E\Phi^{v_i}(2^{-\mu})(\Gamma_{2^{-\mu}}^{j2^{\mu-l}-(i+1)}(G))(q_i) = E\Gamma_{2^{-\mu}}^{j2^{\mu-l}-(i+1)}(G)(q_{i+1})$$

hence we obtain the relation

$$Ee^{-\beta i 2^{-\mu}}\Gamma_{2^{-\mu}}^{j2^{\mu-l}-i}(G)(q_i) \leq Ee^{-\beta i 2^{-\mu}}(l_{\mu,v_i}, q_i)$$
$$+ Ee^{-\beta(i+1)2^{-\mu}}E\Gamma_{2^{-\mu}}^{j2^{\mu-l}-(i+1)}(G)(q_{i+1}).$$

Adding up from $i = 0$ to $j2^{\mu-l} - 1$, we obtain

$$\Gamma_{2^{-\mu}}^{j2^{\mu-l}}(G)(p) \leq J_{p,j2^{-l}}(v(\cdot)). \tag{8.6.11}$$

Now considering the feedback $\hat{v}(p)$ as in Lemma 8.5.7, we may define a particular sequence $\hat{q}_i$, $\hat{v}_i$, starting at $i = 0$, with $\hat{q}_0 = p$, up to $i = j2^{\mu-l}$. We define in this way a control $\hat{v}(\cdot)$ up to $t = j2^{-l}$ (excluded). We extend it between $j2^{-l}$ and 1 by a deterministic quantity (arbitrary). In this way we have defined an element $\hat{v}(\cdot)$ of W_μ. The same argument as above this time yields, instead of (8.6.11), an equality:

$$\Gamma_{2^{-\mu}}^{j2^{\mu-l}}(G)(p) = J_{p,j2^{-l}}(\hat{v}(\cdot))$$

and the result (8.6.9) has been proven. ∎

The representation formula (8.6.9) will allow us to prove additional regularity of $\Sigma(j/2^l)$, which we do not know how to prove directly, namely

Lemma 8.6.5 *The functional $\Sigma(j/2^l)$ belongs to C_1, $\forall l \geq 0$, $0 \leq j < 2^l$.*

Proof

In a way exactly similar to (8.5.10) we see that

$$\left|\frac{J_{p,t}(v(\cdot))}{1+|p|} - \frac{J_{p',t}(v(\cdot))}{1+|p'|}\right| \leq 2\sup_v |l_v||p-p'|\frac{(1-e^{-(\beta-c/2)t})}{\beta-c/2}$$
$$+ e^{-(\beta-c/2)t}\left[\rho_{1,G}(|p-p'|^{1/2})\right.$$
$$\left.+ 2\|G\|_1(|p-p'| + e^{ct/2}|p-p'|^{1/2})\right]$$

for any $v(\cdot)$. Hence, taking $t = j2^{-l}$, we deduce from (8.6.9) that $S_\mu(j/2^l) \in C^1$ and $\rho_{S_\mu(j/2^l)}(\delta)$, the modulus of continuity, is independent of μ. Therefore it carries over to $\Sigma(j/2^l)$, after letting μ tend to $+\infty$. We obtain

$$\rho_{\Sigma(j/2^l)}(\delta) = 2\sup_v |l_v|\delta\frac{(1-e^{-(\beta-c/2)j/2^l})}{\beta-c/2}$$
$$+ e^{-(\beta-c/2)j/2^l}[\rho_{1,G}(\delta^{1/2}) + 2\|G\|_1(\delta + e^{cj/2^l}\delta^{1/2})]. \tag{8.6.12}$$

∎

We can then interpret $\Sigma(j/2^l)(p)$. We have

Lemma 8.6.6 *The following interpretation holds:*

$$\Sigma(j/2^l)(p) = \inf_{v(\cdot)} J_{p,j2^{-l}}(v(\cdot)). \tag{8.6.13}$$

Proof

We first have, from (8.6.9),

$$\inf_{v(\cdot)} J_{p,j2^l}(v(\cdot)) \leq \Sigma(j/2^l)(p). \tag{8.6.14}$$

Next for any admissible $v(\cdot)$ we define, as in the proof of Theorem 8.5.2,

$$v^\mu(t) = \begin{cases} v_0 \in \mathcal{U}_{ad} & \text{if } 0 \leq t < 2^{-\mu} \\ 2^\mu \int_{(i-1)2^{-\mu}}^{i2^{-\mu}} v(s)ds, & \text{if } i2^{-\mu} \leq t < (i+1)2^\mu, \quad i=1,\dots,2^\mu-1 \end{cases}$$

and we may assert that $v^\mu(\cdot) \in W_\mu$, hence

$$J_{p,j2^{-l}}(v^\mu(\cdot)) \geq S_\mu(j2^{-l})(p) \geq \Sigma(j2^{-l})(p). \tag{8.6.15}$$

If $q^\mu(\cdot) = q^{v^\mu(\cdot)}(\cdot)$, then we have, by Proposition 8.4.4,

$$q^\mu(\cdot) \to q^{v(\cdot)}(\cdot) \text{ in } L_Z^2(0,1;V) \cap C(0,1,L^2(\Omega,\mathcal{A},P;H))$$

as $\mu \to \infty$. Therefore

$$J_{p,j2^{-l}}(v^\mu(\cdot)) \to J_{p,j2^{-l}}(v(\cdot)).$$

From (8.6.15) we deduce

$$J_{p,j2^{-l}}(v(\cdot)) \geq \Sigma(j2^{-l})(p),$$

and since $v(\cdot)$ is arbitrary, and in view of (8.6.14), the desired result (8.6.13) obtains.

∎

8.6.4   Proof of Theorem 8.6.1

Let us define for any $t \in [0,1]$,

$$\Sigma(t)(p) = \inf_{v(\cdot)} J_{p,t}(v(\cdot)). \tag{8.6.16}$$

This formula provides a natural extension of (8.6.13). We shall derive from the explicit expression some regularity properties.

Lemma 8.6.7 *For any t, $\Sigma(t) \in C_1$. The function $t \to \Sigma(t)(p)$ is upper semi-continuous for any fixed $p \in H$. If $p \in V = H^1(R^n)$, this function is $C^0([0,1])$.*

Proof

The fact that $\Sigma(t) \in C_1$ has been proven in Lemma 8.6.5; see (8.6.11). Next, as for (8.4.20), we derive the estimate

$$
\begin{aligned}
|J_{p,t}(v(\cdot)) - J_{p,s}(v(\cdot))| \leq {} & \sup_v |l_v| \|p\| \frac{\left|e^{-(\beta - c/2)t} - e^{-(\beta - c/2)s}\right|}{\beta - c/2} \\
& + |e^{-\beta t} - e^{-\beta s}|(1 + |p|e^{cs/2}) \\
& + e^{-\beta t}\Big\{\rho_{1,G}[E|q_p^{v(\cdot)}(t) - q_p^{v(\cdot)}(s)|^2](1 + |p|e^{ct/2}) \\
& + 2\|G\|_1 (1 + |p|e^{ct/2})[E|q_p^{v(\cdot)}(t) - q_p^{v(\cdot)}(s)|^2]^{1/4} \\
& + \|G\|_1 [E|q_p^{v(\cdot)}(t) - q_p^{v(\cdot)}(s)|^2]^{1/2}\Big\}.
\end{aligned}
\tag{8.6.17}
$$

This estimate shows that, for a fixed control $v(\cdot)$, the function $t \to J_{p,t}(v(\cdot))$ is continuous.

This follows from Proposition 8.4.2, in which a frozen control can be replaced by any control. This implies that $t \to \Sigma(t)(p)$ is upper semicontinuous, since whenever $t_m \to t$

$$
\overline{\lim}\, \Sigma(t_n)(p) \leq \overline{\lim}\, J_{p,t_n}(v(\cdot)), \quad \forall v(\cdot)
$$
$$
= J_{p,t}(v(\cdot))
$$

hence

$$
\overline{\lim}\, \Sigma_{t_n}(p) \leq \Sigma(t)(p).
$$

We can say something more when $p \in V$. This amounts to deriving a better estimate for $E|q_p^{v(\cdot)}(t) - q_p^{v(\cdot)}(s)|^2$ in this case. This is where the assumption (8.6.1′) is used. It will allow us to assert that

$$
\text{if } p \in V, \text{ then } E\|q_p^{v(\cdot)}(t)\|_V^2 \leq C_p, \quad \forall t \in [0,1].
\tag{8.6.18}
$$

This follows from differentiating equation (8.4.6). Set $q_\lambda = \partial q / \partial x_\lambda$. We obtain that q_λ is the solution of

$$
\begin{aligned}
dq_\lambda + [Aq_\lambda - G(v_t)q_\lambda]dt &= \left[\frac{\partial}{\partial x_i}(a_{ij,\lambda}q) + \frac{\partial}{\partial x_i}(a_{i,\lambda}(x, v_t)q)\right]dt \\
&= (q_\lambda h_j + q h_{j,\lambda})dz_j. \\
q_\lambda(x,0) &= p_\lambda(x)
\end{aligned}
\tag{8.6.19}
$$

where we have set

$$
a_{ij,\lambda} = \frac{\partial a_{ij}}{\partial x_\lambda}, \quad a_{i,\lambda} = \frac{\partial a_i}{\partial x_\lambda}, \dots
$$

We deduce from (8.6.19) the energy equality

$$
\begin{aligned}
& \frac{d}{dt} E|q_\lambda|^2 + 2E \int a_{ij} \frac{\partial q_\lambda}{\partial x_j} \frac{\partial q_\lambda}{\partial x_i} dx + 2E \int \frac{\partial q_\lambda}{\partial x_i} a_i(x, v_t) q_\lambda dx \\
& + 2E \int a_{ij,\lambda} \frac{\partial q_\lambda}{\partial x_i} \frac{\partial q}{\partial x_j} dx + 2E \int \frac{\partial q_\lambda}{\partial x_i} a_{i,\lambda}(x, v_t) q dx \\
& = \Sigma_j \int (q_\lambda h_j + q h_{j,\lambda})^2 dx.
\end{aligned}
$$

Taking account of the fact that $\|q\|_{L^2(0,1;V)}$ is bounded uniformly in $v(\cdot)$, we deduce easily from the preceding equality that

$$E|q_\lambda(t)|^2 \leq C, \quad \forall t \in [0,1],$$

where C does not depend on $v(\cdot)$. This implies (8.6.18) with a constant C_p depending on the V norm of p.

Define next, for $t > s$,

$$\tilde{q}(t) = q(t) - q(s),$$

then one has

$$d\tilde{q} + A\tilde{q}dt = [-Aq(s) + G(v_t)q]dt + qh \cdot dz$$
$$\tilde{q}(s) = 0.$$

hence for any constant γ

$$\frac{1}{2}\frac{d}{dt}E|\tilde{q}(t)|^2 e^{-\gamma t} + \frac{\gamma}{2}e^{-\gamma t}E|\tilde{q}(t)|^2 + e^{-\gamma t}E\int a_{ij}\frac{\partial \tilde{q}}{\partial x_j}\frac{\partial \tilde{q}}{\partial x_i}dx$$

$$+ e^{-\gamma t}E\int a_i(x,v_t)q\frac{\partial \tilde{q}}{\partial x_i}dx + e^{-\gamma t}E\int a_{ij}\frac{\partial q}{\partial x_j}(x,s)\frac{\partial \tilde{q}}{\partial x_i}(x,t)dx$$

$$= \frac{1}{2}e^{-\gamma t}E\int q^2|h|^2 dx$$

and, for a convenient choice of γ, we deduce thanks to (8.6.18)

$$\frac{d}{dt}E|\tilde{q}(t)|^2 e^{-\gamma t} \leq Ce^{-\gamma t}, \quad t > s$$

hence

$$E|q_p^{v(\cdot)}(t) - q_p^{v(\cdot)}(s)|^2 \leq \frac{C}{\gamma}(e^{\gamma(t-s)} - 1)$$

uniformly in $v(\cdot)$, but with a constant C depending on the V norm of p. Using this estimate in the right hand side of (8.6.17), we get an expression which tends to 0 as $t-s$ tends to 0, uniformly with respect to $v(\cdot)$. It follows that $t \to \Sigma(t)(p)$ is $C^0([0,1])$. ∎

Let us prove that $\Sigma(t)$ belongs to the set $\mathcal{K}$. It satisfies the regularity requirements, as well as $\Sigma(0) = G$. It remains to prove that

$$\Sigma(t)(p) \leq \int_0^{t-s} e^{-\beta\sigma}\Phi^v(\sigma)l_v(p)d\sigma + e^{-\beta(t-s)}\Phi^v(t-s)\Sigma(s)p, \tag{8.6.20}$$
$$\forall s, t, \ 0 \leq s \leq t \leq 1, \quad \forall v, \forall p.$$

From Lemma 8.6.3 (see (8.6.8)) we can assert that (8.6.20) holds for $t = j/2^l$, $s = k/2^l$. Taking $j = [t2^l]$, $k = [s2^l]$, we can then write

$$\Sigma([t2^l]2^{-l})(p) \leq \int_0^{2^{-l}([t2^l]-[s2^l])} e^{-\beta\sigma}\Phi^v(\sigma)l_v(p)d\sigma \tag{8.6.21}$$
$$+ e^{-\beta 2^{-l}([t2^l]-[s2^l])}\Phi^v(([t2^l]-[s2^l])2^{-l})\Sigma(2^{-l}[s2^l])(p).$$

We shall let l tend to $+\infty$, in (8.6.21), assuming that $p \in V$. We first have, since $p \in V$,

$$\Sigma([t2^l]2^{-l})(p) \to \Sigma(t)(p)$$

and clearly

$$\int_0^{2^{-l}([t2^l]-[s2^l])} e^{-\beta\sigma}\Phi^v(\sigma)l_v(p)d\sigma \to \int_0^{t-s} e^{-\beta\sigma}\Phi^v(\sigma)l_v(p)d\sigma.$$

We next consider

$$\Phi^v(2^{-l}([t2^l]-[s2^l]))\Sigma(2^{-l}[s2^l])(p) = E\Sigma(2^{-l}[s2^l])(q_p^v(2^{-l}([t2^l]-[s2^l])))$$

$$= I_l + II_l$$

with

$$II = E\Sigma(2^{-l}[s2^l])(q_p^v(t-s))$$

$$I = E\Big[\Sigma(2^{-l}[s2^l])(q_p^v(2^{-l}([t2^l]-[s2^l]))) - \Sigma(2^{-l}[s2^l])(q_p^v(t-s))\Big]$$

We have

$$|I_l| \leq E\Big[\rho_{\Sigma(2^{-l}([s2^l])}(|q_p^v(2^{-l}([t2^l]-[s2^l])) - q_p^v(t-s)|)(1+|q_p^v(2^{-l}([t2^l]-[s2^l]))|)\Big]$$
$$+ \|\Sigma(2^{-l}[s2^l])\|_1 \times [E|q_p^v(2^{-l}([t2^l]-[s2^l])) - q_p^v(t-s)|^2]^{1/2}.$$

We have from Proposition 8.4.2,

$$E|q_p^v(2^{-l}([t2^l]-[s2^l])) - q_p^v(t-s)|^2 \to 0 \quad \text{as } l \to \infty.$$

Moreover from (8.6.6) applied with $t = j/2^l$, $j = [s2^l]$

$$\|\Sigma(2^{-l}[s2^l])\|_1 \leq \frac{\sup|l_v|}{\beta - c/2}(1 - e^{-(\beta-c/2)2^{-l}[s2^l]}) + \|G\|_1 e^{-(\beta-c/2)2^{-l}[s2^l]}$$

$$\leq C$$

and

$$E|q_p^v(2^{-l}([t2^l]-[s2^l]))|^2 \leq |p|^2 e^{c2^{-l}([t2^l]-[s2^l])}$$

$$\leq C.$$

Collecting results we deduce that $I_l \to 0$, as $l \to \infty$.

Now

$$\Sigma(2^{-l}[s2^l])(q_p^v(t-s)) \to \Sigma(s)(q_p^v(t-s)) \quad \text{a.s.}$$

and

$$|\Sigma(2^{-l}[s2^l])(q_p^v(t-s))| \leq C(1 + |q_p^v(t-s)|).$$

From Lebesgue's theorem, it follows that

$$II_l \to E\Sigma(s)(q_p^v(t-s)), \quad \text{as } l \to \infty.$$

We thus have proven that (8.6.20) holds true whenever $p \in V$. To prove that (8.6.20) holds for any $p \in H$, pick a sequence $p_k \in V$, $p_k \to p$. We have

$$|\Sigma(t)(p_k) - \Sigma(t)(p)| \leq \rho_{\Sigma(t)}(|p_k - p|)(1 + |p_k|) + \|\Sigma(t)\|_1|p_k - p| \to 0.$$

and

$$|\Phi^v(t-s)\Sigma(s)(p_k) - \Phi^v(t-s)\Sigma(s)(p)|$$
$$= |E[\Sigma(s)(q^v_{p_k}(t-s)) - \Sigma(s)(q^v_p(t-s))]|$$
$$\leq E[\rho_{\Sigma(s)}(|q^v_{p_k}(t-s) - q^v_p(t-s)|)(1 + |q^v_{p_k}(t-s)|)]$$
$$+ \|\Sigma(s)\|_1 [E|q^v_{p_k}(t-s) - q^v_p(t-s)|^2]^{1/2}$$
$$\to 0, \quad \text{as } k \to \infty.$$

Therefore (8.6.20) has been proven.

It remains to prove that $\Sigma(t)$ given by (8.6.16) is the maximum element of the set $\mathcal{K}$. Pick $S(t) \in \mathcal{K}$; one has

$$S(\tau) \leq \Gamma_\tau(G)$$
$$S(2\tau) \leq \Gamma_\tau(S(\tau)) \leq \Gamma^2_\tau(G)$$

hence

$$S(2^{-\mu}[t2^\mu]) \leq \Gamma^{[t2^\mu]}_{2^{-\mu}}(G) = S_\mu(t).$$

Applying this inequality with $t = j/2^l$, $\mu \geq l$, yields

$$S(j/2^l)(p) \leq S_\mu(j/2^l)(p)$$

and, letting μ tend to $+\infty$, we obtain

$$S(j/2^l)(p) \leq \Sigma(j/2^l)(p).$$

Suppose $p \in V$, and pick $j = [t2^l]$. We can let l tend to $+\infty$, to deduce

$$S(t)(p) \leq \Sigma(t)(p), \quad \forall p \in V.$$

Since $S(t), \Sigma(t)$ are C_1, using an approximation scheme $p_k \to p$, $p_k \in V$, we can prove that

$$S(t)(p) \leq \Sigma(t)(p), \quad \forall p \in H.$$

The proof of Theorem 8.6.1 has been completed.   ∎

8.7 Non linear semigroup

8.7.1 Definition

We shall define

$$\Psi(t)(G) = \Sigma(t). \tag{8.7.1}$$

We can then assert

Proposition 8.7.1 *The operator* $\Psi(t)$ *maps* C_1 *into* C_1. *We have:*

$$t \mapsto \Psi(t)(G)(p) \in C^0([0,1]), \quad \forall p \in V,$$
$$\text{and is upper semi continuous for } p \in H; \tag{8.7.2}$$

$$\Psi(t+\tau)(G) = \Psi(t)(\Psi(\tau)(G)). \tag{8.7.3}$$

Proof

The properties of $\Sigma(t)$ imply (8.7.2). It remains to prove (8.7.3). Consider

$$\Psi_\tau(t)(G) = \Psi(t+\tau)(G)$$

where τ is fixed as a parameter. We shall show that this is the maximum element of the set $\tilde{\mathcal{K}}$ defined by (8.6.1), with $G = \Psi(\tau)(G)$, which will imply (8.7.3).

Clearly

$$\Psi_\tau(0)(G) = \Psi(\tau)(G)$$

where $\Psi_\tau(t)(G) \in C_1$, $\forall t$, and satisfies the regularity requirements of $S(t)$ in (8.6.1). Also

$$\Psi_\tau(t) \leq \int_0^{t-s} e^{-\beta\sigma}\Phi^v(\sigma)l_v d\sigma + e^{-\beta(t-s)}\Phi^v(t-s)\Psi_\tau(s).$$

Therefore $\Psi_\tau(t)$ belongs to $\tilde{\mathcal{K}}$. Therefore we can assert that

$$\Psi(t+\tau)(G) \leq \Psi(t)(\Psi(t)(G)). \tag{8.7.4}$$

On the other hand, let us define

$$\eta_\tau(s) = \begin{cases} \Psi(s)(G), & \text{if } s \leq \tau \\ \Psi(s-\tau)(\Psi(\tau)(G)), & \text{if } s \geq \tau. \end{cases}$$

We have $\eta_\tau(0) = G$, $\eta_\tau(s) \in C_1$, and the regularity requirements of (8.6.1) are satisfied. Let us now check that

$$\eta_\tau(t) \leq \int_0^{t-s} e^{-\beta\sigma}\Phi^v(\sigma)l_v d\sigma + e^{-\beta(t-s)}\Phi^v(t-s)\eta_\tau(s). \tag{8.7.5}$$

If $0 \leq s \leq t \leq \tau$, this is clear since $\eta_\tau(t) = \Sigma(t)$ and $\eta_\tau(s) = \Sigma(s)$. Suppose next that $s \leq \tau \leq t$, then $\eta_\tau(s) = \Sigma(s)$, and

$$\eta_\tau(t) = \Psi(t - \tau)(\Psi(\tau(G))$$
$$= \Psi(t - \tau)(\Sigma(\tau)).$$

Therefore

$$\eta_\tau(t) \leq \int_0^{t-\tau} e^{-\beta\sigma} \Phi^v(\sigma) l_v d\sigma + e^{-\beta(t-\tau)} \Phi^v(t - \tau)(\Sigma(\tau)).$$

But

$$\Sigma(\tau) \leq \int_0^{\tau-s} e^{-\beta\sigma} \Phi^v(\sigma) l_v d\sigma + e^{-\beta(\tau-s)} \Phi^v(\tau - s) \eta_\tau(s)$$

and (8.7.5) follows easily. Finally, if $\tau \leq s \leq t$, then

$$\eta_\tau(t) = \Psi(t - \tau)(\Psi(\tau)(G)), \quad \eta_\tau(s) = \Psi(s - \tau)(\Psi(\tau)(G))$$

and the property (8.7.5) is also verified. Therefore $\eta_\tau(t)$ belongs to $\mathcal{K}$, which implies

$$\eta_\tau(t) \leq \Psi(t)(G).$$

In particular

$$\eta_\tau(t + \tau) = \Psi(t)(\Psi(\tau)(G)) \leq \Psi(t + \tau)(G)$$

which together with (8.7.4) implies the equality (8.7.3).   ∎

8.7.2   Behaviour at $t \to \infty$

We first notice that the element $\Sigma(t)(G)(p)$, which has been defined on the fixed interval $[0, T]$ and thus may seem to depend on T, is in fact independent of T. Indeed, suppose we denote by $\Sigma_T(t)$ the maximum element of the set $\mathcal{K}_T$, where $\mathcal{K}_T$ represents the set of functions $S(t)$ satisfying (8.6.1) with $0 \leq s \leq t \leq T$. Now pick $T' > T$ and consider $\Sigma_{T'}(t)$. We have

$$\Sigma_{T'}(t) = \Sigma_T(t) \quad \text{for } t \leq T. \tag{8.7.6}$$

Indeed since $\mathcal{K}_{T'} \subset \mathcal{K}_T$ necessarily

$$\Sigma_{T'}(t) \leq \Sigma_T(t), \quad \forall t \leq T.$$

On the other hand let $\eta(t)$ be defined by

$$\eta(t) = \begin{cases} \Sigma_T(t), & \text{for } t \leq T \\ \\ \Psi_{T'}(t - T)(\Sigma_T(T)), & \text{for } t > T \end{cases}$$

where we have denoted

$$\Psi_{T'}(t)(G) = \Sigma_{T'}(t), \quad t \in [0, T'].$$

The element $\eta(t)$ belongs to $\mathcal{K}_{T'}$, hence

$$\eta(t) \leq \Sigma_{T'}(t), \quad \forall t \in [0, T'].$$

In particular

$$\Sigma_T(t) \leq \Sigma_{T'}(t), \quad \forall t \in [0, T]$$

and thus (8.7.6) has been proved.

Therefore we may consider the set $\tilde{\mathcal{K}}$ defined by

$$S(t) \in C_1; \quad t \mapsto S(t)(p) \text{ is upper semicontinuous } \forall p \in H,$$

$$t \mapsto S(t)(p) \text{ belongs to } C^0([0, \infty)) \text{ if } p \in V.$$

$$S(0) = G \tag{8.7.7}$$

$$S(t) \leq \int_0^{t-s} e^{-\beta\sigma} \Phi^v(\sigma) l_v d\sigma + e^{-\beta(t-s)} \Phi^v(t-s) S(s)$$

$$\forall s, t, \ 0 \leq s \leq t, \quad \forall v.$$

This set has a maximum element $\Sigma(t)$, which defines a non linear semigroup in C_1, denoted by $\Phi(t)$. Let us study the limit of $\Psi(t)(G)$ as $t \to \infty$.

We can state the following.

Theorem 8.7.1. *Under the assumptions of Theorem 8.6.1, one has*

$$\Psi(t)(G)(p) \to \Sigma(p) \text{ as } t \to \infty, \forall p \in H, \forall G \in C_1 \tag{8.7.8}$$

where

$$\Sigma(p) = \inf_{v(\cdot)} J_p(v(\cdot))$$

coincides with $\bar{S}(p)$, the maximum element of the set (8.5.4).

Proof

Consider

$$J_p(v(\cdot)) - J_{p,t}(v(\cdot)) = E\left[\int_0^\infty e^{-\beta s}(l_{v_s}, q_p^{v(\cdot)}(s))ds - e^{-\beta t} G(q_p^{v(\cdot)}(t))\right].$$

Therefore

$$|J_p(v(\cdot)) - J_{p,t}(v(\cdot))| \leq \frac{\sup_v |l_v||p|e^{-(\beta-c/2)t}}{(\beta - c/2)} + \|G\|_1|p|e^{-(\beta-c/2)t}.$$

Since this estimate is uniform in $v(\cdot)$, the result (8.7.8) is easily deduced: $\Sigma \in C_1$. Moreover, writing

$$\Sigma(t+\tau) \leq \int_0^t e^{-\beta\sigma} \Phi^v(\sigma) l_v d\sigma + e^{-\beta t} \Phi^v(t) \Sigma(\tau)$$

and letting τ tend to $+\infty$, we deduce that Σ belongs to (8.5.4). Hence $\Sigma \leq \bar{S}$. On the other hand

$$\bar{S} \leq \Psi(t)(\bar{S}),$$

therefore

$$\bar{S} \leq \Sigma,$$

and the equality follows. ∎

9   Existence results for stochastic control problems with partial information

Introduction

In this chapter, we pursue a limited objective. In Chapter 8 we considered the stochastic control problem with full information (8.1.15), (8.1.37), for a stochastic PDE, the Zakai equation. Unfortunately, this problem has no solution in general, i.e., there exists no optimal control. A standard mathematical trick for this type of difficulty is to try to embed the problem into a more general one, by extending the class of admissible controls and the corresponding cost function. This is the idea of relaxation. The relaxed problem is of interest because it has a solution.

There are some natural ideas which lead to this type of result, and some of them are described here. However, the objective is limited, in the sense that we do not try to characterize fully the optimal relaxed controls, particularly in relation to dynamic programming, nor to prove that they are Markov controls. This would imply developments too lengthy to give here.

For related work, I refer the reader to Fleming and Pardoux (1982), Haussmann (1982), Borkar (1983), Davis and Kohlmann (1988), Elkaroui, Huu Nguyen and Jeanblanc-Picque (1988), Elliott and Kohlmann (1982), Fleming and Nisio (1984), Nagase (1988), Bensoussan and Nisio (1990), Nagase and Nisio (1988), where various approaches are considered, in particular the problems mentioned above, which are not treated here, are addressed.

9.1 Notation: setting of the problem

9.1.1 Assumptions

To fix the ideas, we restrict ourselves to the set up of § 8.1.2, in particular considering only controls in the drift term and not in the diffusion term. So we set

$$A_t \phi = - \sum_{i,j} \frac{\partial}{\partial x_i} \left(a_{ij} \frac{\partial \phi}{\partial x_j} \right) \tag{9.1.1}$$

with

$$a_{i,j}(x,t) \text{ bounded, and } a_{ij}\xi_i\xi_j \geq \alpha |\xi|^2, \ \forall \xi \in R^u, \ \alpha > 0, \ \partial a_{ij}/\partial x_i \text{ bounded.} \tag{9.1.2}$$

Also

$$G_t(v)\phi = \sum_i \frac{\partial}{\partial x_i}(a_i(x,v)\phi) \tag{9.1.3}$$

where

$$a_i(x,v,t): \ R^n \times U_{ad} \times (0,\infty) \to R \text{ is Borel and, for almost all } x,$$
$$\text{is continuous in } v,t: \tag{9.1.4}$$

$$U_{ad} \text{ is a compact metric space.} \tag{9.1.5}$$

Moreover we consider

$$h(x,t): \ R^n \times (0,\infty) \to R^m \text{ bounded,}$$
$$\frac{\partial h_i}{\partial t}, \quad \frac{\partial h_i}{\partial x_j}, \quad \frac{\partial h_i}{\partial x_j \partial x_k} \quad \text{bounded.} \tag{9.1.6}$$

Let

$$p_0 \in L^2(R^n). \tag{9.1.7}$$

9.1.2 Functional spaces

Let $M(U_{ad})$ be the set of probability measures on U_{ad}. It is a subset of $R(U_{ad})$, the set of Radon measures on U_{ad}, which is the dual of $C(U_{ad})$, the set of continuous functions with values in U_{ad}. $R(U_{ad})$ is equipped with the weak* topology, and $M(U_{ad})$ is a compact metrizable subset of $R(U_{ad})$.

Now let $R([0,T] \times U_{ad})$ be the space of Radon measures on the compact set $[0,T] \times U_{ad}$, equipped also with the weak* topology. We consider the set $L^\infty(0,T; M(U_{ad}))$ and we embed it into the set of Radon measures $R([0,T] \times U_{ad})$, as follows.

If $\mu_t(dv) \in L^\infty(0, T; M(U_{ad}))$, define $\mu \in R([0, T] \times U_{ad})$ by the formula

$$\langle \mu, f \rangle = \int_0^T \left(\int_{U_{ad}} f(t, v) \mu_t(dv) \right) dt, \quad \forall f \in C([0, T] \times U_{ad}).$$

Again $L^\infty(0, T; M(U_{ad}))$ is a compact metrizable subset of $R([0, T] \times U_{ad})$.

Note that the set U_{ad} is embedded in $M(U_{ad})$ by the map

$$v \mapsto \delta_v$$

which is continuous and one to one. Similarly $L^\infty(0, T; U_{ad})$ is embedded into

$$L^\infty(0, T; M(U_{ad})).$$

We can also extend Borel bounded functions on U_{ad} into functions on $M(U_{ad})$, by setting

$$\tilde{\phi}(\mu) = \int_{U_{ad}} \phi(v) \mu(dv)$$

and clearly

$$\tilde{\phi}(\delta_v) = \phi(v).$$

The notation $\phi \to \tilde{\phi}$ will be used from now on.

To the operation $G_t(v)$ we associate $\tilde{G}_t(\mu)$ defined, by

$$\tilde{G}_t(\mu)\phi = \sum_i \frac{\partial}{\partial x_i}(\tilde{a}_i(x, \mu)\phi).$$

9.1.3   Robust form of Zakai equation

We shall use the formula (8.1.31) to give a functional definition of the solution of the Zakai equation. Specifically, let $z \in C([0, T]; R^m)$ and $\mu_t \in L^\infty(0, T; M(U_{ad}))$; we define $q_t(x) = q_t(x; z, \mu)$ as the function

$$q(x, t) = \nu(x, t) \exp[z(t) \cdot h(x, t)] \tag{9.1.8}$$

where $\nu(x, t)$ is the solution of

$$\frac{\partial \nu}{\partial t} + A_t \nu - \frac{\partial}{\partial x_k}(\tilde{a}_t(\mu_k)\nu) - 2D\nu \cdot aD(h_t \cdot z_t)$$

$$+ \nu_t \left(A_t(h_t \cdot z_t) - D(h_t \cdot z_t)A_u aD(h_t \cdot z_t) + z_t \cdot \frac{\partial h_t}{\partial t} + \frac{1}{2}|h_t|^2 - \tilde{a}_k(\mu_t)\frac{\partial}{\partial x_k}(z_t \cdot h_t) \right)$$

$$= 0$$

$$\nu(x, 0) = p_0(x).$$

$$\tag{9.1.9}$$

Next let $l_t(x, v)$, $k(x)$ be such that

l_t, k are Borel bounded and, for almost all x, l is continuous in v, t;   (9.1.10)

$$l(\cdot, v) \in L^2(R^n) = H, \quad |l(\cdot, v)|_H \le C$$
$$k \in H. \tag{9.1.11}$$

We consider the function

$$F(z(\cdot); \mu(\cdot)) = \int_0^T (\tilde{l}_t(\mu_t), q_t(z, \mu)) dt + (k, q(T; z, \mu)) \tag{9.1.12}$$

which is the pointwise value of the cost function, for any given $z(\cdot)$, $\mu(\cdot)$, later viewed as the realization of a random variable.

9.2   Stochastic optimal control

9.2.1   Properties of the functional F

We begin with the technical result

Proposition 9.2.1 *We make the assumptions (9.1.2), (9.1.4), (9.1.5), (9.1.6), (9.1.7), (9.1.10), (9.1.11). Then the functional F is continuous on the space*

$$\Omega = C(0, T; R^n) \times L^\infty(0, T; M(U_{ad})).\dagger$$

Proof

Let z^n, μ^n be sequences in Ω, such that

$$z^n \to z \text{ in } C(0, T; R^m)$$
$$\mu^n \to \mu \text{ in } L^\infty(0, T; M(U_{ad})) \quad \text{with the topology}$$
$$\text{induced by } R([0, T] \times U_{ad}).$$

Let ν^n be the solution of (9.1.9) corresponding to z^n, μ^n and let q^n be related to ν^n, z^n by (9.1.8).

We shall prove that

$$q^n \to q \text{ in } L^2(0, T; H^1(R^n))$$
$$q^n(t) \to q(t) \text{ in } L^2(R^n), \quad \forall t. \tag{9.2.1}$$

It is clear from (9.1.8) that it is sufficient to prove that

$$\nu^n \to \nu \text{ in } L^2(0, T; H^1(R^n))$$
$$\nu^n(t) \to \nu(t) \text{ in } L^2(R^n), \quad \forall t. \tag{9.2.2}$$

Noting that $\tilde{a}_k(x, \mu^n)$ is bounded, it is easy to check from (9.1.9) that ν^n remains in a bounded subset of $L^2(0, T; H^1(R^n)) \cap C(0, T; L^2(R^n))$. Consider a weakly convergent subsequence, and let $\beta(t) \in C^1(0, T)$, $\beta(T) = 0$, $\beta(0) = 1$, $\phi \in \mathcal{D}(R^n)$. We deduce

---

$\dagger$ Ω is locally compact, metrizable.

from (9.1.9) written for the index n,

$$
-\int_0^T \int \beta' \nu^n(x,t)\phi(x)\mathrm{d}x\mathrm{d}t - \int p_0(x)\phi(x)\mathrm{d}x + \int_0^T \int_{R^n} a_{ij}\frac{\partial \nu^n}{\partial x_j}\frac{\partial \phi}{\partial x_i}\beta\mathrm{d}x\mathrm{d}t
$$

$$
+\int_0^T \int_{R^n} \int_{U_{ad}} a_k(x,v)\nu^n \frac{\partial \phi}{\partial x_k}\beta\mu_t^n(\mathrm{d}v)\mathrm{d}x\mathrm{d}t - 2\int_0^T \int_{R^n} \mathrm{D}\nu^n \cdot a\mathrm{D}(h \cdot z^n)\phi\beta\mathrm{d}x\mathrm{d}t
$$

$$
+\int_0^T \int_{R^n} \nu^n \left[A(h \cdot z^n) - \mathrm{D}(h\cdot^n)a\mathrm{D}(h \cdot z^n) + z^n \cdot \frac{\partial h}{\partial t} + \frac{1}{2}|h|^2\right]\phi\beta\mathrm{d}x\mathrm{d}t
$$

$$
-\int_0^T \int_{R^n} \int_{U_{ad}} \nu^n a_k(x,v)\frac{\partial}{\partial x_k}(z^n \cdot h)\mu_t^n(\mathrm{d}v)\phi\beta\mathrm{d}x\mathrm{d}t
$$

$$
= 0.
$$

(9.2.3)

Since we may assume that $\nu^n \to \nu$ strongly in $L^2(0,T;O)$ for any bounded domain O of R^n, and since the support of ϕ is compact, we have

$$
\left|\int_0^T \int_{R^n} \int_{U_{ad}} a_k(x,v)\frac{\partial \phi}{\partial x_k}\beta\nu^n\mu_t^n(\mathrm{d}v)\mathrm{d}x\mathrm{d}t \right.
$$

$$
\left. -\int_0^T \int_{R^n} \int_{U_{ad}} a_k(x,v)\frac{\partial \phi}{\partial x_k}\beta\nu^n\mu_t(\mathrm{d}v)\mathrm{d}x\mathrm{d}t\right|
$$

(9.2.4)

$$
\leq C\int_0^T \int_{R^n} |\mathrm{D}\phi||\nu^n - \nu|\mathrm{d}x\mathrm{d}t
$$

$$
+\left|\int_0^T \int_{R^n} \int_{U_{ad}} a_k(x,v)\frac{\partial \phi}{\partial x_k}\nu\beta(\mu_t^n(\mathrm{d}v) - \mu_t(\mathrm{d}v))\mathrm{d}x\mathrm{d}t\right|.
$$

But

$$
\int_0^T \int_{U_{ad}} a_k(x,v)\beta(t)\mu_t^n(\mathrm{d}v)\mathrm{d}t \to \int_0^T \int_{U_{ad}} a_k(x,v)\beta(t)\mu_t(\mathrm{d}v)\mathrm{d}t
$$

for almost all x, and remains bounded with respect to n.

By Lebesgue's theorem the second part of the right hand side of (9.2.4) tends to 0. Similarly

$$
\int_0^T \int_{R^n} \int_{U_{ad}} \nu^n a_k(x,v)\frac{\partial}{\partial x_k}(z^n \cdot h)\mu_t^n(\mathrm{d}v)\phi\beta\mathrm{d}x\mathrm{d}t
$$

$$
\to \int_0^T \int_{R^n} \int_{U_{ad}} \nu a_k(x,v)\frac{\partial}{\partial x_k}(z^n \cdot h)\mu_t(\mathrm{d}v)\phi\beta\mathrm{d}x\mathrm{d}t.
$$

It is then possible to take the limit in (9.2.3) to obtain

$$
-\int_0^T \int \beta' \nu(x,t)\phi(x)\mathrm{d}x\mathrm{d}t - \int p_0(x)\phi(x)\mathrm{d}x + \int_0^T \int_{R^n} a_{ij}\frac{\partial \nu}{\partial x_j}\frac{\partial \phi}{\partial x_i}\beta\mathrm{d}x\mathrm{d}t
$$

$$
+\int_0^T \int_{R^n} \int_{U_{ad}} a_k(x,v)\nu\frac{\partial \phi}{\partial x_k}\beta\mu_t(\mathrm{d}v)\mathrm{d}x\mathrm{d}t - 2\int_0^T \int_{R^n} \mathrm{D}\nu \cdot a\mathrm{D}(h \cdot z)\phi\beta\mathrm{d}x\mathrm{d}t
$$

$$
+\int_0^T \int_{R^n} \nu \left[A(h \cdot z) - \mathrm{D}(h \cdot z)a\mathrm{D}(h \cdot z) + z \cdot \frac{\partial h}{\partial t} + \frac{1}{2}|h|^2\right]\phi\beta\mathrm{d}x\mathrm{d}t
$$

$$
-\int_0^T \int_{R^n} \int_{U_{ad}} \nu a_k(x,v)\frac{\partial}{\partial x_k}(z \cdot h)\mu_t(\mathrm{d}v)\phi\beta\mathrm{d}x\mathrm{d}t = 0
$$

which is the weak formulation of (9.1.9).

It remains to prove strong convergence.

Write

$$\theta^n(x,t) = A_t(h_t \cdot z_t^n) - D(h_t \cdot z_t^n)aD(h_t \cdot z_t^n) + z_t^n \cdot \frac{\partial h_t}{\partial t}$$
$$+ \frac{1}{2}|h_t|^2 - \tilde{a}_k(\mu_t^n)\frac{\partial}{\partial x_k}(z_t^n \cdot h_t)$$

and for $\beta > 0$ we have

$$\frac{\partial}{\partial t}(\nu^n e^{-\beta t}) + A_t(\nu^n e^{-\beta t}) + \beta \nu^n e^{-\beta t} - \frac{\partial}{\partial x_k}(\tilde{a}_k(\mu_t^n)\nu^n e^{-\beta t})$$
$$- 2D(\nu^n e^{-\beta t}) \cdot aD(h_t \cdot z_t^n) + \nu_t^n e^{-\beta t}\theta^n = 0.$$

Define similarly

$$\theta(x,t) = A_t(h_t \cdot z_t) - D(h_t \cdot z_t)aD(h_t \cdot z_t) + z_t \cdot \frac{\partial h_t}{\partial t}$$
$$+ \frac{1}{2}|h_t|^2 - \tilde{a}_k(\mu_t)\frac{\partial}{\partial x_k}(z_t \cdot h_t)$$

and

$$\frac{\partial}{\partial t}(\nu e^{-\beta t}) + A_t(\nu e^{-\beta t}) + \beta \nu e^{-\beta t} - \frac{\partial}{\partial x_k}(\tilde{a}_k(\mu_t)\nu e^{-\beta t})$$
$$- 2D(\nu e^{-\beta t}) \cdot aD(h_t \cdot z_t) + \nu_t e^{-\beta t}\theta = 0.$$

We consider the quantity

$$X_t^n = \frac{1}{2}|\nu^n - \nu|_H^2 e^{-2\beta t} + \int_0^t \int_{R^n} a_{ij}\frac{\partial}{\partial x_j}(\nu^n - \nu)\frac{\partial}{\partial x_i}(\nu^n - \nu)e^{-2\beta s}dxds$$
$$+ \beta \int_0^t e^{-2\beta s}|\nu^n - \nu|_H^2 ds$$
$$+ \int_0^t \int_{R^n} (\tilde{a}_k(\mu_t^n)\nu^n - \tilde{a}_k(\mu_t)\nu)\frac{\partial}{\partial x_k}(\nu^n - \nu)e^{-2\beta s}dxds$$
$$- 2\int_0^t \int_{R^n} (D\nu^n \cdot aD(h_s \cdot z_s^n) - D\nu \cdot aD(h_s \cdot z_s))(\nu^n - \nu)e^{-2\beta s}dxds$$
$$+ \int_0^t \int_{R^n} (\nu^n\theta^n - \nu\theta)(\nu^n - \nu)e^{-2\beta s}dxds = 0.$$

We note

$$X_t^n = X_t^{n,1} + X_t^{n,2} + X_t^{n,3}$$

where

$$X_t^{n,1} = \frac{1}{2}|\nu^n|_H^2 e^{-2\beta t} + \int_0^t \int_{R^n} a_{ij}\frac{\partial}{\partial x_j}\nu^n \frac{\partial}{\partial x_i}\nu^n e^{-2\beta s}dxds$$
$$+ \beta \int_0^t e^{-2\beta s}|\nu^n|_H^2 ds + \int_0^t \int_{R^n} \tilde{a}_k(\mu_s^n)\nu^n \frac{\partial}{\partial x_k}\nu^n e^{-2\beta s}dxds$$
$$- 2\int_0^t \int_{R^n} D\nu^n \cdot aD(h_s \cdot z_s^n)\nu^n e^{-2\beta s}dxds + \int_0^t \int_{R^n} (\nu_s^n)^2\theta^n e^{-2\beta s}dxds$$

$$X_t^{n,2} = -(\nu^n, \nu)e^{-2\beta t} - 2\int_0^t \int_{R^n} a_{ij}\frac{\partial \nu^n}{\partial x_j}\frac{\partial \nu}{\partial x_i}e^{-2\beta s}dxds$$

$$- 2\beta \int_0^t e^{-2\beta s}(\nu^n, \nu)ds - \int_0^t \int_{R^n} \tilde{a}_k(\mu_s)\nu\frac{\partial}{\partial x_k}\nu^n e^{-2\beta s}dxds$$

$$- \int_0^t \int_{R^n} \tilde{a}_k(\mu_s^n)\nu^n\frac{\partial}{\partial x_k}\nu e^{-2\beta s}dxds$$

$$+ 2\int_0^t \int_{R^n} D\nu^n \cdot aD(h_s z_s^n)\nu_s e^{-2\beta s}dxds$$

$$+ 2\int_0^t \int_{R^n} D\nu \cdot aD(h_s z_s)\nu_s^n e^{-2\beta s}dxds - \int_0^t \int_{R^n} \nu^n \theta^n \nu e^{-2\beta s}dxds$$

$$- \int_0^t \int_{R^n} \nu\theta\nu^n e^{-2\beta s}dxds$$

$$X_t^{n,3} = \frac{1}{2}|\nu|_t^2 e^{-2\beta t} + \int_0^t \int_{R^n} a_{ij}\frac{\partial \nu}{\partial x_j}\frac{\partial \nu}{\partial x_i}e^{-2\beta s}dxds + \beta \int_0^t e^{-2\beta s}|\nu|_H^2 ds$$

$$+ \int_0^t \int_{R^n} \tilde{a}_k(\mu_s)\nu\frac{\partial \nu}{\partial x_k}e^{-2\beta s}dxds - 2\int_0^t \int_{R^n} D\nu \cdot aD(h_s z_s)\nu e^{-2\beta s}dxds$$

$$+ \int_0^t \int_{R^n} \nu^2 \theta e^{-2\beta s}dxds = X_t^3 = \frac{1}{2}|p_0|^2.$$

But

$$X_t^{n,1} = \frac{1}{2}|p_0|^2.$$

Noting that among other things,

$$\tilde{a}_k(\mu_s^n)\nu^n \to \tilde{a}_k(\mu_s)\nu \text{ in } L^2(0, T; H) \text{ weakly}$$

$$\nu^n \theta^n \to \nu\theta \text{ in } L^2(0, T; H) \text{ weakly}$$

We can check

$$X_t^{n,2} \to -|\nu|^2 e^{-2\beta t} - 2\int_0^t \int_{R^n} a_{ij}\frac{\partial \nu}{\partial x_j}\frac{\partial \nu}{\partial x_i}e^{-2\beta s}dxds - 2\beta \int_0^t e^{-2\beta s}|\nu|^2 ds$$

$$- \int_0^t \int_{R^n} \tilde{a}_k(\mu_s)\nu\frac{\partial \nu}{\partial x_k}e^{-2\beta s}dxds - \int_0^t \int_{R^n} \tilde{a}_k(\mu_s)\nu\frac{\partial \nu}{\partial x_k}e^{-2\beta s}dxds$$

$$+ 4\int_0^t \int_{R^n} D\nu \cdot aD(h_s z_s)\nu_s e^{-2\beta s}dxds - 2\int_0^t \int_{R^n} \nu^2 \theta e^{-2\beta s}dxds$$

$$= -2X_t^3 = -|p_0|^2.$$

Therefore

$$X_t^n \to 0.$$

But, choosing β sufficiently large, we have

$$|\nu^n(t) - \nu(t)|^2 e^{-2\beta t} + \alpha \int_0^t \|\nu^n(s) - \nu(s)\|^2 e^{-2\beta s}ds \leq cX_t^n$$

and the strong convergence is proven (i.e. (9.2.2) is proven).

Therefore (9.2.1) is also proven. We can then complete the proof of the proposition. Indeed

$$F(z^n(\cdot); \mu^n(\cdot)) = \int_0^T \int_{U_{ad}} \int_{R^n} l_t(x,v) q^n(x,t) \mu_t^n(dv) dx dt$$

$$+ \int_{R^n} k(x) q^n(x,T) dx.$$

But

$$\left| \int_0^T \int_{U_{ad}} \int_{R^n} l_t(x,v)(q^n(x,t) - q(x,t)) \mu_t^n(dv) dx dt \right|$$

$$\leq \left(\int_0^T \int_{R^n} (q^n - q)^2 dx dt \right)^{1/2} \int_0^T \int_{U_{ad}} \int_{R^n} (l_t(x,v))^2 \mu_t^n(dv) dx dt$$

$$\leq C |q^n - q|_{L^2(0,T;H)} \to 0$$

from the second assumption (9.1.11). Moreover

$$\int_0^T \int_{U_{ad}} \int_{R^n} q(x,t) l_t(x,v) \mu_t^n(dv) dx dt \to \int_0^T \int_{U_{ad}} \int_{R^n} q(x,t) l_t(x,v) \mu_t(dv) dx dt.$$

If q were continuous in t, for almost all x, then this would follow from the convergence of μ^n to μ. Since we can find a sequence of smooth functions converging to q in $L^2(0,T;H)$, the result obtains.

Clearly the desired result follows.   ∎

9.2.2   Probabilistic set up

Consider Ω, provided with its Borel σ-algebra as a probability space, denoted by $\mathcal{A}$. An element ω will be a pair $(z(\cdot), \mu(\cdot))$.

We define

$$\mathcal{A}_t = \sigma(z(s), \mu(s), s \leq t).$$

The canonical process $z_t(\omega), \mu_t(\omega)$ is as usual $\omega(t)$. Now we provide $(\Omega, \mathcal{A})$ with probabilities P such that

$$z_t \text{ is a standard Wiener process for the system } (\Omega, \mathcal{A}, \mathcal{A}_t, P). \tag{9.2.5}$$

Let $\mathcal{P}_{ad}$ be the set of probabilities on $(\Omega, \mathcal{A})$ such that (9.2.5) holds. It is a closed subset of $\mathcal{P}$, where $\mathcal{P}$ denotes the set of probabilities on (Ω, A), equipped with the weak convergence

$$P^n \to P \text{ if } \int f(\omega) dP^n \to \int f(\omega) dP \text{ whenever } f \text{ is continuous and bounded.} \tag{9.2.6}$$

Remark 9.2.1 Since Ω is locally compact metric, one can consider the set of Radon measures on Ω, $R(\Omega)$, i.e. the vector space of the linear forms on $C_k(\Omega)$ (continuous functions with compact support), equipped with the *vague convergence*

$$\Lambda^n \to \Lambda, \quad \text{if} \quad \langle \Lambda^n, f \rangle \to \langle \Lambda, f \rangle \quad \forall f \in C_k(\Omega).$$

The set of probabilities $\mathcal{P}$ is a subset of $R(\Omega)$, and the weak topology (9.2.6) is stronger than the vague topology. The weak topology is metrizable by Prokhorov's metric (see Prokhorov 1956 or Ikeda and Watanabe 1981).   ∎

Lemma 9.2.1 $\mathcal{P}_{ad}$ *is a compact subset of* $\mathcal{P}$, *hence a compact metric space.*
Proof
Clearly $\mathcal{P}_{ad}$ is closed. To show that it is compact, it suffices to show that Prokhorov's criterion holds, namely

$$\forall \delta > 0, \exists K_\delta \text{ compact in } \Omega, \text{ such that}$$
$$P(K_\delta) \geq 1 - \delta, \quad \forall P \in \mathcal{P}_{ad}. \tag{9.2.7}$$

Now let

$$K = \left\{ z, \mu \mid z \text{ is Hölder with exponent } 0 < \gamma < \tfrac{1}{2} \right\}$$

which is a compact subset of Ω, by the Ascoli theorem.
But by the definition of $\mathcal{P}_{ad}$

$$P(K) = 1, \quad \forall P \in \mathcal{P}_{ad}.$$

Hence the desired result. ∎

For any $P \in \mathcal{P}_{ad}$, we define

$$J(P) = \int F(\omega) \mathrm{d}P(\omega). \tag{9.2.8}$$

The functional $J(P)$ is well defined, thanks to
Lemma 9.2.2 $F(\omega) \in L^2(\Omega, \mathcal{A}, P)$, $\forall P \in \mathcal{P}_{ad}$, *and* $E|F|^2 \leq C_0$, *where* C_0 *does not depend on the particular* P *in* $\mathcal{P}_{ad}$.
Proof
Since on $(\Omega, \mathcal{A}, \mathcal{A}_t, P)$ with $P \in \mathcal{P}_{ad}$, z is a Wiener process, we can check by stochastic calculus that q defined by (9.1.8) satisfies the Zakai equation

$$\mathrm{d}q + A_t q - \sum_k \frac{\partial}{\partial x_k}(\tilde{a}_k(x, \mu_t)q) = qh \cdot \mathrm{d}z$$
$$q(x, 0) = p_0 \tag{9.2.9}$$

and as is easily seen

$$E|q(t)|_H^2 \leq C$$

where the constant C does not depend on the probability P, provided P is in $\mathcal{P}_{ad}$.
On the other hand, from (9.1.12)

$$|F(\omega)|^2 \leq C \int_0^T |q(t)|_H^2 \mathrm{d}t + C|q(T)|_H^2$$

where C does not depend on ω. And thus

$$\int |F(\omega)|^2 \mathrm{d}P(\omega) \leq C_0 \tag{9.2.10}$$

with a constant C_0 independent of the choice of P in $\mathcal{P}_{ad}$. ∎

9.3   Existence of a solution

9.3.1   Optimization problem

The optimization problem that we consider is stated as follows: find

$$\inf_{P \in \mathcal{P}_{ad}} J(P).$$

We can assert

Theorem 9.3.1 *Under the assumptions of Proposition 9.2.1, there exists P^* in $\mathcal{P}_{ad}$ such that*

$$J(P^*) = \inf_{P \in \mathcal{P}_{ad}} J(P). \qquad (9.3.1)$$

Proof
Since $\mathcal{P}_{ad}$ is compact, it is sufficient to prove that $J(P)$ is continuous. But let $P^n \to P$, then

$$\int F(\omega) dP^n(\omega) = \int \frac{F(\omega)}{1 + \epsilon |F(\omega)|} dP^n + \epsilon \int \frac{|F|F}{1 + \epsilon |F|} dP^n.$$

Since $F(\omega)$ is continuous, we have

$$\int \frac{F(\omega)}{1 + \epsilon |F(\omega)|} dP^n \to \int \frac{F(\omega)}{1 + \epsilon |F(\omega)|} dP, \ \forall \epsilon.$$

On the other hand

$$\left| \epsilon \int \frac{|F|F}{1 + \epsilon |F|} \right| dP^n| \le \epsilon \int |F|^2 dP^n \le C\epsilon.$$

These properties suffice to imply

$$\int F(\omega) dP^n(\omega) \to \int F(\omega) dP(\omega)$$

and the continuity is proved.   ∎

9.3.2   Wide sense admissible controls

We want to show here the connections between the problem (9.3.1) and the original problem corresponding to the set up of § 8.1.2 where, to summarize, one starts with

$$(\Omega, \mathcal{A}, P, z_t)$$

where z_t is a standard Wiener process, and an admissible control v_t is adapted to $\mathcal{Z}^t = \sigma(z(s), s \leq t)$ with values in U_{ad}.

In fact, it is necessary to generalize this set up slightly in order to come close to problem (9.3.1). This generalization is quite natural, as will be shown and leads to the concept of *wide sense* admissible control, introduced by Fleming and Pardoux (1982). It goes as follows. Consider a probability space $(\Omega, \mathcal{A}, \mathcal{A}_t, P)$ where $\mathcal{A}_t$ is a filtration on Ω, and two processes z_t, v_t which are adapted to $\mathcal{A}_t$, and

$$z_t \text{ is a standard Wiener process (with respect to } \mathcal{A}_t\text{); } v_t \text{ takes values in } U_{ad}. \quad (9.3.2)$$

For such a set up, we can solve the Zakai equation

$$\mathrm{d}q + A_t q \mathrm{d}t - \sum_k \frac{\partial}{\partial x_k}(a_k(x, v_t)q) = qh \cdot \mathrm{d}z \qquad (9.3.3)$$

$$q(0) = p_0$$

as usual and define

$$J(v(\cdot)) = E\left[\int_0^T (l_t(v_t), q_t)\mathrm{d}t + (k, q_T)\right] . \dagger \qquad (9.3.4)$$

We can always take the probabilistic set $(\Omega, \mathcal{A}, \mathcal{A}_t, P)$ as in § 9.2.2, provided $P \in \mathcal{P}_{ad}$. We say that $P \in \mathcal{P}_{ad}$ corresponds to a wide sense admissible control whenever there exists $v_t(\omega)$ with values in U_{ad}, such that a.s. for P

$$a_k(x, t, v_t) = \tilde{a}_k(x, t, \mu_t), \quad \forall t \in [0, T], \quad \text{a.e. } x$$

$$l_t(x, v_t) = \tilde{l}_t(x, \mu_t).$$

In that case, it is clear that

$$J(P) = J(v(\cdot)). \qquad (9.3.5)$$

We shall define

$$\bar{\mathcal{P}}_{ad} = \text{subset of } \mathcal{P}_{ad} \text{ corresponding to wide sense admissible controls.}$$

Clearly

$$\inf_{P \in \bar{\mathcal{P}}_{ad}} J(P) \geq \min_{P \in \mathcal{P}_{ad}} J(P). \qquad (9.3.6)$$

With additional convexity assumptions we can have equality in (9.3.6). Let us make this precise.

Define

$$b(x, t, v) = \begin{pmatrix} a_1(x, t, v) \\ \vdots \\ a_n(x, t, v) \\ l(x, t, v) \end{pmatrix}$$

$\dagger$ The notation is not precise. It is the whole set $(\Omega, \mathcal{A}, \mathcal{A}_t, P, z_t, v_t,)$ which is a wide sense admissible control. To simplify we only write v_t. We can fix $\Omega, \mathcal{A}, \mathcal{A}_t$, and it is the pair $P \in \mathcal{P}_{ad}, v_t$ which has to be chosen eventually.

which takes values in R^{n+1}, and make the following assumptions:

b is uniformly continuous on $R^n \times [0, T] \times U_{ad}$ with values in R^{n+1};   (9.3.7)

the set $B = \{b(\cdot\,; v),\ v \in U_{ad}\}$ is a convex subset

of $C(R^n \times [0, T]; R^{n+1})$ with the topology of the sup norm.   (9.3.8)

Note that it follows from (9.3.7) that the map $v \to b(\cdot; v)$ is continuous from U_{ad} to $C(R^n \times [0, T]; R^{n+1})$, hence B, being the continuous image of a compact set, is compact. Therefore B is compact and convex.

Let $\mu \in M(U_{ad})$, and using the notation of § 9.1.2

$$\tilde{b}(\cdot\,; \mu) = \int_{U_{ad}} b(\cdot\,; v)\mu(dv).$$

Since B coincides with its closed convex hull, $\overline{co}\, B$, it follows that $\tilde{b}(\cdot\,; \mu) \in B$, $\forall \mu \in M(U_{ad})$.

Therefore we can define a set-valued map

$$D(\mu):\ M(U_{ad}) \to U_{ad}$$

$$D(\mu) = \left\{ v \in U_{ad} | b(\cdot\,; v) = \tilde{b}(\cdot\,; \mu) \right\}$$

Lemma 9.3.1 *For any μ, $D(\mu)$ is compact, and the multivalued map $\mu \to D(\mu)$ has a closed graph.*

Proof

To prove that $D(\mu)$ is compact, it is enough to show that it is closed. But if $v_n \in D(\mu)$ and $v_n \to v$ in U_{ad}, then $b(\cdot\,; v_n) \to b(\cdot\,; v)$ in $C(R^n \times [0, T]; R^{n+1})$.

But, by definition

$$b(\cdot\,; v_n) = \tilde{b}(\cdot\,; \mu)$$

hence also

$$b(\cdot\,; v) = \tilde{b}(\cdot\,; \mu)$$

which proves that $v \in D(\mu)$.

Next suppose that $\mu_n \to \mu$ in $M(U_{ad})$ (recall that it is equipped with the weak topology, which is the same as that induced by the weak* topology of Radon measures on U_{ad}). Suppose also that $v_n \in D(\mu_n)$ converges to v; let us show that $v \in D(\mu)$.

But

$$\tilde{b}(x, t, \mu_n) \to \tilde{b}(x, t, \mu), \quad \forall x, t.$$

On the other hand

$$b(x, t, v_n) = \tilde{b}(x, t, \mu_n), \quad \forall x, t$$

and

$$b(x, t, v_n) \to b(x, t, v)$$

hence

$$b(\cdot\,; v) = \tilde{b}(\cdot\,; \mu)$$

which proves that $v \in D(\mu)$.   ∎

The properties of Lemma 9.3.1 suffice to imply that there exists a Borel selection (see Stroock and Varadhan 1979)

$$S(\mu) : M(U_{ad}) \to U_{ad}$$

such that S is Borel and

$$S(\mu) \in D(\mu), \quad \forall \mu.$$

To $\mu_t(\omega)$ we can associate $v_t(\omega) = S(\mu_t(\omega))$. Therefore $\forall P \in \mathcal{P}_{ad}$; the process v_t constructed in this way is such that (9.3.5) holds. Hence $\mathcal{P}_{ad} = \mathcal{P}$.

We can then state

Theorem 9.3.2 *We make the assumptions of Proposition 9.2.1, and (9.3.7), (9.3.8). Then there exists an optimal wide sense admissible control.* ∎

References

Preface

Bensoussan, A. and Runggaldier, W. (1987), An approximation method for stochastic control problems with partial observation of the state, *Acta Applican. Math.* **10**, 145–70, Reidel.

Legland, F. (1989), Refined and high order time discretization of non linear filtering equations, Proceedings of the 28th Control and Decision Conference IEEE.

Chapter 1

Kalman, R.E. (1960), A new approach to linear filtering and prediction problems, *J. Basic Eng.*, 35–45.

Kalman, R.E. and Bucy, R.S. (1961), New results in linear filtering and prediction theory, *J. Basic Eng.*, 95–108.

Chapter 2

Bensoussan, A. and Lions, J.L. (1978), *Application des Inequations Variationnelles en Contrôle Stochastique*, Dunod, Paris.

Bismut, J.M. (1977), On optimal control of linear stochastic equations with a linear–quadratic criterion, *SIAM J. Contr. Optim.* **15(1)**.

Faurre, P. (1972), Realisations Markoviennes de processes alétories, rapport Laboria, IRIA.

Fleming, W. and Rishel R. (1975), *Optimal Deterministic and Stochastic Control*, Springer–Verlag, Berlin.

Gikhman, I.I. and Skorokhod, A.V. (1972), *Stochastic Differential Equations*, Springer–Verlag, Berlin.

Holt, C.C., Modigliani, F., Muth J.F. and Simon, H.A. (1960), *Planning Production, Inventories and Work Force*, Prentice Hall.

Liptser, R.S. and Shiryaev A.N. (1977), *Statistics of Random Processes*, two volumes, Springer–Verlag, New York.

Messulam, P. (1983), Contrôle stochastique avec informations partielles, *Thesis*, Université de Paris Dauphine.

Wonham, W.M. (1968a), On a matrix Riccati equation of stochastic control, *SIAM J. Cont.*, **6**, 312–26.

Wonham, W.M. (1968b), On the separation theorem of stochastic control, *SIAM J. Cont.* **6(2)**. et décideurs multiples, Thèse de 3e-cycle, Université Paris Dauphine.

Chapter 3

Bensoussan, A. and Van Schuppen, J.H. (1985), Optimal control of partially observable stochastic systems with an exponential-of-integral performance index, *SIAM J. Cont. Opt.* **23(4)**.

Gikhman, I.J. and Skorokhod, A.V. (1972), see Chapter 2.

Jacobson, D.H. (1973), Optimal stochastic linear systems with exponential criteria and their relation to deterministic differential games, *IEEE Trans. Auto. Control* **AC-18**, 124–31.

Kumar, P.R. and Van Schuppen, J.H. (1981), On the optimal control of stochastic systems with an exponential-of-integral performance index, *J. Math. Anal. App.* 312–32.

Speyer, J.L. (1976), An adaptive terminal guidance scheme based on an exponential cost criterion with application to homing missile guidance, *IEEE Trans. Auto. Control* **AC-21**, 371–5.

Speyer, J.L., Deyst, J. and Jacobson, D.H. (1974), Optimization of stochastic linear systems with additive measurement and process noise using exponential performance criteria, *IEEE Trans. Auto. Control* **AC-19**, 358–66.

Whittle, P. (1974), Risk sensitive linear quadratic gaussian control, *Adv. Appl. Prob.* **13**, 764–77.

Whittle, P. (1982, 1983) *Optimization over Time: Dynamic Programming and Stochastic Control*, two volumes, Wiley, New York.

Chapter 4

Baras, J.S., Blankenship, G. and Hopkins, W. (1983), Existence, uniqueness and asymptotic behavior of solutions of a class of Zakai equations with unbounded coefficients, *IEEE Trans. Auto. Control* **ACH 2(2)**, 203–14.

Bensoussan, A. (1972), Filtrage Optimal des Systēmes Linéaires, Dunod, Paris.

Bensoussan, A. (1982), *Stochastic Control by Functional Analysis Methods*, North–Holland, Amsterdam.

Kallianpur, G. (1980), *Stochastic Filtering Theory*, Springer–Verlag, New York.

Bensoussan, A. (1987), On a general class of stochastic partial differential equations, *J. Hydrology Hydraulics* **1(4)**, 297–303.

Bensoussan, A. and Lions, J. L. (1978), see Chapter 2.

Chaleyat-Maurel, M., Michel, D. and Pardoux, E. (1989), Un theorēme d'unicité pour l'equation de Zakai, to appear in *Stochastics*.

Clark, J.M.C. (1978), The design of robust approximations to the stochastic differential equations of non linear filtering, in *Communication Systems and Random Process Theory*, J. Skwirzaynski, ed., Sijthoff Noordhoff.

Davis, M.H.A. (1981), On a multiplicative transformation arising in nonlinear filtering, *Z. Wahrschein. verw. Geb.* **54**, 125–39.

Doss, H. (1977), Liens entre equations differentielles stochastiques et ordinaires, *Ann. Inst. H. Poincaré* **13**, 99–125.

Fleming, W.H. and Mitter, S.K. (1982), Optimal control and pathwise nonlinear filtering of non degenerate diffusions, *Stochastics* **8(1)**, 63–77.

Fujisaki, M. Kallianpur, G. and Kunita, H. (1979), Stochastic differential equations for the nonlinear filtering problem, *Osaka J. Math.* **9(1)**, 19–40.

Haussmann, V.G. and Pardoux, E. (1985), Time reversal of diffusions, *Stochastics* **23**, 241–75.

Kailath, T. (1968), An innovation approach to least square estimation, Parts I, II, *IEEE Trans. Auto. Control* **AC-13**, 164, 660.

Kailath, T. (1970), The innovations approach to detection and estimation theory, *Proc. IEEE* **58**, 680–95.

Kallianpur, G. and Striebel, C. (1968), Estimation of stochastic systems: arbitrary systems process with additive white noise observation errors, *AMS* **39**, 785–801.

Krylov, N.V. and Rozovskii, B.L. (1978), On conditional distributions of diffusion processes, *Math USSR Izvestya* **12**, 336–56.

Kunita, H. (1982), Stochastic partial differential equations connected with non linear filtering, in *Nonlinear Filtering and Stochastic Control*, S.Mitter and A. Moro, ed, Lecture Notes in Mathematics No. 972, Springer-Verlag.

Kunita, H. and Watanabe, S. (1967), On square integrable martingale, *Nagoya Math. J.* **30), 209-45.**

Kurtz, T. and Ocone, D. (1988), Unique characterization of conditional distributions in non linear filtering, *Annals Prob.* **16(1)**, 80–107.

Kushner, H.J. (1967), Dynamical equations for optimal non linear filtering, *J. Diff. Equations,* **3**, 179–90.

Liptser, R.S. and Shiryaev A.N. (1977), see Chapter 2.

Pardoux, E. (1979), SPDEs and filtering of diffusion processes, *Stochastics* **3**, 127–67.

Pardoux, E. (1982), Equations du filtrage non lineaire, de la prédiction et du lissage, *Stochastics* **6**, 193–231.

Picard, J. (1987), Méthods de perturbations pour les equations differentielles *Stochastiques et le Filtrage non linéaire, Thēse, Université de Provence.*

Stroock, D. and Varadhan, S.R.S. (1979), *Multidimensional Diffusion Processes*, Springer–Verlag, Berlin.

Sussmann, H. (1978), On the gap between deterministic and stochastic ordinary differential equations, *Ann. Prob.* **6**, 19–41.

Szpirglas, J. and Mazziotto, G. (1978), Modèle general de filtrage non lineaire et equations differentielles stochastiques associées, *Compte R. Acad. Sci. Paris* **286**, 1067–70.

Zakaï, M. (1969), On the optimal filtering of diffusion processes, *Z. Wahrschein verw. Geb.* **11**, 230–43.

Chapter 5

Bensoussan, A. (1988), *Perturbation Methods in Optimal Control*, Dunod-Gauthier Villars, Paris.

Bensoussan, A. and Blankenship, G.L. (1986), Nonlinear filtering with homogenization, *Stochastics* **17**, 67–90.

Kushner, H.J. (1986), Approximate and limit results for non linear filters with wide bandwidth observation noise, *Stochastics* **16**, 65–96.

Kushner, H.J. (1989), Functional occupation measures and ergodic cost problems for singularly perturbed stochastic systems, preprint, CICS, p. 136, MIT.

Pardoux, E. (1982), see chapter 4.

Picard, J. (1986a), Filtrage de diffusions vectorielles faiblement bruitées, *Proceedings of the 7th International Conference on Analysis and Optimization of Systems*, Lecture Notes in Control and Information Systems. No 83, Springer–Verlag.

Picard, J. (1986b), Nonlinear filtering of one-dimensional diffusions in the case of a high signal to noise ratio, *SIAM J. App. Math.*

Picard, J. (1987), see Chapter 4.

Chapter 6

Beneš, V.E. (1981), Exact finite-dimensional filters for certain diffusions with non linear drift, *Stochastics* **5** 65–92.

Beneš, V. and I. Karatzas (1983), Estimation and control for linear, partially observable systems with non-gaussian initial distribution, *Stoch. Proc. Applic.* **14**, 233–48.

Haussmann, U.G. and Pardoux, E. (1986), A conditionally almost linear filtering problem with non gaussian initial condition, University of British Columbia, Department of Mathematics.

Hazewinkel, M. and Markus, S.I. (1982), On Lie algebras and finite dimensional filtering, *Stochastics* **7**, 29–62.

Liptser, R.S. and Shiryaev, A.N. (1977), see Chapter 2.

Makowski, A.M. (1986), Filtering formulae for partially observed linear systems with non-gaussian initial conditions, *Stochastics* **16**, 1–24.

Shukhman, J. (1985), Explicit filters for linear systems and certain nonlinear systems with stochastic initial condition, Department of Electrical Engineering, Technion, Haifa.

Sussmann, H.J. (1982), Approximate finite-dimensional filters for some nonlinear problems, *Stochastics* **7**, 183–204.

Zeitouni, O. and Bobrovski, B.Z. (1984), Another approach to the equation of non-linear filtering and smoothing with applications, Department of Electrical Engineering, Technion, Haifa.

Chapter 7

Beneš, V.E. (1975), Composition and invariance methods for solving some stochastic control problems, *Adv. App. Prob.* **7**, 299–329.

Beneš, V.E. (1976), Full 'bang' to reduce predicted miss is optimal, *SIAM J. Contr. Optim.* **14**, 62–83.

Beneš, V.E. and Karatzas, I. (1982), Examples of optimal control for partially observable systems. Comparison, classical and martingale methods, *Stochastics*.

Bensoussan, A. (1982), see Chapter 4.

Bensoussan, A. and Lions, J. (1978), see Chapter 2.

Christopeit, N. and Helmes, K. (1982), Optimal control for a class of partially observable systems, *Stochastics* **8**, 17–38.

Christopeit, N. and Heimes, K. (1983), *Syst. Contr. Lett.* **3**, 105–12.

Clark, J.M.C. and Davis, M.H.A. (1979), On 'predicted miss' stochastic control problems, *Stochastics* **2**, 197–209.

Coddington, E. and Levinson, N. (1955), *Theory of Ordinary Differential Equations*, MacGraw Hill, New York.

Ekeland, I. and Temam, R. (1976), *Convex Analysis and Variational Problems*, North–Holland, Amsterdam.

Haussmann, U.G. (1981), Some examples of optimal control, or: The stochastic maximum principle at work, *SIAM Review* **23**, 292–307.

Haussmann, U.G. and Zhang, Qing (1988), Optimal control of diffusions with small observation noise, preprint.

Ikeda, N. and Watanabe, S. (1981), *Stochastic Differential Equations and Diffusion Processes*, North–Holland, Amsterdam.

Krylov, N.V. (1980), *Controlled Diffusion Processes*, Springer Verlag, New York.

Ladyzenskaya, O. A., Solonnikov, V.A. and Ural'tseva, N.N. (1967), *Linear and Quasi Linear Equations of Parabolic Type*, American Mathematical Society.

Liptser, R.S. and Shiryaev, A. N. (1977), see Chapter 2.

Qing Zhang, (1988) Controlled Partially Observed Diffusions, PhD Thesis, Brown University.

Ruzicka, J. (1977), On the separation principle with bounded coefficients, *App. Math. Optim.* **3**, 243–61.

Yamada, T. and Watanabe, S. (1971), On the uniqueness of solutions of stochastic differential equations, *J. Math. Kyoto Univ.* **11**, 155-7.

Chapter 8

Baras, J., Elliott, R.J. and Kohlmann, M. (1987), The partially observed stochastic minimum principle, University of Alberta, Edmonton.

Bensoussan, A. (1982), see Chapter 4.

Bensoussan, A. (1983), Maximum principle and dynamic programming approaches of the optimal control of partially observed diffusions, *Stochastics* **9**, 169–222.

Bensoussan, A. and Nisio, M. (1990), Nonlinear semigroup arising in the control of diffusions with partial observation, *Stochastics* **30**, 1–45.

Bensoussan, A. and Robin, M. (1982), On the convergence of discrete time dynamic programming equation for general semi group, *SIAM J. Contr. Optim.* **20(5)**, 722–46.

Borkar, V.S. (1982), *Existence of Optimal Controls for Partially Observed Diffusions*, TIFR Centre, Bangalore.

Cannarsa, P. and G.A.D. Prato (1988), Nonlinear optimal control with infinite horizon for distributed parameter systems and stationary Hamilton–Jacobi equations, Preprints, Scuola Normale Supercore, Lisa.

Davis, M.H.A. (1984), *Lectures on Stochastic Control and Nonlinear Filtering*, TIFR, Springer–Verlag, Berlin.

Davis, M.H.A. and Kohlmann, M. (1988), On the nonlinear semigroup of stochastic control under partial observations, unpublished manuscript.

Fleming, W.H. (1982), Nonlinear semi group for controlled partially observed diffusions, *SIAM J. Contr. Optim.* **20**, 286–301.

Haussmann, U.G. (1987), The maximum principle for optimal control of diffusions with partial information, *SIAM J. Contr. Optim.* **25**, 341–61.

Kunita, H. and Watanabe, S. (1957), On square integrable martingales, *Nagoya Math J.* **30**, 209–12.

Kwakernaak, H. (1981), A minimum principle for stochastic control problems with output feedback, *Syst. Contr. Lett.* **1(1)**.

Lions, P.L. (1988), Viscosity solutions of fully nonlinear second-order equations and optimal stochastic control in infinite dimensions, *Acta Math.* **161**.

Nisio, M. (1976a), On a nonlinear semigroup attached to stochastic optimal control, *RIMS, Kyoto University* **13**, 513–37.

Nisio, M. (1976b), On stochastic optimal controls and envelopes of Markovian semi groups, *Proceedings of the International Symposium, Kyoto*, pp. 297–325.

Chapter 9

Bensoussan, A. and Nisio, M. (1990), see Chapter 8.

Borkar, V.S. (1983), Existence of optimal control for partially observed diffusions, *Stochastics* **11**, 103–41.

Davis, M.H. and Kohlmann, M. (1988), see Chapter 8.

Elkaroui, N. Huu Nguyen, D. and Jeanblanc-Picqué, M. (1988), Existence of an optimal Markovian filter for the control under partial observations, *SIAM J. Contr. Optim.* **26(5)**.

Elliott, R.J. and Kohlmann, M. (1982), On the existence of partially observable controls, *Applied Math. Opt.* **3**, 41–66.

Fleming, W. H. and Nisio, M. (1984), On stochastic relaxed control for partially observed diffusions, *Nagoya Math. J.* **93**, 71–108.

Fleming, W.H. and Pardoux, E. (1989), Optimal control for partially observed diffusions, *SIAM J. Contr. Optim.* **20**, 261–85.

Haussmann, U.G. (1982), On the existence of optimal controls for partially observed diffusions, *SIAM J. Contr. Optim.* **20**, 385–407.

Ikeda, N. and Watanabe, S. (1981), see Chapter 7.

Nagase, N. (1988), On the existence of optimal control for controlled stochastic partial differential equations, preprint.

Nagase, N. and Nisio, M. (1988), Optimal controls for stochastic partial differential equations, preprint.

Prokhorov, Y.V. (1956), Convergence of random processes and limit theorems in probability theory, *Theory Prob. Appl.* **1**, 157–214.

Stroock, D. and Varadhan, S.R.S. (1979), see Chapter 4.

Index

Printed in the United States
By Bookmasters